palgrave advances in the european reformations

Palgrave Advances

Titles include:

Patrick Finney (*editor*)
INTERNATIONAL HISTORY

Jonathan Harris (*editor*)
BYZANTINE HISTORY

Marnie Hughes-Warrington (*editor*)
WORLD HISTORIES

Helen J. Nicholson (*editor*)
THE CRUSADES

Alec Ryrie (*editor*)
EUROPEAN REFORMATIONS

Jonathan Woolfson (*editor*)
RENAISSANCE HISTORIOGRAPHY

Forthcoming:

Jonathan Barry (*editor*)
WITCHCRAFT STUDIES

Katherine O'Donnell (*editor*)
IRISH HISTORY

Richard Whatmore (*editor*)
INTELLECTUAL HISTORY

Palgrave Advances
Series Standing Order ISBN 1–4039–3512–2 (Hardback) 1–4039–3513–0 (Paperback)
(*outside North America only*)

You can receive future titles in this series as they are published by placing a standing order. Please contact your bookseller or, in the case of difficulty, write to us at the address below with your name and address, the title of the series and the ISBN quoted above.

Customer Services Department, Macmillan Distribution Ltd, Houndmills, Basingstoke, Hampshire RG21 6XS, England

palgrave advances in the european reformations

edited by

alec ryrie

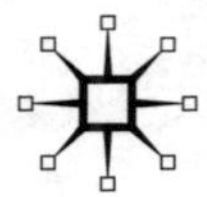

First published 2006 by
PALGRAVE MACMILLAN
Houndmills, Basingstoke, Hampshire RG21 6XS and
175 Fifth Avenue, New York, N.Y. 10010
Companies and representatives throughout the world

PALGRAVE MACMILLAN is the global academic imprint of the
Palgrave Macmillan division of St Martin's Press LLC and of
Palgrave Macmillan Ltd.
Macmillan® is a registered trademark in the United States,
United Kingdom and other countries. Palgrave is a registered
trademark in the European Union and other countries.

ISBN-13 978–1–4039–2041–6 hardback
ISBN-10 1–4039–2041–9 hardback
ISBN-13 978–1–4039–2042–3 paperback
ISBN-10 1–4039–2042–7 paperback

This book is printed on paper suitable for recycling and
made from fully managed and sustained forest sources.

A catalogue record for this book is available
from the British Library.

Library of Congress Cataloging-in-Publication Data
Palgrave advances in the European reformations / edited by Alec Ryrie.
p. cm. — (Palgrave advances)
Includes bibliographical references and index.
ISBN 1–4039–2041–9 (cloth) — ISBN 1–4039–2042–7 (paper)
1. Reformation. I. Ryrie, Alec. II. Series.

BR309.P34 2005
270.6—dc22

205051191

10 9 8 7 6 5 4 3 2 1
15 14 13 12 11 10 09 08 07 06

Printed and bound in Great Britain by
Antony Rowe Ltd, Chippenham and Eastbourne

contents

notes on contributors

David Bagchi is Lecturer in the History of Christian Thought in the Department of Theology at the University of Hull, UK. He has particular interests in Luther's theology, pre-Tridentine Catholic polemical theology, and theological pamphlet literature. He is the author of *Luther's Earliest Opponents* (1991) and co-editor, with David Steinmetz, of *The Cambridge Companion to Reformation Theology* (2004).

Craig D'Alton lectures in Reformation History in the Melbourne College of Divinity, Australia, and is a research fellow of the Department of History at the University of Melbourne. His publications include articles on heresy in Tudor England. He is currently working on a book on the writings of Thomas More.

Michael Driedger is an Associate Professor of History and Liberal Studies at Brock University in St Catharines, Ontario. His research is on the transformation of early modern radical Protestantism, and his most recent book is *Obedient Heretics: Mennonite Identities in Lutheran Hamburg and Altona during the Confessional Age* (2002).

Bruce Gordon is Reader in Modern History at the University of St Andrews, UK, and Deputy Director of the St Andrews Reformation Studies Institute. His recent publications include *The Swiss Reformation* (2002) and (ed., with Emidio Campi) *Architect of Reformation: an Introduction to Heinrich Bullinger, 1504–1575* (2004).

Trevor Johnson is Senior Lecturer in History at the University of the West of England, Bristol. The author of a number of essays on the Catholic Reformation, he has edited (with Bob Scribner) *Popular Religion in Germany and Central Europe, 1400–1800* (1996).

William Monter is Professor Emeritus of History at Northwestern University, USA, where he taught from 1963 to 2002. His most recent books are *Judging the French Reformation: Heresy Trials by Sixteenth-Century French Parlements* (1999) and *Frontiers of Heresy: The Spanish Inquisition from the Basque Lands to Sicily* (1990, paperback 2002).

Graeme Murdock is Senior Lecturer in Modern History at the Centre for Reformation and Early Modern Studies, University of Birmingham, UK. Recent publications include *Calvinism on the Frontier, 1600–1660: International Calvinism and the Reformed Church in Hungary and Transylvania* (2000); with Maria Crăciun and Ovidiu Ghitta (eds), *Confessional Identity in East-Central Europe* (2002); and *Beyond Calvin: the Intellectual, Political and Cultural World of Europe's Reformed Churches* (2004).

Andrew Pettegree is Professor of Modern History at the University of St Andrews, UK, and the Founding Director of the St Andrews Reformation Studies Institute. His authored and edited books on sixteenth-century and Reformation history include *Emden and the Dutch Revolt* (1992), *The Reformation World* (2000) and *Europe in the Sixteenth Century* (2002). His latest book, *Reformation and the Culture of Persuasion,* was published by Cambridge University Press in 2005.

Judith Pollmann is Senior Lecturer in Early Dutch Modern History at the University of Leiden. She is the author of *Religious Choice in the Dutch Republic: the reformation of Arnoldus Buchelius (1565–1641)* (1999). She is currently working on a book on Catholic laypeople in the sixteenth-century Netherlands.

Penny Roberts is Senior Lecturer in History at the University of Warwick, UK. She has written extensively on the social, religious and political history of sixteenth-century France, primarily focusing on the religious wars and the Huguenots. Her publications include *A City in Conflict: Troyes during the French Wars of Religion* (1996) and 'Royal Authority and Justice during the French Religious Wars', *Past and Present,* 184 (2004).

Alec Ryrie is Senior Lecturer in Modern History at the Centre for Reformation and Early Modern Studies, University of Birmingham, UK. His publications on the early Reformations in England and Scotland include *The Gospel and Henry VIII* (2003) and *The Origins of the Scottish Reformation* (2006). He is currently researching early Protestant practices of prayer.

Philip M. Soergel is Associate Professor of History at University of Maryland, College Park. He is the author of *Wondrous in His Saints*

(1993), as well as numerous articles on the history of the Protestant and Catholic Reformations in Germany. Currently, he is completing a monograph on the uses of miracles and natural wonders among early-modern Protestants.

Merry Wiesner-Hanks is Professor of History and Director of the Center for Women's Studies at the University of Wisconsin-Milwaukee. Her books include *Women and Gender in Early Modern Europe* (2nd edn, 2000), *Christianity and Sexuality in the Early Modern World: Regulating Desire, Reforming Practice* (2000) and *Gender in History* (2002).

glossary

adiaphora	Literally, 'things indifferent': practices which are neither forbidden by nor commanded in the Bible
confession	(1) A statement of belief, and, by extension, those people and territories who subscribe to it: eg. the Lutheran Confession of Augsburg gave rise to 'confessional', i.e. orthodox, Lutherans (2) Auricular confession: the practice of confessing one's sins to a priest
confessionalisation	From 'confession' (sense 1): the process by which people and territories were divided into clearly defined and clearly separated religious groups during the Reformation period
consistory	In Reformed Protestantism, a church court which regulates the morals and behaviour of church members
Eucharist	The most neutral term for the sacrament known to Catholics as the Mass, and to Protestants as Holy Communion or the Lord's Supper
evangelical	Literally, 'gospeller': often applied to early Protestants before clear confessional distinctions emerged. Also used as a synonym for 'Lutheran'
Huguenots	French Protestants
monarchomach	A term for political theories justifying regicide or armed resistance to a tyrannical monarch
Nicodemism	A pejorative term for the practice of holding a secret allegiance to one religion while outwardly

conforming to another, usually to escape persecution

Reformed	With a capital R: the broad theological tradition arising from Zwingli, Calvin and others, often (misleadingly) referred to as Calvinism
Tridentine	Relating to the Council of Trent (Latin *Tridentium*), the reforming Catholic council which sat intermittently from 1545 to 1563
Utraquist	A Bohemian church which broke with Rome in the fifteenth century, which insisted that the laity should receive both bread and wine in the Eucharist (i.e. in both kinds, *sub utraque specie*)

introduction: the european reformations

alec ryrie

The Reformation is the ghost at the feast of European history. It should be safely distant, dead and buried, but it stubbornly refuses to disappear. Indeed, it presses itself inconveniently onto our attention. It marks the period which historians know, awkwardly, as 'early modern': an appropriately uneasy term for a period which is neither close enough for reliable empathy, nor remote enough for any kind of objectivity.

Several things are immediately apparent about the state of Reformation history in the early twenty-first century. First, the field is in rude health, whether measured by the sheer volume of both scholarly and popular writing on the subject over the past generation, or by the intellectual energy of its historians. Secondly, it is criss-crossed with battle-lines, old ones between Catholic and Protestant historians persisting beneath more recent disputes. But thirdly, those battle-lines are fragmenting as they proliferate. There are fewer certainties in Reformation historiography than there were a generation ago: generalisations are being dissolved into complexity and definitions are disintegrating into confusion. This volume attempts a critical survey of past and current scholarship in thirteen key areas of Reformation research. It is intended both as a guide to those complexities and confusions for students and scholars; and as a discussion of the 'state of the question' as we consider where Reformation history might go next.

It is a truism that all histories are the products of their own time, but this can always be seen more clearly with hindsight. It is only with the Cold War era firmly behind us that we can see how far it coloured Reformation history. Its most obvious effect was entirely malign: western historians were denied access to the archives of the Eastern Bloc, nor could historians behind the Iron Curtain make free use of them. The result was an attenuated, impoverished history of the Reformation. Events east of the Elbe could not be entirely forgotten (after all, Luther's Wittenberg was in East Germany), but the remarkable parts played in the European Reformation by Poland, by Hungary and above all by Bohemia were driven to the periphery. In Chapter 2 of this volume, Graeme Murdock surveys these 'forgotten' Reformations, and how, since 1989, they have begun to be remembered.

However, Cold War mentalities affected western historical preoccupations as well. It was easy to read the Reformation as an analogue of the Cold War, in which two competing ideologies struggled for continental dominance. It directed our attention to the great-power rivalry and diplomacy that accompanied the Reformation – a topic which, as William Monter comments in this volume, is now thoroughly out of fashion.[1] It seemed natural to view Calvin's Geneva, or the Society of Jesus, as directing international webs of influence in the fashion of the Comintern or the CIA. Robert Kingdon's justly famous study of the Genevan-sponsored missions to France in the late 1550s might have been written about Communist infiltrators.[2] And of course, historians on both sides of the Iron Curtain who were influenced by Marx's methods had very specific interests in the Reformation. The Marxist tradition is predisposed to see religious disputes as a cloak for socio-economic grievances; East German historiography, in particular, sometimes saw the Lutheran Reformation as little more than a warm-up act for the 1525 Peasants' War. Marxist history also focused on the radical political ideologies of the Reformation period, especially those of Calvinist rebels in France, the Netherlands and the British Isles. Those historians inclined to trace a 'European revolutionary tradition' could find its origins in the sixteenth and seventeenth centuries.[3]

Cold War *détente*, in the 1970s, fostered a different atmosphere. There was a flurry of studies of the summitry of the Reformation: the repeated (and failed) attempts by a number of leading theologians to negotiate religious compromises.[4] The political priorities of the 1970s also helped to produce a consciously even-handed approach in a field previously known for its partisan rancour. Historians increasingly viewed the Protestant and Catholic Reformations as parallel forces rather than opposing ones, and became interested in how they managed to coexist (a theme discussed in Trevor Johnson and Philip M. Soergel's essays in this volume). In particular, German historians – for whom the issue of Cold War coexistence was inescapable – elaborated the 'confessionalisation' hypothesis, first formed in the 1950s. This described how both Protestant and Catholic princes used religion as a state-building tool. The fact that princes enforced religious uniformity onto their territories, and in so doing bolstered their own authority, was seen as more important than any differences between the religions which they used to this end. It is a powerful model, as far as it goes; but it has always worked much better for Protestant than for Catholic territories, and its applicability beyond the Holy Roman Empire is doubtful. Moreover, we are now more aware that religious communities, and especially minorities such as Anabaptists, might 'confessionalise' themselves: the process could create as much division as unity.[5]

The post-Cold War era no doubt has its own, equally distinctive historical preoccupations, but as we are still in the midst of them, they are not yet clearly visible. Certainly, some early twenty-first-century political preoccupations – terrorism, state attempts to fight it, and religious fundamentalisms of all kinds – find an echo in Reformation studies. No doubt we will continue using the present and the past to try to explain one another. However, a more obvious characteristic of post-Cold War Reformation history (indeed of western historiography in general) is the disintegration of neat intellectual certainties and divisions. This is partly, of course, the mere by-product of research. Confident generalisations are a hallmark of ignorance: the more we know, the more complex the picture appears. Moreover, for budding historians, established truisms represent career opportunities. Academia actively rewards those who replace simplicity with complexity, even if doing so does not always produce a net gain in understanding. Yet the mood of restless questioning, and of impatience with answers, which is often labelled 'postmodernity' is also a part of this story. It has left us certain of much less than we were a generation ago, even as our knowledge has increased.

This is most obviously visible in our terminology. Almost all of the inherited historical shorthand used to describe sixteenth-century European religion has been eaten through with holes. 'The Reformation' – an eighteenth-century German coinage – was once thought of as a clear ideological package. Historians would regularly speak of towns or principalities choosing to adopt 'the Reformation'. Now we are more comfortable fragmenting religious change into 'Reformations' – Lutheran, Calvinist, Catholic and Radical Reformations, national Reformations, princely and popular Reformations, first and second Reformations, Reformations of life, of manners, of buildings, even of the dead. Nor is the timescale clear. This volume focuses on the traditional 'Reformation century', from the 1510s to the 1620s, but we are increasingly aware of the longer context: one recent survey has stretched the 'long' Reformation from the fourteenth to the eighteenth century, and another has even claimed that the Reformation is not yet over.[6] These various Reformations were obviously not unconnected to one another, but they cannot be reduced to a single phenomenon, nor traced to a single hammer-blow on the Wittenberg door.

Likewise, the sharp definition of the religious 'confessions' themselves has been questioned. Textbooks still include maps professing to show which territories subscribed to which confessions at which point, but we are increasingly uneasy about such clear demarcations. We are more inclined now to remember that 'Calvinism', 'Catholicism' and 'Lutheranism' were all broad churches, and that 'Anabaptism' was so diverse as scarcely to exist as a single entity. We are also more conscious that while a state can be drawn in one colour on a map, the Dutch Republic's Calvinist pink (for example) conceals a kaleidoscope of religious coexistence, and that Netherlandish 'Calvinism' owed as much to Zurich as to Geneva. The most awkward feature of such maps – the designation of post-Reformation England as 'Anglican', as if this were a third, distinct Protestant confession – is vanishing from our historiography entirely.[7]

These confessions did have a real existence, but they were fuzzy at the edges. While we are no longer so interested in attempts at formal ecumenical *détente*, we are looking at those who existed between the cracks, outside the tight boundaries which some theologians and politicians were attempting to draw. The study of 'tolerance' and of 'moderation' is fashionable.[8] And rather than negotiated theological compromises, we are now more interested in the unheroic ways in which private individuals negotiated the murderous religious landscapes of the sixteenth century. Those who gave their private allegiance to one religion, while (for the sake of safety or neighbourliness) conforming to another, were condemned by contemporaries as 'Nicodemites' or 'church papists', and have traditionally been discounted as fainthearts who do not deserve a place in Reformation history. Several recent, and less condescending studies have suggested that these people may have had a more distinctive and a more significant part to play.[9] History is not predominantly made by heroes and villains, but by ordinary people trying to preserve their safety, their prosperity and (when possible) their consciences in a complex and unpredictable world.

The heroes and villains remain important, however. They were few in number, but they fired the imaginations of whole societies. Almost all participants in the Reformation disputes understood that those disputes were more than passing political squabbles. They were located in the wider story of Christian history, which was itself framed by the apocalyptic struggle between Christ and Antichrist. We have always known that the Reformation era thought of itself in these stirring terms, but it is only recently that historians have emphasised the need to understand this world-view from within. Catholics and Protestants, conservatives and evangelicals, were battling for ownership of the Christian story. Those who could place themselves most credibly in that story stood to gain legitimacy and, thereby, support. The Reformation era's battles to control history have attracted considerable interest.[10] In particular, we have paid more attention recently to the competing attempts to package the events of the sixteenth century itself as apocalyptic history. These attempts were made predominantly in the time-honoured Christian fashion, through telling the stories of martyrs and their persecutors. The rich martyrologies of the Reformation era are currently a focus for renewed and fruitful study – less for the facts they relate than for the mentalities to which they testify.[11]

The focus on the histories and martyrologies of the Reformation era is a part of one of the most striking features of the last quarter-century of Reformation historiography: the revival of ideas in general and of theology in particular. Most current approaches emphasise not merely the power of religious belief in this period, but the importance of specific doctrines. Where surveys of the Reformation might once have sat light to religion and concentrated on politics, society and the economy, more recent treatments invariably give detailed and generous treatment to theology.[12] This perhaps surprising shift is due to two apparently opposing historical forces. On one side is the postmodern 'linguistic turn', which has reawakened our interest in the power of words and ideas to define human reality, sometimes to the point at which the grim physicality of most of history is evaded altogether. On the other side is the continued interest in the Reformation era from Christians – both from believers, and from former or lapsed believers who still find this great crisis of their religion congenial or fascinating.

In the largely secularised world of western academia, a surprising number of Reformation historians retain some kind of personal religious interest in their subject. This has of course always been the case. For centuries, as most of the essays in this volume demonstrate, Reformation historians wrote from openly partisan viewpoints, either lavishing attention and sympathy on their own confessional forebears, or polemically denigrating the other side. Obviously enough, this produced bad history, but it had its redeeming features: much of this scholarship was meticulous and, in places, insightful. Openly partisan histories are now rarer, but religious axes of various kinds are still being ground on all sides. Alongside those who retain a clear confessional commitment are Christians of a more ecumenical bent who are interested in tracing moderations, cross-currents or continuities behind the stark divisions of the sixteenth century. The lapsed believers, too, have agendas to pursue, whether of nostalgia for or anger with the traditions to which they once belonged. (Nor, of course, are those of other faiths, or of none, 'neutral'.)

Historians across this spectrum will argue the strengths of their particular vantage points. In 1968 the English Catholic historian J. J. Scarisbrick claimed that Catholics who had embraced the spirit of the Second Vatican Council were best placed to understand the Reformation. By contrast, Diarmaid MacCulloch suggests that religious history written by believers will always be suspect – 'certain aspects that might be construed as embarrassments might not be dwelt on, certain questions might not be raised' – and argues that his own position, as a former Anglican, provides an ideal mixture of insight and objectivity.[13] Yet none of these claims to priority seem well borne out, for if we examine the most compelling and influential histories of the Reformation period, there is in fact no clear pattern. Passionate partisanship can provide revealing empathy or blinkered bigotry; the outsider's detachment can penetrate to the heart of the matter, or miss the point entirely. No one stance can be relied on to produce value-free history; even if it could be written, such history would scarcely be worth reading. Readers of Reformation history should be aware, however, that even when the surface seems calm, religious tensions of some kind are often bubbling away beneath. It is usually wise to read the 'Acknowledgements' pages of books for health warnings; and to be aware of the biases and insights that we ourselves bring to the subject.

This old tradition of religious and theological history has crossed with the new, 'postmodern' preoccupations with language and the fluidity of meaning; and the match has produced some intriguing fruit. One of the most valuable aspects of these new histories is their interdisciplinarity. Theology has not only been made respectable again, but integrated into the political, social, economic and – above all – the cultural history of the period. As a result, some topics have prospered – 'popular religion', for example.[14] Likewise, studies of mass violence in the religious wars have made it clear how genuinely and specifically religious such violence was. And the history of gender in the Reformation is now not merely about whether the Reformation was 'good for women' (a near-meaningless question), but about the unpredictable ways in which the various doctrinal systems and systems of gender roles interacted with one another.[15] Other subjects have faded: the Protestant 'theories of resistance' now seem much less important, after-the-fact justifications of actions forced onto the reformers by political necessity. (They may, however, have a stubborn importance in explaining the forms in which rebellious Protestant movements established governments.) Other, old subjects have been given a new lease of life. Max Weber's long-unfashionable suggestion of a connection between Protestantism and capitalism seems to be returning cautiously to favour.[16]

Above all, our renewed interest in belief now focuses less on magisterial theology than on the oddballs and the also-rans of the period; and less on theologians in general than on how their ideas were received and interpreted in the wider population. Scepticism about neatly-defined 'confessions' has made us aware that in so far as they existed, they only coalesced slowly and unsteadily across the period. Our interest in the first generation of the Reformation now focuses on the halting processes by which myriad overlapping religious possibilities were collapsed into distinct confessions. In keeping with current historiographical fashion, the course and the specific outcomes of this process are regarded as 'contingent': it was not inevitable that the Reformation's battle-lines would be drawn the way they eventually were. Some religious movements which did not survive were for a time realistic contenders, such as the evangelical Catholicism which was fashionable in Germany, Italy and Scotland in the 1540s and 1550s. This perception helpfully undermines the still-pervasive idea of 'progress'. It can also create a damaging illusion of chance: we are now, perhaps, prejudiced not merely against 'inevitabilism' but against any suggestion of deep, underlying causes at all.

What, then, are the currently urgent themes in Reformation research? No two historians' answers would be the same – in itself a healthy sign. At the time of writing my own list would headline three themes, touching on Protestantism, Catholicism and society. For Protestantism, following on from the emphasis on the contingency and plurality of the Reformation, we need to continue to detach sixteenth-century Protestantism from eighteenth-century rationalism. The late Bob Scribner powerfully challenged Weber's claim that Protestantism 'disenchanted' the world, and a recent volume has drawn our attention to Protestant 'superstition', but we are not yet free of the association between Protestantism and secular individualism.[17] In particular, we need to understand not merely the beliefs of Protestants, but how those beliefs were lived out. The prayer and worship of sixteenth-century Protestants has been studied by liturgical and musical historians, but needs to be taken more seriously as a part of the Protestant life, and placed properly in its political and social context. If we do not understand what Protestants thought they were doing when they were praying, we are unlikely to understand much of the rest of what they did. This is perhaps one approach to the deeper and more baffling question: quite why were so many Europeans attracted to Protestantism at all?

Our understanding of the devotional life of Catholicism is far more advanced, but other areas of Catholic history remain under-researched. In particular, it is becoming clear that the outcome of the Reformation struggles in any given region was significantly determined by the Catholic response. Yet why was that response so ineffectual in some areas and so powerful in others? German Catholicism cracked, and Scandinavian Catholicism crumbled, in the face of the Lutheran onslaught. Elsewhere, Catholicism mobilised to meet the Protestant threat in the parishes (in Ireland, or much of central Europe) or on the battlefield (in France). There is a generational pattern to this: from the 1570s, at the latest, Catholicism was rallying, although it seems unlikely that this can be ascribed to the work of the Council of Trent (which was completed in 1563). How and why Catholicism defeated the Protestant challenge, or was defeated by it, remains one of the major unanswered questions of Reformation history. And this, too, leads to a deeper and more baffling question: what drove the longer movement of Catholic reform which predated, inspired, opposed and (arguably) outlasted the Protestant Reformation?

Part of the answer may lie in my third theme: the role of lay elites. The nobility, gentry and educated laity do not fit neatly into either 'top-down' or 'bottom-up' models of religious change. In particular, neither confessional, nor Enlightenment, nor Marxist histories of the Reformation have had much place for the European nobility, a class still often viewed as thuggish throwbacks to a more barbaric age. The 'confessionalisation' hypothesis links the Reformation to the rise of absolutism in the seventeenth and eighteenth centuries, when the remaining powers of Europe's aristocracies were broken. Yet from the so-called Knights' War in Germany, through the religious wars in Scotland, France and the Netherlands, to the estate-by-estate patchwork of Reformations in Poland and the Habsburg lands, nobles were amongst the earliest converts to and the most vigorous defenders of the Protestant causes. We do not know why, although if princes could use 'confessionalisation' to bolster their authority, so could nobles. If this is so, then perhaps (again) it was Protestantism which was the 'reactionary', 'medieval' force, and (post-Tridentine?) Catholicism which was the motor of change.[18] Likewise, other lay elites, in particular the educated laity and those in courtly circles, often proved decisive in either destroying or sustaining religious traditions.[19] Women in these elites were sometimes particularly influential. We are familiar with the parts these people played, but there has been little attempt systematically to examine or compare their roles.

Until we understand how and why these people acted as they did, we will not understand the Reformations which they made.

The agenda for Reformation historians has shifted, and will continue to shift. The essays in this volume represent some of the major themes of recent research. Part One looks at six of the most important national or regional Reformations. Other recent collections have followed similar divisions,[20] reflecting the diversity and specificity of the different Reformations. The list is, inevitably, incomplete. Scandinavia is here given short shrift; more importantly, the Mediterranean lands where Protestantism struggled to obtain a foothold are neglected. They are excluded partly because the relevant themes are dealt with elsewhere in the volume and partly from considerations of space, but also because the historiographies of these national Reformations are relatively thin. The six regions covered have provided the richest and most controversial discussions over the past few decades, and this looks set to continue.

Part Two moves to look at seven wider themes which have dominated recent Reformation research. Some of these are obvious (Catholicism, Anabaptism, gender); others less so (printing, violence). It is a shifting and overlapping list. Others might have been added: in particular the arts (discussed in this volume's companion on the Renaissance) and witchcraft, a subject whose whirlwind of historiography deserves a volume to itself. Needless to say, no effort has been made to ensure that the contributors keep off one another's patches, much less to ensure that they do not disagree. All thirteen contributors are active researchers in their fields, and as well as surveying those fields they have distinct opinions to offer. This book does not, then, offer a new interpretation of the European Reformation. It does, perhaps, suggest where someone looking to build such an interpretation might begin.

notes

1. See below, pp. 286–7.
2. Robert M. Kingdon, *Geneva and the Coming of the Wars of Religion in France, 1555–1563* (Geneva, 1956).
3. One example amongst many: Michael Walzer, *The Revolution of the Saints: a Study in the Origins of Radical Politics* (London, 1966).
4. Notably Peter Matheson, *Cardinal Contarini at Regensburg* (Oxford, 1972); Donald Nugent, *Ecumenism in the Age of the Reformation: the Colloquy of Poissy* (Cambridge, 1974).
5. See Chapters 1, 2, 3, 4 and 10 below.
6. Peter G. Wallace, *The Long European Reformation* (Basingstoke, 2004); Felipe Fernández-Armesto and Derek Wilson, *Reformations: a Radical Interpretation of Christianity and the World (1500–2000)* (New York, 1996).

7. Indeed, it is finally disappearing from the maps, too: compare the maps of Europe's confessional divisions in 1600 in Andrew Pettegree (ed.), *The Reformation World* (London, 2000), p. xvi, and in Diarmaid MacCulloch, *Reformation: Europe's House Divided, 1490–1700* (London, 2003), p. 488.
8. Ole Peter Grell and Bob Scribner (eds), *Tolerance and Intolerance in the European Reformation* (Cambridge 1996); Luc Racaut and Alec Ryrie (eds), *Moderate Voices in the European Reformation* (Aldershot, 2005); Thierry Wanegffelen, *Ni Rome ni Genève: Des fidèles entre deux chaires en France au XVIe siècle* (Paris, 1997).
9. Notably Peres Zagorin, *Ways of Lying: Dissimulation, Persecution and Conformity in Early Modern Europe* (Cambridge, MA, 1990); Alexandra Walsham, *Church Papists: Catholicism, Conformity and Confessional Polemic in Early Modern England* (Woodbridge, 1993); Andrew Pettegree, 'Nicodemism and the English Reformation', in his *Marian Protestantism: Six Studies* (Aldershot, 1996).
10. For example, Bruce Gordon (ed.), *Protestant History and Identity in Sixteenth-Century Europe*, 2 vols (Aldershot, 1996).
11. See, above all, Brad S. Gregory, *Salvation at Stake: Christian Martyrdom in Early Modern Europe* (Cambridge, MA, 1999).
12. For a classic twentieth-century survey which favoured politics over theology, see G. R. Elton, *Reformation Europe* (London, 1963). For more recent examples of the restoration of theology to the story, see Euan Cameron, *The European Reformation* (Oxford, 1991); Carter Lindberg, *The European Reformations* (Oxford, 1996); MacCulloch, *Reformation*.
13. J. J. Scarisbrick, 'Post-Vatican II Catholicism – some historical perspectives', *Dublin Review*, 202 (1968), 117–28; MacCulloch, *Reformation*, p. xxv; MacCulloch quoted in *The Spectator*, 16 December 2004.
14. See Chapter 11, below.
15. See Chapter 12, below.
16. Max Weber, *The Protestant Ethic and the Spirit of Capitalism*, trans. T. Parsons (London, 1930); for a recent treatment of this subject, see Margaret C. Jacob and Matthew Kadane, 'Missing, Now Found in the Eighteenth Century: Weber's Protestant Capitalist', *American Historical Review*, 108 (2003), 20–49.
17. Robert Scribner, 'The Reformation, Popular Magic and the Disenchantment of the World', *Journal of Interdisciplinary History*, 23 (1993), 475–94; Helen Parish and William G. Naphy (eds), *Religion and Superstition in Reformation Europe* (Manchester, 2002).
18. An idea pursued in R. Po-chia Hsia, *The World of Catholic Renewal 1540–1770* (Cambridge, 1998).
19. For one recent discussion of this, see Elaine Fulton, *Catholic Belief and Survival in Late Sixteenth-Century Vienna* (Aldershot, 2005).
20. Andrew Pettegree (ed.), *The Early Reformation in Europe* (Cambridge, 1992); Robert Scribner, Roy Porter and Mikuláš Teich (eds), *The Reformation in National Context* (Cambridge, 1994); Pettegree, *Reformation World*.

part one
national reformations

1
germany and the lutheran reformation

david bagchi

The Lutheran Reformation in Germany has long been recognised as a complex phenomenon that cannot adequately be explained by any one historical discipline. Religious and intellectual, social and cultural, political and economic approaches all have some relevance to the task in hand. In previous decades, there was undoubtedly a tendency towards the polarisation and fragmentation of these approaches; but more recently historians have fostered a genuine multidisciplinarity. An example of the way in which the approaches complement each other is provided by what is usually identified as the starting point of the German Reformation, Luther's nailing up of the ninety-five theses against indulgences. Earlier iconography depicted this moment as a deeply significant one in which the hammer blows of a lone, brave monk on the church doors reverberated not only through the castle-church of Wittenberg but through the entire western Church. The current scholarly consensus, however, paints a more complex picture in more subtle hues.

Historians of education have shown that there was nothing extraordinary or dramatic about Luther's nailing up the theses. It was an invitation to a theological disputation, and the church door was Wittenberg university's main noticeboard. The posting was a routine academic exercise, which would have attracted as much (or indeed as little) attention as a notice advertising a staff seminar would on any university noticeboard today. In fact, the incident was so routine that Luther himself made no reference to it in later life, beyond marking All Hallows' Eve as the day on which he attacked indulgences. This omission led one scholar famously to deny that the posting ever happened, and indeed an influential view is that the posting (as opposed to Luther's letter to the bishops attacking the untoward pastoral effects of over-enthusiastic indulgence preaching) took place later, perhaps in mid-November.[1] The precise date of the posting of the theses is less important to modern scholarship than the fact that they circulated widely, far beyond the academic community. The Latin theses could not easily have sparked a popular movement. Social historians would rightly point out that it was only when they were circulated outside the academic milieu Luther intended, and were translated into German against his will, that the Reformation as a movement can be said truly to have begun.

Another widely-held assumption, challenged this time by doctrinal historians, is that the *Ninety-five Theses* express Luther's characteristic belief that Christians are justified (that is, put in a right relationship with God) by grace alone, through faith and not through good works such as the purchase of indulgences. Had he not been steeled by such a mighty conviction, the older account would have it, how could a lone monk have dared to speak out against this pan-European trade? But even a cursory examination reveals that there is no mention in these theses of justification or of faith. Their vocabulary is that of late medieval penitential theology: contrition, penance, merit. Admittedly, a distinctive theological note is sounded in the climactic final two theses; not the doctrine of justification by faith alone but what Luther would soon call 'the theology of the cross'. This was the idea, which owed much to the German mystical tradition in which Luther was immersed, that Christians should seek out or at least not avoid suffering, and so follow their master in the way of the cross. Indulgences were a 'let-off' (the literal meaning of *Ablaß*, the German word for indulgences) from penance that no conscientious Christian could countenance. In itself, Luther's idea was not new, though this was the first time it had been used to mount a critique of indulgences. Nor was the criticism of indulgences new or unusual: almost contemporaneously with Luther, such pillars of the church establishment as the Sorbonne and the papal theologian Cardinal Cajetan were also making warning noises about the weak theological foundations upon which most indulgence preaching was based.

a unified reformation message?

The case of the *Ninety-five Theses* is an obvious one to use to illustrate the way in which recent historical approaches have worked together to challenge accepted interpretations of the Lutheran Reformation in Germany. It is time now to consider the historiographical issues on which these approaches are currently focused. One of the most important controversies in recent years has been over the content of the early Reformation message. To what extent should we think of a unified message that in all essentials followed Luther's own teachings, and to what extent one that differed fundamentally in its emphases and concerns, from one preacher and one community to another? Our answer to this will depend on our answer to a number of other related questions: how central we think the person and teachings of Luther himself were to the early German Reformation as a whole; whether we regard that Reformation as a unified and coherent movement; how effective we estimate the mass media of the day to have been at producing a uniform message; and whether we think that uniformity was intended and achievable.

In the 1960s, Franz Lau first characterised the years 1520 to 1525 (that is, the period before the Peasants' War) as a time of 'rank growth' (*Wildwuchs*), when all manner of opinions found expression, relatively unaffected at this time by official censorship.[2] Twenty years later, on the basis of his research into pamphlets containing summaries of individual preachers' sermons, Bernd Moeller reported that, on the contrary, a basic consensus existed among evangelical preachers on a range of issues. Common to all were half-a-dozen points: an emphasis on justification by faith; a radical rejection of good works as a means of salvation; criticism of the Church and of the clergy; a definition of the true Church which included some notion of the priesthood of all believers; the importance of love expressed through care for the poor; opposition to the tyrannical use of force and a corresponding emphasis on suffering and the way of the cross. Moeller concluded that not only was there a basic consensus among German urban preachers in the early years of the Reformation, but that this consensus largely followed Luther's own lead.[3] Since Moeller's essay, Lau's idea of *Wildwuchs* has been re-asserted by Susan Karant-Nunn in the service of the confessionalisation thesis.[4] For her, the 1520s represent the brief period of theological freedom which preceded the strict enforcement of confessional orthodoxy (whether Roman, Lutheran, or Reformed) by the civil powers.[5] Moeller's thesis, she argues, had been overly influenced by the fact that his research related to the imperial cities, where the need for conformity was already keenly felt. Her own findings, from the territories of Saxony and Thuringia, point to a much greater heterogeneity.

It is of course impossible for the non-specialist to decide between Moeller's and Karant-Nunn's arguments. It might however be pointed out that the debate hitherto has focused rather narrowly on the content of preaching, at the expense of the wider spectrum of media that urban (and rural) populations of this period would have experienced. Such a self-denying ordinance is less defensible when we consider that the sermon-summaries originally considered by Moeller took the form of printed pamphlets, not live sermons. If we broaden the focus to include pamphlets as such, some interesting results emerge. The doyen of modern pamphlet-research in Germany, Hans-Joachim Köhler, took a random sample of 3,000 religious pamphlets published in German-speaking lands in the 1520s. From these, he selected 356 as representative of the larger sample in terms of language, date, and place of publication. He found that almost two-thirds of these pamphlets included some appeal to the Scripture principle (over 70 per cent for those published in the period 1520–26), and that the doctrine of justification similarly dominated the Protestant pamphlets in the selection.[6] Köhler's results, which are based on pamphlets published throughout the German-speaking lands and not just from the imperial cities, therefore appear to strengthen Moeller's case. Similarly, in an extremely detailed and nuanced study, Thomas Hohenberger has examined all the religious pamphlets published in Germany in the years 1521–22, and concluded that they demonstrate 'all the essential contours of the justification message as communicated by Luther through the mass media'.[7]

However, we should be careful of deducing from this evidence that evangelical pamphleteers and preachers in 1520s Germany regarded themselves as clones of Luther. On the contrary, a feature of their rhetoric was that they were followers of Christ, or followers of the gospel, not disciples of Luther, who was a mere man. A good example of this tendency was the Bavarian writer Argula von Grumbach, who wrote under her maiden name of von Stauffen.[8] Although her literary career shows some parallels with Luther's own, intentionally or not, she fiercely asserted her independence of the Wittenberger. Her teaching was, she claimed, based on Christ and not on any fallible human being. There was of course a rhetorical advantage in taking this stance; but it meant that writers such as von Stauffen would have had no hesitation in taking leave of Luther wherever they perceived his teaching and the Gospel's to conflict. In these circumstances, one would expect a free-for-all rather than the toeing of any party line. 'The wonder is,' as Karant-Nunn herself points out, 'that as many of Luther's ardent followers managed to agree with him as did, even before his faith was defined, its limits surveyed and staked out'.[9]

justification: the heart of the reformation message?

The weight of evidence seems to point to a convergent, if not identical, Reformation message centred on a common understanding of justification by faith alone. Here two questions have dominated recent scholarship: what were the sources of Luther's new understanding of justification, and how (if at all) could ordinary, theologically untrained, people have grasped its doctrinal complexities? A definition of the Reformation doctrine of justification which has commanded broad agreement in recent years is that provided by Berndt Hamm.[10] One advantage of his definition is that it applies not just to Luther's doctrine but to those of all the major players of the early Protestant movement, including Melanchthon, Zwingli, and Calvin. Another is that it offers a clear demarcation of the Protestant doctrine both from medieval doctrines on the one hand, and from the Roman Catholic doctrine, as established by the Council of Trent, on the other.

Hamm believes that what is most distinctive about the Reformation idea is the notion of God's unconditional acceptance of the sinner. Unlike the various medieval schemes of salvation, which all presupposed to a greater or lesser extent human cooperation with God in the form of some token effort or disposition to love God, the Reformation scheme had God forgive and accept sinners as they are. This was not because the reformers regarded sin with less seriousness. On the contrary, they (and especially Luther) believed it was more serious that any medieval theologian suspected: sin for Luther was a state of permanent rebellion from God, so that one could lead an outwardly sinless life (that is, committing no actual crimes against God or one's neighbour) but still be in a sinful state. The difference was rather that the reformers took the idea of God's grace more seriously. The medieval systems had in general confused the concept of grace (literally, 'that which is freely given') with merit (literally, 'that which is deserved'), so that grace was seen as the bestowal of merit which cancelled out sin and freed believers to do good works and build up merit ('created grace') of their own. For the reformers, grace was synonymous with the merciful God. Justification by grace means that God covers the sinner with an alien or external righteousness, that is, with God's own righteousness. Underneath, in himself or herself, the sinner remains a sinner. But in the sight of God, the sinner is resplendent with no less than divine righteousness. This was how Luther could call the Christian 'at one and the same time both righteous and a sinner'.

The ethical implications of this doctrine are obvious. Freed from the necessity of doing good works to put one in a right relation with God (for God has already done that), the believer is able to do good works simply to help other people, as an expression of love. The Protestant believer can have absolute certainty of this right relationship (another dissimilarity with both medieval and post-Tridentine systems) through faith in God's promise of salvation as revealed through and wrought by Jesus Christ. Hamm concludes that 'justification by faith alone' does not therefore simply replace one condition, that of good works, with another, that of faith. For the reformers, faith was not a human quality or virtue but something passive and receptive, a divine gift that re-orientates us from trusting in our own righteousness to trusting in God's promises.

It is important to remember that neither Luther nor the other reformers claimed any originality for their ideas on justification. Luther himself attributed his re-discovery of it to his reading of the Bible and St Augustine. The correlation of the two is important, because Luther read the Bible through the lens of the fifth-century African bishop. A major debate among the first Christians was the extent to which non-Jewish converts to Christianity were bound by the Jewish law. Both sides of the debate are represented in the pages of the New Testament, but the earliest and predominant voice in it is that of St Paul, who strongly opposed the imposition of the law of Moses on Gentiles, who were, in his opinion, 'justified' (brought into a right relationship with God) by an act of sheer grace on God's part. It was the Pauline position that was appropriated and adapted by Augustine in his debate with the British monk Pelagius. Pelagius had asserted that the way to salvation lay simply in doing what divine law commanded, and avoiding what it prohibited. Augustine, a late convert to Christianity, knew from his own experience that no-one, purely by their own powers, could do what was pleasing to God. This had not been God's intention; but Adam's and Eve's disobedience in the garden of Eden had so flawed humanity that thereafter it would always seek its own way rather than its Creator's. In fact, the human race had chosen the way of death rather than life, and just as physical mortality was a consequence of the first couple's disobedience, so all men and women are now headed for destruction of the soul. By grace, God has decided to save some – to rescue some timbers, as it were, from the burning house. Why God should save some and not others is not therefore an injustice to be questioned but a wholly unmerited mercy to be praised. In this debate the western Church, at the Second Council of Orange (529), sided with Augustine against Pelagius.

It might seem therefore that Luther's own insistence that justification was purely a result of divine grace, not human activity ('works-righteousness', as he dubbed it), was fully in line with the western theological tradition stemming from Augustine and tracing its inspiration ultimately to Paul. That was certainly the earliest evangelicals' understanding of the matter: they regarded the period after Augustine as a decline from the true faith, at least in the matter of justification. Until relatively recently, Protestant historians of the Reformation tended to take the same view.[11] Medieval theology, particularly that inspired by the thirteenth-century Dominican Thomas Aquinas, who had tried to marry Augustine's thought with that of the pagan philosopher Aristotle, had lost the plot, Christianly speaking, and Reformation theology was therefore the restoration of biblical and patristic truths. It was therefore important to accentuate the differences between Reformation teaching and what had gone immediately before, portraying the Reformation as a return to a golden age, and isolating medieval theology as an illegitimate but temporary departure from the common tradition. The rhetoric of discontinuity was also adopted by Roman Catholic historians. Their problem was to explain how the Middle Ages, the 'age of faith', could possibly have given birth to the Reformation. The solution was to identify a period of transition between the 'high' Middle Ages and the Reformation itself. This was the 'late' Middle Ages, in which the ontological certainties of Aquinas were replaced by the corrosive scepticism of Ockham, and a vibrant Christianity by the mere externals of religiosity. The language associated with this period was one of decline, of gathering gloom at the end of the day, of autumnal chill.[12]

In the last forty years or so, these confessionally-inspired depictions of decline and discontinuity have been replaced by a more positive evaluation of the late medieval period. Roman Catholic historians have affirmed the continuing vigour of a recognisable Catholicism well into the era of Catholic Reformation (or 'early modern Catholicism'),[13] while Protestants have been willing to seek the sources for Reformation teaching as much in the medieval tradition as in the repristination of a biblical message.[14]

This process can clearly be seen in relation to the doctrine of justification, so that Luther's breakthrough is no longer regarded as a complete discontinuity with the late-medieval tradition, but as a development of at least some of its strands. Aquinas's (or 'Thomist') thought represented a strand which Luther seems to have had little sympathy with or knowledge of.[15] But as an undergraduate at Erfurt university he had been exposed to the thought of William of Ockham and his followers, and Luther was happy to call himself an 'Ockhamist' throughout his life.[16] Ockhamists strongly opposed the Thomist belief that God was in a sense obliged to reward his creatures whenever they demonstrated any 'divine' quality, such as when they acted out of love (love being 'essential' to God, that is, part of God's being – so I John 4:8). Ockhamists insisted on the contrary that the Creator was wholly free with regard to the creature. If God rewarded human love, it was because God had freely decided to reward it; but God was free, in principle, to have established hatred as the precondition of reward, or even to have established no precondition whatever. This potentially radical line of thinking made Ockhamists in practice highly conservative: after all, if there were an infinite number of ways of ordering the moral universe available to God, the one which God chose must literally be the best of all possible worlds. But it has long been recognised that Luther the self-confessed Ockhamist could well have been influenced by the idea of the sovereign freedom of God that is not necessarily bound to the Church's system of rewards and punishments.

This brings us to the second historiographical issue in relation to this topic: how could ordinary Christians have understood the doctrinal complexities of the new thinking on justification? Historians of doctrine perhaps have themselves to blame that this question has been raised to the level of importance it has. Over the last 120 years, they have pored over every word of Luther's early theological works in their search for the origins and development of Luther's teaching on justification. Delving deeply into the niceties of Augustinian, Thomist, Scotist, and Ockhamist thought, they have given the impression that only a German theology professor could have produced such a formula, and only a German theology professor could understand it. More recent studies have however emphasised the extent to which Luther's theology was attuned to that of popular devotion in the early sixteenth century. We have already alluded to the importance of Luther's theology of the cross. This was in effect the transposition into academic terms of a phenomenon we can still see in the visual art and literature of late medieval northern Europe: its Christ-centredness, and more specifically its cross-centredness. The ubiquitous 'doom' – the depiction of Christ from a rainbow judging between the saved and the damned – was giving way to another image of Christ, his suffering in his human nature, with and for his fellow humans. This image, whether a physical one or one preached in sermons, carried powerful messages about God's love in Christ: that God was on the side of the suffering, not the secure; the poor, not the wealthy; the mean and the sinful, not the great and the good. Above all, Christ in judgement was being superseded in popular devotion by the Christ of sorrows, and the vengeful by a loving God. These are the accents that can be heard in the *Ninety-five Theses*, alongside the more obviously incendiary ones, and these are the themes explored in countless of Luther's early vernacular pamphlets. It is easy to forget that Luther's contemporary reputation in German-speaking lands was sealed by his German devotional works in the tradition of consolation literature, not his Latin polemical ones.[17] The starting point of Luther's justification doctrine, the notion of a God who calls the heavy-laden and takes their burden of sin, was certainly not foreign to the averagely serious Christian of his day.

Another feature of Luther's justification doctrine, which distinguishes it not only from Thomist and Ockhamist systems but also from Augustine's own thinking, is the question of certainty, and this was also an idea that ordinary Christians would have been all too familiar with. Late medieval theologians were agreed that salvation was an uncertain business. Echoing Augustine's belief that salvation was granted only to those given the grace of persevering to the end, it was understood that the Christian was like a traveller (*viator*) en route to the heavenly city, unsure of safety unless and until the destination was reached. Any other attitude would be to presume dangerously on God's grace. No-one could be sure that they would be saved. In practice, however, the Church offered as much certainty as it could. The most common use to which the printing press was put before the Reformation was the mass-production of indulgence letters – guarantees that the punishment for sins previously forgiven would not have to be suffered in Purgatory. In an important study, Moeller showed that, in the period from about 1480 to the outbreak of the Reformation, all the recorded measures of piety indicate unprecedented religious activity throughout the Holy Roman Empire. The building and enlargement of churches; the endowing of masses for the dead; the cult of the saints and the rise of new cults, such as that of St Anne; pilgrimages; and of course the sale of indulgences – all these increased massively. The sinister side of late medieval religion was also more in evidence, with an increase in the number of witch-burnings, anti-Jewish manifestations, and obsessive self-flagellation.[18]

How should we interpret this evidence? Does it suggest a near-frantic desire to win God's favour, stirred up by the uncertainty and fear generated by late medieval theology? That is what Moeller himself and other scholars such as Ozment concluded.[19] Or does it suggest that people had the religion they wanted and were taking full advantage of it? Rather than assume that late medieval religion was burdensome, perhaps we should regard it as serving a function and meeting a need. That is what other scholars, such as Duggan, have argued.[20] In either case, it was Luther's attack on the indulgence system that first touched a nerve with a wider public, and enabled his more positive utterances to gain a hearing.

the spread of the reformation in germany

Although in recent years the importance of Luther to the early development of the Reformation has been re-asserted, the importance of the lesser figures (the 'reformers in the wings') remains a lively aspect of Reformation studies.[21] This applies in particular to two of Luther's earliest allies, Karlstadt and Melanchthon. In earlier studies dismissed as a rather comical character, Karlstadt has come to be seen as a key player. It was he who seems to have revived interest in Augustine's anti-Pelagian writings at Wittenberg, precipitating the educational reforms there in which Luther was at first a junior partner. And it was he who gave intellectual weight to a more radical interpretation of the gospel that rejected the elements of clericalism, sacramentalism, ritualism, and, above all, collaboration with secular power that Luther famously retained. More constructively, he was the first to address in a practical way the social implications of Reformation theology. Having renounced his ecclesiastical and academic positions, Karlstadt's became an isolated voice, but a powerful one nonetheless, as the popularity of his writings show.[22]

Philip Melanchthon was one of the most controverted figures of the early German Reformation. Variously depicted in the earlier literature as either Lutheranism's saviour or its betrayer, he has continued to attract a deal of scholarship devoted to his multi-faceted contribution to German intellectual life in general and to the Reformation in particular.[23] The relationship between Melanchthon and Luther was an odd one. Luther was clearly in awe of his younger colleague and often deferred to his judgement, even in theological matters. In most cases, Melanchthon was also more staid and conservative than Luther: staid in his horror at Luther's sometimes foul and intemperate polemic; conservative in his understanding of justification, in which he was prepared to stress far more emphatically than Luther the necessity of good works. Whether these traits were part of Melanchthon's personality or a result of his thoroughly humanist training has been a matter of conjecture, but his humanism was clearly his defining characteristic. It gave him an international orientation and an international appeal that Luther lacked. Catholic contemporaries were of the opinion that he could have been a second Erasmus, but for his damning association with evangelical reform. His broad and largely irenic sympathies made him suspect with hard-line Lutherans, and even Luther thought him more open to Catholic overtures than he should have been at the 1530 Diet of Augsburg, when hammering out the basic evangelical statement of faith, the Augsburg Confession. His role in toning down that confession's article on justification secured his reputation as a traitor to the Lutheran cause; but equally it has been argued that without such concessions, and without Melanchthon's willingness to concede on non-decisive points (see below), Lutheranism as a religious force in mid-sixteenth century Germany could have expired.

Just as important as what Luther actually said and preached was the way in which his message was interpreted by other interest groups in the Empire, from peasants to city-dwellers to princes. We should perhaps think not so much of his message being misinterpreted by others, as of the whole Luther affair crystallising a number of pre-existing grievances and aspirations: though Luther's message was vital for determining the course and character of the German Reformation, he was as important for what he represented as for what he said. For village life, Blickle has argued, the early evangelical movement struck chords with the aspirations of these largely autonomous communities.[24] Over the preceding two centuries, they had developed codes governing not only their civil life, but also to an appreciable degree their religious life. At one level, therefore, Luther's 1523 pamphlet *That a Christian Assembly or Congregation (Gemeinde) Has the Right and Power to Judge all Things and to Call, Appoint, and Dismiss Teachers, Established and Proven by Scripture* was an expression of his theology of the priesthood of all believers. But when the villagers of Wendelstein in Franconia put it into practice, stating explicitly that the pastor was an appointee of the *Gemeinde* (as both church congregation and village commune), not a lord over them, it was at another level the continuation of a movement already under way.

Rural congregations were among the first to adopt the Reformation message and to adapt it to their own horizon of expectations. This included the translation of a theological message into a social and political revolution, the first 'liberation theology' of the modern era. German peasants worked under different systems. In some regions they were free, and could even accumulate some capital. In others, they were little better than serfs. The early sixteenth century saw an increase of exactions on these peasants and the private appropriation of land considered common since time immemorial, reducing still further the peasants' sources of food and fuel. The Peasants' War began in 1524. Starting in the Black Forest area in the south-west, with outbreaks of passive resistance, it spread north and east, and its character became much more violent. Armed mobs of peasants roamed the countryside, attacking the castles of nobles – and bishops' palaces, for the Church was at least as important a landowner as the nobility. Monks, nuns, and priests became the target of casual violence which drew on a tradition of anticlericalism. By 1525, the peasants of Switzerland and Austria had joined those of central and southern Germany. Popular revolution had come to the heart of the Holy Roman Empire. But it lasted little more than a year: once they were organised, it was a fairly straightforward job for the landed interests, with modern weapons and trained soldiers at their disposal, to defeat the peasant armies in open warfare.[25]

There has been some disagreement over the role that religious motivation played in the Peasants' War. Some interpretations, particularly those from a Marxist viewpoint emanating from the old East Germany, have tended to follow Engels in seeing it as a wholly secular event, the first in a series that would continue with the French and Russian revolutions: the peasants' aspirations may have been expressed in religious terms, but that was simply the vocabulary available to them at the time to establish a purely social agenda. More recent studies have, on the contrary, pointed to the adoption of the evangelical message as the keynote of the 1525 revolt which distinguished it from the numerous, less successful revolts of late fifteenth- and early sixteenth-century Germany.[26] Historians are still divided over the content and extent of 'anticlericalism', and the degree to which it represented a religious or a social protest.[27]

the princely reformation

Whatever its causes and nature, the Peasants' War was a failure for its peasant protagonists. One result of this failure was to drive many of the surviving rebels out of the mainstream of reform (which was believed to have sold its soul for princely patronage) into the tightly-knit, often apocalyptic, sects which comprised the so-called radical Reformation. Another result of the Peasants' War was to alert the German princes to the fact that the Reformation was not simply a quarrel between clerics, or a useful means of keeping the church in check, but was a double-edged sword. So from 1525 we see a new stage of reform, the so-called princely or magisterial Reformation.

Luther's famous *Address to the Christian Nobility of the German Nation* in 1520 had been an appeal to the new Emperor to initiate the reform of the church. In one sense, this was an expression of Luther's belief in the priesthood of all believers. All Christians are equally charged, by virtue of their baptism, to preach the gospel, and if the clergy fail in their duty then the laity are obliged to step in. But, in another sense, it was recognition of an existing duty to protect the church to which all Christian princes, kings, and governors were liable. Princely protection of and control over the church in German lands had over the years been extended in a piecemeal way at the expense of religious houses and even bishops, in just the same way as ecclesiastical privileges had been assumed by town councils and village communities.[28] This was not simply a power struggle with the church – though it often was that – but also a programme that stemmed from the sincere belief among princes that they had been divinely entrusted with providing for the spiritual as well as the physical welfare of their subjects.

If the aims of the evangelical reformers coincided with the aspirations of secular powers, it was not immediately obvious. Luther's *Address to the Christian Nobility* went unheeded by its addressees, and the early success of the Reformation was due entirely to popular, not noble, support. Even Luther's own prince, Frederick of Electoral Saxony, never backed him publicly. Indeed, the first princely intervention of the Reformation era was that of Frederick's relative, George of Ducal Saxony, who banned the printing and sale of Luther's works and commissioned others to write refutations.[29] Luther developed his famous, or infamous, doctrine of the 'two kingdoms' as a direct result of the duke's actions. Often seen as a blueprint for state intervention in church affairs and the start of a trail that leads to Lutheran complicity with the Third Reich, it actually establishes clear limits on a prince's right to interfere with his subjects' consciences, and in this sense can be seen as retrogressive. Luther himself did little to promote the princes' programme of ecclesiastical encroachment.

It was only after the shock of the Peasants' War that princes start to take control of the Reformation in their lands, so that it could proceed in an orderly way without risk of unrest. This is known as 'the princely Reformation'. While Luther himself reluctantly conceded that princes had the right to act as bishops in a situation where there were no real bishops, his followers promoted with enthusiasm the divine right of kings to control the church routinely. Finally, in the 1580s, the so-called 'Second Reformation' had the effect of radicalising the German Reformation along more Calvinistic lines, introducing a more thoroughgoing reform of Christian lives as well as Christian doctrine. The stage was set for what some historians regard as 'confessionalisation', the process by which denominational allegiance was used as an element in the building of new, independent states.[30] While princes had in the pre-Reformation period already assumed a degree of control over the church, now the 'social disciplining' associated with the new denominations gave them control over the beliefs of all their subjects.[31]

Several echoes of the progress of the princely Reformation in Germany can be heard in the Reformations that took place during the sixteenth century in Scandinavia, along with several discordant notes that make a comparison between the two areas a fruitful exercise. Of the three kingdoms, Denmark (with its dependent territories such as Schleswig-Holstein) was most similar to German lands, being urbanised to the same degree and with an active humanist network linking the Danish cities to developments elsewhere. Here evangelical ideas, assisted by close trading links with German cities, were circulating as early as 1520, and Malmø had become a fully reformed city within ten years. The spread of Lutheranism was held up by civil war, but with the victory of King Christian III in 1536, and his invitation to Luther's lieutenant Johannes Bugenhagen to reform the church, the formation of a Lutheran state was well under way.[32]

In the kingdoms of Norway and Sweden (which included Finland), the Reformations took different courses. Lutheranism was imposed upon Norway by the Danish crown in 1537, against a conservative opposition which remained strong. Norway was at this time largely rural, and evangelical ideas had not circulated to any extent. Only after seventy years of evangelical indoctrination and pastoral care could Norway be described as Lutheran by conviction as well as by royal decree. The situation in Sweden shows perhaps more parallels with the English than with the German Reformation. The riksdag of Västerås in 1527 secured the independence of the Swedish church from Rome, but it remained firmly Catholic. King Gustav Vasa exploited Swedish nationalist sentiment and flirted with evangelical ideas, but never fully embraced them: his interest seems to have been more in the Swedish church's funds than in its faith. Commitment to Lutheranism was strong amongst sections of the Stockholm bourgeoisie, but so weak in the countryside that Gustav faced a series of peasant rebellions in which religious change was cited as a grievance. The progress of Reformation was slow and in its nature highly conservative. It was only when evidence of a Jesuit conspiracy at court came to light that anti-popish sentiment was able to consolidate widespread support for Lutheranism, against royal opposition.[33]

the near-death of lutheranism

The takeover of the Reformation by the German princes, together with its arguable connection with state-building, gives the impression of steady and ineluctable progress. In fact, nothing was further from the truth in the case of German Lutheranism, which very nearly died as a movement shortly after Luther's own death. Its eventual survival and expansion has been described, by a hard-nosed historian not noted for his sympathy with Protestantism, as 'little less than a miracle' and as 'one of the most striking instances of success against what appear to be overwhelming odds to be found in history'.[34] Three forces were acting to bring about the destruction of Lutheranism. First, there were the Catholic armies of the Emperor Charles V and his allies, which defeated the Protestant

'Schmalkaldic' league just months after Luther's death. Without princely patronage, there was the distinct possibility that Lutheranism might go the way of Anabaptism, or else be absorbed back into Catholicism. The second force was the Swiss Reformation, and more particularly Genevan Calvinism, which in 1549 had allied with Zurich through the Zurich agreement (*Consensus Tigurinus*). The Reformed movement had for a long time influenced the cities of south-west Germany. It presented itself as a more thorough form of Protestantism than evangelicalism, and might have appealed on those grounds.[35]

The third force working to crush Lutheranism after Luther's death, and arguably the most potent, was Lutheranism itself. Luther had left very little in the way of balanced, thoughtful, and considered opinions on any theological issue. His words were more often forged in the heat of controversy, tendentious and over-emphatic, and even misleading if abstracted from the specific context for which they were intended. His opponents were quick to do exactly this, and delighted in pointing out his apparent self-contradictions. Among his followers, this facilitated the growth of parties which could each appeal to his own words in support of their beliefs. It happened even during Luther's life, so that his most heated exchanges were not with the Romanists, but with people who claimed to be acting in accordance with his own principles. But after his death, these disputes proliferated and became even more serious: did Luther allow some place for good works in salvation, or none at all? Was he accommodating towards ecclesiastical practices which could not be justified from Scripture, or did he take a hard line? These and many other differences could not be resolved by perusing his writings, which often seemed contradictory and inconclusive; so they persisted and grew, dividing theologians, universities, and even territories into different camps. By 1560, it was likely that Lutheranism would simply disappear under the pressure of these destructive forces, internal and external.[36]

Following the military defeat of the Protestant princes at the battle of Mühlberg in 1547, the Emperor Charles V imposed a broadly Catholic settlement, the Augsburg Interim (May 1548), which would regulate religious matters in the Empire until a Council could settle the questions permanently. (The Council of Trent had begun its long, drawn-out course in 1545.) The Interim allowed clerical marriage and communion in both kinds, but obliged Protestant territories to re-adopt a large number of Catholic practices and ceremonies, including direction by Catholic bishops, the sevenfold sacramental system, belief in transubstantiation, the cult of the saints, and many minor rites and ceremonies.[37]

There was a good deal of resistance to this new agreement. In southern Germany, where the Emperor was stronger, he enforced it by banishment and execution on a large scale – some four hundred were killed; but in the north he could not enforce it. Nevertheless, in Saxony, there was a move to implement it by Moritz, duke of Saxony. Although he was a Protestant, he had changed sides during the war and as a reward had received from the Emperor the confiscated lands of Electoral Saxony. He decided to implement the Interim through an interim of his own, the Leipzig Interim of 1548.[38] Some senior Wittenberg theologians, such as Nicholas Amsdorf and Matthias Flacius Illyricus, left Saxony rather than submit to it. But Melanchthon supported it. He based his decision on a distinction he made between the essentials (he used the Greek word *diaphora*) and the non-essentials (*adiaphora*) of the faith. External rites such as prayers for the dead or the canonical hours are not concerned with the heart of the gospel message, and so can be tolerated if need be. To argue the opposite, that external matters permit of *no* latitude whatever, is to set up a new legalism and a new popery.[39]

Melanchthon's readiness to compromise was anathema to those hard-line Lutherans who had left Wittenberg, such as Flacius Illyricus. Flacius's position was essentially twofold: first, that it was dangerous to make concessions in a time of persecution – all ceremonies cease to be *adiaphora* (no matter how unimportant they may be) when they are made compulsory; secondly, that ceremonies are never neutral, but embody the beliefs that underlie them – the sacrificial character of the Catholic Mass, for instance, is expressed in the sacrificial imagery of its ceremonial, and one cannot deny one while retaining the other.[40]

The stand-off between Melanchthon and Flacius was eventually overtaken by events. In 1552, Duke Moritz finally decided to assert his independence of the Emperor, and the new treaty of Passau made the Leipzig Interim a dead letter. The treaty was confirmed by the Diet of Augsburg in 1555, which concluded a religious settlement which lasted until 1648. But although the adiaphoristic controversy was now at an end, Melanchthon came under increasing suspicion from the hard-line theologians (now removed to the university of Jena), and northern Lutheranism was split between the Philippists and the Gnesio-Lutherans on a number of different matters, notably predestination and the real presence of Christ in the Eucharist. (The term 'Gnesio-Lutheran', from the Greek word for 'real' or 'genuine', was first introduced in the eighteenth century but is still used by scholars today for convenience. Its use carries no suggestion of partiality.) Melanchthon was suspected of toning down both doctrines; in the first favouring a larger place for the exercise of free will, along lines favoured by Erasmus in his debate with Luther, and in the second preferring the Swiss idea that Christ's body was truly present in the Eucharist only to believers. The Gnesio-Lutherans bitterly opposed Melanchthon and stuck to what they regarded as Luther's own view on both points. A favourite nickname for the Philippists was 'the crypto-Calvinists', and a favourite slogan was 'Better to be Catholic than Calvinist'. It was against this background that Shakespeare was able to speak of 'spleeny Lutherans'; and this certainly represents the low point of Lutheranism in Germany.

The first indications of a rapprochement between the Philippists and Gnesio-Lutherans came in 1567 when the chancellor of the University of Tübingen, Jakob Andreae, drew up five articles which he thought the two sides could agree on. They achieved some support, but it was not until the 1570s that the political climate was right for another attempt. In 1574–75, Andreae and Martin Chemnitz drew up a list of eleven articles known as the Swabian Concord, which created peace in northern Saxony. These were enlarged into the Formula of Torgau (1576), and finally were further enlarged into the Formula of Concord of 1577, published as the *Book of Concord* in 1580.[41] The Formula of Concord was not universally recognised (Danish Lutherans, for example, refused it and still held to the Augsburg Confession of 1530), but it did eventually gain majority assent in Germany. It gave Lutheranism an identity which was comprehensive and unambiguous, and it marked the end of the danger that Lutheranism might have been swallowed up by an intellectually stronger and more satisfying Calvinism. Thanks to the Formula of Concord, Lutherans remain to this day the largest denomination in the unified Germany (though they are geographically concentrated in the north and east), despite the massive advances made by Calvinism and Catholicism.[42]

The near-death experience of the Lutheran church had a profound effect on the nature of Lutheranism. Lutheran theology became a strict orthodoxy, systematised to an extreme and regulated by the disciplines of Aristotelian reason and logic, which Luther had rejected and which Melanchthon had reintroduced. Lutheranism embraced neo-scholasticism with as much enthusiasm as did Reformed theology and Roman Catholicism. The memory of Luther himself suffered a double fate. On the one hand, his words became used as a quarry for proof-texts, as if they were an authoritative commentary on Scripture – the very thing Luther himself had opposed in the excessive veneration paid to the medieval schoolmen in the theology of his younger days. On the other hand, the mystical and poetic elements of Luther's theology were driven underground, becoming part of the tradition of Lutheran pietism, but separate from academic theology.

success and failure of the german reformation

The debates between Lutheran theologians were one thing; but how successful were Protestant ministers, and the princely and urban authorities who supported them, at indoctrinating the populace as a whole? This has been a live question for historians since the publication of Gerald Strauss's *Luther's House of Learning* in 1978. Indoctrination had clearly been the aim of Luther and his followers. At first, Luther had believed that, once the Word was allowed to go forth untrammelled, people would be taught directly by God; but early setbacks with Karlstadt and the peasants convinced him of the need to eradicate centuries of ignorance and false, papal teaching by a programme of education. Central to this was the establishment of schools, using the resources of dissolved chantries and other religious foundations no longer needed, to create a cadre of learned ministers, who in turn would bring knowledge and understanding of the gospel to their parishioners. The first visitation of the Saxon church in 1528 revealed a parlous situation, which Luther summarised in his preface to his *Small Catechism* of 1529: 'the common people, especially in the villages, know absolutely nothing about Christian doctrine, and unfortunately many pastors are practically unfit and incompetent to teach ... They live just like animals and unreasoning sows.' However, later visitations throughout Protestant Germany show that the situation did not improve for the generality of the population, despite an increasingly well-trained clergy and the provision of catechisms and other learning aids for the laity. For example, in the duchy of Grubenhagen, where the Reformation had been introduced in 1532, the visitations of 1579, 1580, and 1610 show the same story repeated: few adults could remember their catechism.[43]

Of course, the picture was not this bad everywhere. In larger cities, the schoolchildren showed that they had memorised and understood all that was expected of them. And we should remember that the visitation panels might have had reasons for exaggerating the poor state of religious knowledge; perhaps because they wished to prise additional funding from the princes or city authorities.[44] But the overall picture presented by Strauss has generally been confirmed by subsequent studies. Why was this?

The visitation records show that parishioners were not only ignorant of the catechism, but to a great extent absent altogether from church. Even where the pastor was knowledgeable and motivated (as was increasingly the case after the first Saxon visitation), he could not teach, or reinforce what was already taught, if his flock never came to lessons or sermons. The successive returns for Grubenhagen, for instance, show that individuals who knew their catechisms as children had forgotten them by the time they grew up. The problem was not so much that they were unteachable, rather that they were irreligious.

The visitations normally contain a litany of what parishioners spent their time doing instead of going to church: drinking, gambling, fornicating, and so on. Witchcraft, sorcery, and soothsaying figured frequently in these lists, and one explanation for the failure of Lutheranism to capture hearts and minds is its concentration on the abstract Word at the expense of the tangible and immediate helps that the medieval church tolerated, or even provided. In other words, Protestantism had 'disenchanted' the world, by denying religious power to sacramentals and other artefacts.[45] The story now appears to be more complicated than this, in that popular religion in Protestant lands continued to be 'enchanted': the only difference was that the charms used were no longer the sacramentals of the Catholic church (salt, holy water, blessed bread), but words from Scripture used as powerful incantations and 'in effect, a Protestant form of magic'.[46] This reminds us that Lutheran visitation reports, depressing as their compilers found them, were evidence of no less success than the comparable visitation reports of Reformed or Catholic authorities. The failure of the German Reformation cannot be seen in isolation from the failure of all denominations to bring their adherents up to the required standard of Christian discipleship.

further reading

Good surveys of the subject include C. Scott Dixon, *The Reformation in Germany* (Oxford, 2002) and R. W. Scribner, *The German Reformation* (Basingstoke, 1986). C. Scott Dixon (ed.), *The German Reformation: The Essential Readings* (Oxford, 1999) is an invaluable introduction to the debates on the subject. See also Eric Lund (ed.), *Documents from the History of Lutheranism, 1517–1750* (Minneapolis, 2002).

On Luther, see R. Marius, *Martin Luther: the Christian Between God and Death* (Cambridge, MA, 1999); Martin Brecht, *Martin Luther: His Road to Reformation, 1483–1521* (Philadelphia, 1985); Mark U. Edwards, *Printing, Propaganda, and Martin Luther* (Berkeley, 1994).

On other leading figures, see the essays in Carter Lindberg (ed.), *The Reformation Theologians* (Oxford, 2002) and in David Bagchi and David

C. Steinmetz (eds), *The Cambridge Companion to Reformation Theology* (Cambridge, 2004); also David C. Steinmetz, *Reformers in the Wings: From Geiler von Keysersberg to Theodore Beza*, 2nd edn (Oxford, 2001); Ronald J. Sider, *Andreas Bodenstein von Karlstadt* (Leiden, 1974); Sachiko Kusukawa, *The Transformation of Natural Philosophy: the Case of Philip Melanchthon* (Cambridge, 1995).

On late medieval religion in Germany, see Gerald Strauss (ed.), *Pre-Reformation Germany* (London, 1972); Steven E. Ozment, *The Reformation in the Cities: the Appeal of Protestantism to Sixteenth-Century Germany and Switzerland* (London, 1975); Lawrence G. Duggan, 'Fear and Confession on the Eve of the Reformation', *Archiv für Reformationsgeschichte*, 75 (1984), 153–75.

On the debates over the pre-1525 period, see especially Susan Karant-Nunn, 'What was Preached in German Cities in the Early Years of the Reformation? *Wildwuchs* versus Lutheran Unity', in Phillip N. Bebb and Sherrin Marshall (eds), *The Process of Change in Early Modern Europe* (Athens, OH, 1988); and Peter A. Dykema and Heiko A. Oberman (eds), *Anticlericalism in Late Medieval and Early Modern Europe* (Leiden, 1993).

On the rural Reformation and the Peasants' War, see Peter Blickle, *Communal Reformation: the Quest for Salvation in Sixteenth-Century Germany* (London, 1992); Tom Scott and Bob Scribner (eds), *The German Peasants' War: a History in Documents* (Atlantic Highlands, NJ, and London, 1991); Bob Scribner and Gerhard Benecke (eds), *The German Peasants' War of 1525: New Viewpoints* (London, 1979).

On Scandinavia: Ole Peter Grell (ed.), *The Scandinavian Reformation* (Cambridge, 1995).

On Lutheranism after Luther, see especially the work of Robert Kolb: *Confessing the Faith: Reformers Define the Church, 1530–1580* (St Louis, MO, 1991) and *Luther's Heirs Define His Legacy: Studies on Lutheran Confessionalization* (Aldershot, 1996). On the debates about 'success', see Geoffrey Parker, 'Success and Failure During the First Century of Reformation', *Past and Present*, 139 (1992), 43–82.

notes

1. Erwin Iserloh, *The Theses Were Not Posted* (London, 1966). For evidence for a mid-November posting, see Martin Brecht, *Martin Luther: His Road to Reformation, 1483–1521* (Philadelphia, 1985).
2. Franz Lau and Ernst Bizer, *A History of the Reformation in Germany to 1555* (London, 1969).
3. Bernd Moeller, 'Was wurde in der Frühzeit der Reformation in den deutschen Städten gepredigt?', *Archiv für Reformationsgeschichte*, 75 (1984), 176–93. This

seminal article is helpfully translated in C. Scott Dixon (ed.), *The German Reformation: the Essential Readings* (Oxford, 1999).

4. On the confessionalisation thesis, see below, n. 30 and pp. 95, 204–5, 223–4.
5. Susan Karant-Nunn, 'What was Preached in German Cities in the Early Years of the Reformation? *Wildwuchs* versus Lutheran Unity', in Phillip N. Bebb and Sherrin Marshall (eds), *The Process of Change in Early Modern Europe: Essays in Honor of Miriam Usher Chrisman* (Athens, OH, 1988); idem, 'Clerical Anticlericalism in the Early German Reformation: an Oxymoron?', in Peter A. Dykema and Heiko A. Oberman (eds), *Anticlericalism in Late Medieval and Early Modern Europe* (Leiden, 1993), esp. pp. 521, 533. The case for the essential heterogeneity of Reformation pamphlets as such is stated by R. Po-chia Hsia, 'Anticlericalism in German Reformation Pamphlets: a Response' in Dykema and Oberman (eds), *Anticlericalism.*
6. Hans-Joachim Köhler, 'Erste Schritte zu einem Meinungsprofil der frühen Reformationszeit', in Volker Press and Dieter Stievermann (eds), *Martin Luther. Probleme seiner Seit, Spätmittelalter und Frühe Neuzeit* (Stuttgart, 1986), p. 259.
7. Thomas Hohenberger, *Lutherische Rechtfertigungslehre in den reformatorischen Flugschriften der Jahre 1521–1522* (Tübingen, 1996), p. 396.
8. Peter Matheson (ed.), *Argula von Grumbach: a Woman's Voice in the Reformation* (Edinburgh, 1995).
9. Karant-Nunn, '*Wildwuchs*', p. 92.
10. Berndt Hamm, 'Was ist reformatorische Rechtfertigungslehre?', *Zeitschrift für Theologie und Kirche*, 83 (1986), 1–38; translated in Dixon (ed.), *German Reformation.*
11. A good example can be found in James Atkinson's extensive introductions in *Luther: Early Theological Works*, Library of Christian Classics 16 (London, 1962).
12. Etienne Gilson, *History of Christian Philosophy in the Middle Ages* (London, 1980); Johann Huizinga, *The Autumn of the Middle Ages* (Chicago, 1996).
13. John W. O'Malley, *Trent And All That: Renaming Catholicism in the Early Modern Era* (Cambridge, 2000). See also below, Chapter 9.
14. See especially Heiko A. Oberman, *The Harvest of Medieval Theology* (Cambridge, MA, 1963); David C. Steinmetz, *Luther in Context* (Bloomington, IN, 1986), and idem, *Calvin in Context* (New York, 1995), among many other works by these writers.
15. Denis Janz, *Luther and Thomas Aquinas: the Angelic Doctor in the Thought of the Reformer* (Stuttgart, 1989).
16. David Bagchi, 'Sic et Non: Luther and the Scholastic Traditions', in Carl R. Trueman and R. Scott Clark (eds), *Protestant Scholasticism: Essays in Reassessment* (Carlisle, 1999).
17. Mark U. Edwards, *Printing, Propaganda, and Martin Luther* (Berkeley, 1994).
18. For a convenient English translation, see Bernd Moeller, 'Religious Life in Germany on the Eve of the Reformation', in Gerald Strauss (ed.), *Pre-Reformation Germany* (London, 1972).
19. Steven E. Ozment, *The Reformation in the Cities: the Appeal of Protestantism to Sixteenth-Century Germany and Switzerland* (London, 1975).

20. Lawrence G. Duggan, 'Fear and Confession on the Eve of the Reformation', *Archiv für Reformationsgeschichte*, 75 (1984), 153–75.
21. For example, David C. Steinmetz, *Reformers in the Wings: From Geiler von Keysersberg to Theodore Beza*, 2nd edn (Oxford, 2001).
22. Ronald J. Sider, *Andreas Bodenstein von Karlstadt: the Development of his Thought, 1517–1525* (Leiden, 1974); Alejandro Zorzin, *Karlstadt als Flugschriftenautor* (Göttingen, 1990); Hans-Peter Hasse, *Karlstadt und Tauler: Untersuchungen zur Kreuzestheologie* (Gütersloh, 1993).
23. See esp. Ralph Keen (ed.), *A Melanchthon Reader* (New York, 1988); Sachiko Kusukawa, *The Transformation of Natural Philosophy: the Case of Philip Melanchthon* (Cambridge, 1995); idem, 'Melanchthon', in David Bagchi and David C. Steinmetz (eds), *The Cambridge Companion to Reformation Theology* (Cambridge, 2004); and Timothy Wengert, *Human Freedom, Christian Righteousness: Philip Melanchthon's Exegetical Dispute with Erasmus of Rotterdam* (Oxford, 1998).
24. Peter Blickle, *Communal Reformation: the Quest for Salvation in Sixteenth-Century Germany* (London, 1992).
25. For a useful account of the course of the Peasants' War, and for a wealth of contemporary documentation, see Tom Scott and Bob Scribner (eds), *The German Peasants' War: a History in Documents* (Atlantic Highlands, NJ, and London, 1991).
26. See Bob Scribner and Gerhard Benecke (eds), *The German Peasants' War of 1525: New Viewpoints* (London, 1979), for an introductory essay on the basic historiographical questions on this issue during the existence of the German Democratic Republic (East Germany). For what might be termed a typical GDR position, see the article by Max Steinmetz reprinted there; for a typical 'Western' approach of the time, emphasising the importance of political and religious factors over social and economic ones, see the article by Heiko Oberman, 'The Gospel of Social Unrest': also in his *The Dawn of the Reformation: Essays in Late Medieval and Early Reformation Thought* (Edinburgh, 1986).
27. Hans-Jürgen Goertz, *Pfaffenhass und gross Geschrei: die reformatorischen Bewegungen in Deutschland, 1517–1529* (Munich, 1987); Dykema and Oberman (eds), *Anticlericalism*.
28. Peter Dykema, 'The Reforms of Count Eberhard of Württemberg', in Beat Kümin (ed.), *Reformations Old and New: Essays on the Socio-Economic Impact of Religious Change, c. 1470–1630* (Aldershot, 1996).
29. Edwards, *Printing, Propaganda, and Martin Luther*.
30. For an early formulation of the confessionalisation thesis, see Wolfgang Reinhard, 'Pressures Towards Confessionalization? Prolegomena to a Theory of the Confessional Age', in Dixon (ed.), *German Reformation*.
31. For an exploration of this concept, see R. Po-chia Hsia, *Social Discipline in the Reformation: Central Europe, 1550–1750* (London, 1989).
32. The most important recent study of this topic is Ole Peter Grell (ed.), *The Scandinavian Reformation: From Evangelical Movement to Institutionalisation of Reform* (Cambridge, 1995). A useful digest of scholarship and issues up to 1990 is provided by Trygve R. Skarsten in William S. Maltby (ed.), *Reformation Europe: A Guide to Research II*, Reformation Guides to Research, vol. 3 (St Louis, MO, 1992), pp. 215–34.

33. Ole Peter Grell, 'Scandinavia', in Andrew Pettegree (ed.), *The Reformation World* (London, 2000); for the comparison with England, Alec Ryrie, 'The Strange Death of Lutheran England', *Journal of Ecclesiastical History*, 53 (2002), 64–92.
34. T. M. Parker, 'Protestantism and Confessional Strife', in R. B. Wernham (ed.), *The Counter-Reformation and the Price Revolution, 1559–1660* (Cambridge, 1968), pp. 83, 89.
35. On this, see below, Chapter 3.
36. For a convenient account of the internecine quarrels, together with a useful selection of documents in English, see Eric Lund (ed.), *Documents from the History of Lutheranism, 1517–1750* (Minneapolis, 2002). Robert Kolb has done much to elucidate the thorny theological issues: *Confessing the Faith: Reformers Define the Church, 1530–1580* (St Louis, MO, 1991); idem, *Luther's Heirs Define His Legacy: Studies on Lutheran Confessionalization* (Aldershot, 1996); and idem, 'Confessional Lutheran Theology', in Bagchi and Steinmetz (eds), *The Cambridge Companion to Reformation Theology*.
37. For the abbreviated English text, see Lund (ed.), *Documents*, pp. 162–4.
38. *Ibid.*, pp. 165f.
39. *Ibid.*, p. 187.
40. On Flacius, see Oliver K. Olson's chapter in Carter Lindberg (ed.), *The Reformation Theologians: an Introduction to Theology in the Early Modern Period* (Oxford, 2002).
41. On the genesis of the Book of Concord, see Robert Kolb and James A. Nestingen (eds), *Sources and Contexts of the Book of Concord* (Minneapolis, 2001).
42. Eric W. Gritsch, *Fortress Introduction to Lutheranism* (Minneapolis, 1994).
43. Gerald Strauss, 'Success and Failure in the German Reformation', *Past and Present*, 67 (1975), 30–63; reprinted in Dixon (ed.), *German Reformation*.
44. See Geoffrey Parker, 'Success and Failure During the First Century of Reformation', *Past and Present*, 139 (1992), 43–82 for a summary of objections.
45. Max Weber first explicated the theory of 'the disenchantment of the world' by a combination of classically-inspired rationalism and the Christian antipathy to magic. See his *General Economic History* (1961), p. 265, discussed in Keith Thomas, *Religion and the Decline of Magic* (London, 1971).
46. Robert W. Scribner, 'The Reformation, Popular Magic, and the "Disenchantment of the World"', *Journal of Interdisciplinary History*, 23 (1993), 475–94; reprinted in Dixon (ed.), *German Reformation*, p. 275.

2
central and eastern europe

graeme murdock

The Reformation in Central and Eastern Europe formed a vital element of efforts to reform the Latin Church during the early modern period. Movements in favour of reform in the region long pre-dated the emergence of Evangelical churches in the German lands during the 1520s. Within the enormous territory covered by the Polish-Lithuanian Commonwealth, the Austrian provinces, the lands of the Bohemian crown and the medieval Hungarian kingdom, religious divisions already existed by 1520. This space was shared between Roman Catholics, Eastern Orthodox, Utraquists, Bohemian Brethren and Jews. Catholic social and political dominance was subject to new and substantial challenges during the sixteenth century from Evangelical (Lutheran), Reformed, Anabaptist and anti-Trinitarian preachers who attracted support from nobles and in towns. Compromises were reached over religious rights between monarchs and their noble estates during the middle decades of the century in many territories, and legal protection was offered to alternative confessions. However, these hard-won guarantees of religious liberty were soon almost everywhere undermined by Catholic rulers. By 1620, the Protestant character of many parts of Central and Eastern Europe had been terminally curtailed by the recovery of Catholic institutions and by the force of Catholic royal arms. Although the range of other religious confessions played an increasingly marginal part in Polish, Lithuanian, Austrian, Bohemian, Moravian, Hungarian and Transylvanian society, it was not, however, as marginal as Catholic propagandists cared to suggest.

Although the nature and impact of the Reformation differed considerably across this vast region, it is possible to outline an approximate chronology of the advance and retreat of the challenge to Catholic dominance within Central European societies. During the early and middle decades of the sixteenth century a range of reformers questioned the veracity of Catholic doctrine and the value of Catholic religious practices. From the 1550s a variety of confessions attracted sufficient support to press for legal recognition alongside the Catholic church, and Catholic rulers were compelled to reach compromises with estates of divided religious loyalties. These religious settlements gave way at the beginning of the seventeenth century with the re-assertion of Catholic supremacy after violent confrontations within the Habsburg monarchy. This chapter will consider how historians have understood these key features of the success and failure of the Reformation in Central and Eastern Europe. It will begin with a survey of the changing approaches which have been adopted by historians who have studied the Reformation in the region, before focusing on current understanding of how ideas in favour of reform spread and why they were so widely supported. It will then review the reasons now offered by historians as to why a range of religions received legal recognition within Central and Eastern European states, and why this was achieved without any major outbreaks of violence between those of different faiths. It will finally consider study of the revival of Catholic fortunes in the region as well as the nature and extent of Protestant resistance and survival. Rather than attempt to offer comprehensive coverage of the whole region on each issue, particular debates will be highlighted which exemplify these broader themes.

historiographical background

Historians who study the Reformation in Central and Eastern Europe, no less than any other part of the Continent, face the responsibility of writing about religious leaders who continue to inspire loyalty among many to the present day. Historians also chart the emergence of churches and religious institutions whose history is of much more than academic concern to current members, who often strongly identify with co-religionists from previous generations. The force of such confessional loyalty heavily coloured early histories of the Reformation. The first studies of the monarchs and activists who promoted or resisted waves of reform in the Church were marked by vicious attacks and passionate defences of their views and actions. Even up to the Second World War most histories of church life in Central Europe were written by historians drawn from within the religious community which they studied. However, the number of overtly confessional histories which simply set out to justify one particular strand of reform steadily decreased over time. Indeed, many of the works written by church historians during the late nineteenth and early twentieth centuries have stood the test of time very well.[1] Historians who are themselves Catholic, Lutheran, Reformed, Anabaptist or Unitarian have certainly grown more and more sensitive to the charge of having allowed private sympathies to affect their analysis of sixteenth-century religious life. In recent decades more historians have striven to provide non-confessional histories of the Reformation. In addition, an increasing number of historians have declared themselves personally neutral in Christian squabbles, but any claim that a lack of personal religious convictions inevitably better places an author to write about the history of Christianity seems at best rather dubious.

Debates about the importance of reform in Central and Eastern Europe, as elsewhere, have also been significantly affected by national historiographical traditions. Most Reformation historians have chosen to write about the religious history of their own states or ethno-linguistic groups. While in many ways this is perfectly understandable, it has often resulted in isolated debates about the nature and impact of religious reform. Accounts of the Reformation have also been structured according to the shifting borders of the region and, for example, both Slovakia and Slovenia have recently been provided with their own histories of the Reformation.[2] As well as existing within the context of national histories, studies of the Reformation have also been affected by nationalist history-writing. The success or failure of reform has been ascribed a place in the development or lack of development of a particular nation, and sixteenth-century religious heroes and villains have been treated as patriots or traitors. This has been particularly important within histories of Polish and Austrian Catholicism, but the Hussite movement in Bohemia and the Reformed church in Hungary have also both been given the mantle of national religions. However, historians have steadily moved away from viewing the acceptance of reform as either an aberrant period from a nation's natural Catholicism, or as a key hallmark of a nation's nascent identity. Some tangled debates which connect religious affiliation and national identity still persist, particularly in disputed territories within the region. For example, sharply divided opinions have continued between Romanian and Hungarian historians about the impact of the Reformation among the Orthodox community in Transylvania, which will be discussed below.

During the second half of the twentieth century, studies of the Reformation in Central Europe were also profoundly affected by the Communist regimes which came to power across much of the region. Communist authorities implemented measures which limited the range of historical research which was able to be published. Authors were also compelled at the very least to provide some sort of Marxist gloss to their work, whatever their private opinions. It is therefore difficult to know whether it is more appropriate to describe historians of the post-war generations in Central Europe as Marxist historians of the Reformation or rather as Reformation historians living under Communist regimes. However, the imprint of Marxist ideas certainly challenged the dominance of confessional and national narratives about the history of religion in the region. The best of this writing provided important assessments about the significance of social and economic problems within sixteenth-century feudal monarchies in explaining the spread of ideas about religious reform.[3] The focus of attention also turned from the role of monarchs and clergy in the Reformation towards study of changes in the practice of popular religion. However, Marxist historiography has continued to struggle to provide credible accounts of the social and political significance of religious ideas within sixteenth-century communities.

From the other side of the Iron Curtain, Western historians were presented with almost insurmountable difficulties in conducting research within Communist states after the Second World War. This led to the engagement of Eastern European countries with Reformation ideas being increasingly overlooked in the work of Western historians. As Western scholars neglected the importance of the Reformation in somewhere they called Eastern Europe, some historians and intellectuals within the region were turning to the Reformation, among other historical processes, as evidence for the existence of somewhere they preferred to call Central Europe. Their attempts to break down the barriers between Central Europe and the West challenged long-standing Western constructions of Eastern European history and identity. This image of Eastern Europe had in part been developed by travellers from England and France during the eighteenth century who conceived of the eastern lands of their continent as a complementary backward half of Europe against which the Enlightened values of the West could be defined.[4]

The power of such ideas about a wild, barbarian Eastern Europe continued to be felt in the twentieth century. At the end of the Second World War, the classical education of many Western politicians seemed to inform their determination to save an imagined Greece for Western civilisation while loyal allies such as Poland were more readily abandoned to Soviet domination. In response to this perception of Western attitudes, advocates of a Central European space claimed that their region was spiritually and historically of the West and had participated in the key historic processes, such as the Reformation, which characterised the development of modern Europe. How, therefore, could the land of Jan Hus be relegated as belonging to some amorphous Eastern Bloc?[5] The history of Central Europe's Reformation thus became one element within wider moral claims for recognition of the region's essential European identity, and more practically has even been deployed to bolster recent efforts to gain entry to key European institutions. It should also be noted that these arguments about Central Europe implied that lands which failed to experience movements such as the Reformation ought to be definitively and conclusively labelled as Eastern Europe. Partly as a result, historians in Romania and elsewhere have adopted the term East-Central Europe to describe their perception of their countries' position between Russia and Central Europe.

The history of the Reformation in a region described here as Central and Eastern Europe has therefore been written within confessional, nationalist, Marxist and regional perspectives. The current generation of Reformation historians from Central and Eastern Europe have been joined by an increasing number of writers drawn from outside the region. Recent accounts of the progress of religious reform have highlighted the significance of patterns of intellectual communication across the Continent, and have looked for both common features and differences between the impact of the Reformation in Central and Western Europe. These efforts still face a degree of resistance from some Western historians who have failed to appreciate the altered picture of intellectual and religious life during the sixteenth century which inclusion of Central and Eastern European lands demands. For those historians prepared to engage at least in part with the region's cultural and religious history, a huge range of opportunities exists to move forward our understanding of the Reformation period in fundamental ways, and a survey of the current state of research is outlined below.

the spread of reform

What factors explain the spread of a range of reform ideas across Central and Eastern Europe? The region certainly did not produce any outstanding, innovative theologians during the early sixteenth century. Aside from the native Hussite tradition and some anti-Trinitarian writers in Hungary and Poland, movements of reform were rather inspired by the spread, translation and synthesis of confessions, catechisms, tracts and sermons by German-speaking authors, and above all by Martin Luther, Philip Melanchthon and Heinrich Bullinger. German-speaking towns from the Baltic coast to the eastern Carpathians were among the first to be impacted by ideas about reform. Luther's ideas spread quickly from Saxony to neighbouring lands in Lusatia and Silesia, and attracted support among German-speaking Catholics in northern and western Bohemia. The Evangelical cause also gained support among many Utraquist clergy, who had received their education at Prague or in German universities. There were clear similarities between the Utraquist church (which already dominated Bohemian society) and Lutheran ideas. There was a common concern to reform the practice of the sacraments, agreement about the importance of the Bible, about the centrality of Christ in the scheme of salvation, and about the need for moral and spiritual renewal.[6] A fresh Evangelical spur to reform also spread to Moravia during the 1520s, with German-speaking Catholics and Utraquists again among the first to support Evangelical preachers. Alongside the growth in support for Lutheranism, some Moravian towns and nobles also offered protection to Anabaptist refugees after the Peasants' War. In the Austrian provinces, reform-minded preachers were influenced by the humanist training available at Vienna and by strong connections with German universities. Support for the Evangelical cause spread from the 1520s both among the nobility and in towns. During the 1530s and 1540s, preaching which proclaimed the need for doctrinal, ritual and institutional reforms also reached German-speaking towns in Upper Hungary, Transylvania, Greater Poland, and Royal and Ducal Prussia.

In assessing the reasons which explain how ideas about religious reform were communicated across Central and Eastern Europe, historians have noted the importance of the presence of German-speakers throughout the region and of close commercial connections with the Empire. While important academic centres in the region such as the universities at Prague and Vienna were evidently significant, strong links with other universities in the Empire were also crucial in the education of reform-minded clergy. Historians have studied these patterns of clerical education, and considered the influence of particular academic centres, including the universities at Wittenberg, Basel and Heidelberg.[7] Historians have also examined the role of towns in Central Europe as centres of reform and communication. However, the limited size and lack of autonomy of many towns meant that they were less significant than in parts of Western Europe. Rather, it was the reception of reform movements among landed magnates and gentry which was to prove decisive in ensuring their success. This has focused attention on the importance of those preachers and chaplains who converted members of noble families. In addition, historians have noted the linkages between aristocrats who embraced reform, and the influence of magnates over lesser nobles. This seems to have been important for instance in Lithuania, where many followed the example set by the chancellor of the grand duchy, Mikołaj Radziwiłł.

It has certainly proved more straightforward to explain how the message of reform spread across Central Europe, than to assess why nobles and people in towns and the countryside seemed so ready to listen to reformers. The degree of popular enthusiasm for changes to sacramental and salvation theology is difficult to gauge. However, there was clearly a widespread perception of profound failure in both the spiritual and social sphere by the clergy and institutions of the Church. Reformers gained a ready reception for their desire to improve the standards of training for clergy so as to enable them to perform key functions of preaching, teaching and administering the sacraments. Reformers also offered a commitment to use the vernacular in worship, to improve standards of education among the laity, and to grant ordinary people greater access to the world of printed literature. There is also evidence of lay enthusiasm for the message of moral renewal, which in some places was tinged with apocalyptic expectation. This was particularly the case in Hungary, where the progress of Muslim arms and collapse of the Christian monarchy in the 1520s provided dramatic evidence of the failure of the spiritual powers of the Catholic church. The political environment for religious reform was certainly also important. Nobles who left the Catholic church may well have been personally convinced about the need for reform. However, it is clear that nobles who adopted a different religion from that of their monarch were also making a political demonstration which challenged royal power.

In assessing the impact of different reform initiatives, it has also been difficult for historians to explain why some communities who initially supported Evangelical preachers remained loyal to the insights of Luther and Melanchthon, while other communities in the region later changed to adopt subsequent waves of reform. The reasons behind this differing reception of reform among nobles and urban communities have been the subject of close analysis. In particular, it has long been noted that German-speaking towns often remained Lutheran while many Hungarian- and Polish-speaking towns and nobles did not. This attachment to Lutheranism among German-speakers in Central and Eastern Europe seemed to early historians of those communities to be a natural feature of their early modern history. Identifying German-speakers in the region with a distinctively German religion was used to bolster claims of close cultural and social connections with the German lands and to mark a degree of separation from the other peoples of Central Europe. More recently, however, it has been acknowledged that support among Polish- and Hungarian-speakers for the Reformed theology of Heinrich Bullinger can hardly be seen as a rejection of German intellectual influence.

Historians have therefore looked again at the nature of pre-Reformation piety in German-speaking communities in the region for suggestions as to why they so readily and conclusively adopted Evangelical styles of worship and piety.[8]

The spread of Reformed religion in Poland, Bohemia, Hungary and Transylvania certainly reflected some sort of second Reformation, during the middle decades of the sixteenth century, among communities which had already developed a taste for reform. The attraction of Reformed doctrine on the sacraments, Reformed anxiety to abandon all forms of idolatry, and a clear commitment to enforce rigorous standards of moral discipline were all significant. Heinrich Bullinger's 1566 *Second Helvetic Confession* proved influential in shaping the doctrine of Reformed churches across the region. For example, in 1567 the founding Debrecen synod of the Hungarian Reformed church accepted Bullinger's confession alongside a local *Confessio Catholica* by reformers Péter Méliusz Juhász and Gergely Szegedi.[9] However, Reformed churches were weakened during the 1560s by a split between Trinitarians and anti-Trinitarians. The emergence of anti-Trinitarian preaching, influenced by refugee Italian intellectuals including Giorgio Biandrata, emerged first in Poland and then spread to Hungary and Transylvania.[10]

While Western theologians, universities, and refugee intellectuals were therefore crucial agents of reform in Central Europe, the direction of influence did not only run from west to east. One prominent figure who influenced the direction of the Reformed cause across the Continent was the Pole, John a Lasco (Jan Łaski). Lasco worked in a number of key Western centres of Reformed religious life. During the early 1540s Lasco was invited to lead reform in the church at Emden in East Friesland, and he set up a consistory in the town to enforce moral discipline. Lasco then moved to lead the strangers' church in Edwardian London, before returning to Emden in 1553. Lasco's most significant contribution to the development of the Reformed tradition in the middle decades of the sixteenth century was the influential church order, or *Forma ac Ratio*, which he provided for the London strangers' community. Lasco was then appointed superintendent of the Reformed church in Poland from 1556, and he coordinated efforts involving Heinrich Bullinger to attempt to persuade the Polish king, Sigismond Augustus, to support reform.[11]

Reform was rejected at the Polish and Habsburg courts, and reform measures were also resisted by some nobles and communities despite the best efforts of Protestant activists. For example, during the 1540s the German-speaking, Saxon towns of Transylvania adopted Evangelical reforms under the direction of Johannes Honter at Kronstadt (Brassó). As early as 1544 these Evangelical towns then attempted to sponsor reform among Orthodox Romanian-speakers in neighbouring rural communities. A translation of Luther's *Shorter Catechism* was published in Romanian, and a further edition of this catechism was printed around 1560. The Evangelical magistrates of the Saxon towns clearly felt some responsibility for spreading reform to their Orthodox neighbours, and paid for the cost of translating and publishing these catechisms into Romanian. There is no surviving evidence about the impact of these texts, but they cannot be dismissed as an isolated or insignificant moment of interaction between the Protestant and Orthodox worlds. In addition, the Saxon magistrates introduced measures intended to enforce changes to the pattern of ritual and forms of worship in local Orthodox churches. Hungarian Reformed nobles and clergy also took up the cause of promoting reform among Transylvania's Romanians. Measures were proposed at the Transylvanian diet, Romanian schools were set up, and a separate bishopric was established for Romanian Reformed communities. These reform efforts were renewed during the early seventeenth century through institutional reforms, attempts to have Orthodox services conducted in the vernacular, and through the publication of Psalters and catechisms in Romanian.

These efforts to reform the Orthodox church in Transylvania, marked by institutional changes, the publication of vernacular texts and by attempts to promote education among Romanians, largely failed. Only a small number of Romanian communities in south-western Transylvania adopted Reformed religion. This failure has sparked a lively debate among historians, influenced by rival national perspectives. While Romanian historians have recognised the production of books in the vernacular as an important achievement in the development of Romanian printed literature, many writers have expressed suspicions that Hungarian- and German-speaking Protestants' efforts to reform Orthodoxy in fact marked an attack against the identity of Romanians in Transylvania. Romanian historians have stressed the profound differences between Orthodoxy and Protestantism, and praised the Orthodox community's resistance to the imposition of alien creeds. Meanwhile, Hungarian historians have described brave but doomed attempts by Protestants to spread literacy and education among backward Romanian peasants. They have tended to blame the recipients for the failure of reform rather than the message or its means of transmission. Recently some historians have tried to move this debate forward by looking again at the impact of Protestantism among Orthodox communities in Transylvania and in Moldavia, not to support rival national claims but to examine changing styles of piety and popular religiosity.[12]

toleration of confessional differences

By the middle decades of the sixteenth century religious life in Central and Eastern Europe had fractured, with the emergence of a range of different churches. However, the region was not to be afflicted by any major outbreaks of state-sponsored or communal religious violence. There were victims of anti-heresy laws, and some popular religious riots were directed against rival clergy and church buildings, including, for example, an assault on the Reformed church at Cracow in 1574.[13] However, laws were agreed across the region which offered rights of worship to Lutheran, Reformed and other churches, and there was a relatively peaceful pattern of coexistence between different confessional communities. Some historians have been drawn to praise a Central European spirit of reasonable compromise over religious differences, but the region was certainly not home to any modern sense of religious moderation or tolerance. Other writers have pointed out that some parts of Central Europe had prior experience of trying to deal with the political and practical problems raised by confessional divisions, which may have helped to prepare the ground for the compromises which were made during the mid-sixteenth century. In Bohemia, for example, monarchs had been required since 1485 to promise on their election to uphold the freedoms of both the Catholic and Utraquist churches. This arrangement between a Catholic monarchy and Utraquist nobles in Bohemia also highlights the importance of politics in the series of religious settlements reached during the mid-sixteenth century. Catholic monarchs were very reluctant to concede rights of free practice to other religions, but they were compelled to do so because of the power of their estates. The legal rights granted to Lutheran, Reformed and other churches were therefore only secured and upheld because of the support of sympathetic nobles. Aristocratic loyalty remained vital to Protestant churches, whose fate became linked with the political fortunes of estates across the region.

In Bohemia, although the Utraquist church was protected by law, the Bohemian Brethren (who inherited the legacy of the more radical Taborite group of Hussites) had not been offered any legal protection. Ferdinand I was also fully entitled to challenge the spread of Evangelical preaching in Bohemia. During the 1540s, Ferdinand ordered the expulsion of Lutheran preachers from within the Utraquist church and forced the Bohemian Brethren into exile in Poland and Moravia. However, in the wake of the Augsburg settlement of 1555, Bohemian nobles claimed the right to reform religion in their own lands and sheltered Lutheran preachers in their parishes. When the Habsburg court tried to revive Catholic interests in Bohemia, Protestant nobles pressed their clergy to agree to a common confession. In 1575 different groups of Evangelicals, reformist Utraquists and the Bohemian Brethren agreed to the *Confessio Bohemica*, which followed the version of the Augsburg Confession drawn up by Philip Melanchthon. Nobles then successfully pressed Maximilian II into granting them the right to worship according to this *Confessio Bohemica*. In Moravia, nobles had also asserted their right to reform religion on their lands in the 1550s, and many supported Lutheran, Bohemian Brethren and Anabaptist preachers.

This pattern of events in Bohemia and Moravia was followed elsewhere. In Poland, Austria and Royal Hungary noble patrons supported local reform initiatives and claimed the right to appoint reform-minded clergy. Estates then defended aristocratic privileges in religious affairs against any attempts by monarchs or the clergy hierarchy to enforce anti-heresy laws. In the face of any evidence of Catholic pressure to erode their assumed religious rights, nobles demanded legal guarantees of Protestant freedoms. In the Austrian provinces, the authorities were unable to prevent the spread of reform among many noble families. In 1568 the need for financial support from the estates forced Maximilian to accept the free practice of Lutheranism on nobles' estates in Lower Austria, and by 1576 Evangelical nobles were also granted formal rights to worship freely in Vienna and across Inner Austria. In Poland, Protestant churches and schools were also protected on the lands of sympathetic noble patrons, while towns were conceded the right to choose Evangelical preachers. In April 1570 the Evangelical, Reformed and Bohemian Brethren churches in Poland agreed to terms of mutual recognition and to meet in common synods through the Union of Sandomierz. In 1573 the estates won legal protection for the churches of this Union under the terms of the Warsaw Confederation. This Confederation recognised the rights of nobles in matters of religion and declared as its aim the preservation of peace between 'those who differ in religion and rite'. Meanwhile in Royal Hungary, nobles' rights of patronage over local churches simply went unchallenged for most of the sixteenth century, and only some royal free towns were granted formal rights to appoint Evangelical preachers.

The most interesting response to the political and social challenges posed by confessional division came in the Transylvanian principality. During the middle decades of the sixteenth century the Transylvanian diet of Hungarian nobles, Saxon magistrates and Szekler lords passed a series of laws which established the rights of three new religions to worship freely: the Lutheran, Reformed, and anti-Trinitarian churches.[14] In April 1558 Lutherans across the principality were given equal rights alongside the Catholic church. In June 1564 the diet offered rights to worship to a church for the Saxons (German-speaking Lutherans) and to a church for the Hungarians (Reformed). Both sides were only allowed to introduce their faith where local people accepted their preaching, and in cases where a Reformed minister ousted a Lutheran preacher no mockery or offensive behaviour was permitted. As in Poland, the diet claimed to be trying to act 'for the peace of the realm', and banned the use of force to compel individuals to convert to any church. However, by 1566 the collapse in the number of Catholic representatives at the diet led to the decision to expel any clergy who did not want to convert from 'the Pope's learning'. The balance of confessional forces in the diet shifted again in the late 1560s with the spread of anti-Trinitarianism. In 1568 the diet in effect extended rights to anti-Trinitarian preachers by deciding that

> ministers should everywhere preach and proclaim the Gospel according to their understanding of it, and if their community is willing to accept this, good. If not, however, no-one should be compelled by force if their spirit is not at peace, but a minister retained whose teaching is pleasing to the community. Therefore, no-one should harm any superintendent or minister, nor abuse anyone on account of their religion, in accordance with previous laws, and no-one is permitted to threaten to imprison or deprive anyone of their position because of their teaching, because faith is a gift from God which comes from listening, listening to the word of God.[15]

Transylvanian law forbade any behaviour which insulted the religion of others or could lead to public disorder. Church buildings were supposed to belong to the confessional group with majority support in each parish, and churches were to be peaceably shared between different communities where necessary. Even nobles with rights to select parish ministers on their lands were not permitted to introduce a minister of a different religion to that of the local community, although patterns of religious loyalty still tended to reflect the rights of patronage of the social elite.

The laws on religion passed in Transylvania, as elsewhere in Central Europe, were a political compromise rather than any principled charter for religious tolerance. The particular nature of Transylvania's laws reflected the more limited power of elected rulers within the principality, the structure of its diet, the social diversity of the ruling elite, and the spread of a variety of strands of Reformation thought. However, concern had already grown by the early 1570s about the extent of clergy's freedom to preach according to conscience. In particular, anti-Trinitarian ministers who began to question Christ's divinity were the target of laws against preachers who went to 'criminal excess'. In 1572 under the new Catholic voivode (prince), and later Polish king, István Báthory, the diet passed a strict law against any doctrinal innovation among followers of the three legal religions. The following decades under Báthory princes saw some attempts to revive Catholicism and to establish rights for Catholic clergy in the principality. In 1579 Jesuits were allowed to teach at agreed sites in Transylvania, although their work was subject to frequent interference from the estates, and they were expelled from the principality in 1588. However, by April 1595 the diet proclaimed that in Transylvania the four 'received religions, that is, the Catholic or Roman, Lutheran, Calvinist and Arian, can be kept everywhere freely'. In practice, the four churches were not, however, treated equally. From the beginning of the seventeenth century Calvinists benefited from the patronage of a series of Reformed princes, while after only four years of residence the diet expelled a restored Catholic bishop, and the Jesuits were expelled from the principality again in 1607 and 1610.

Nevertheless, all of the princes elected to rule Transylvania until 1691 agreed to uphold the constitutional status of all four received religions in the principality. The question of how to interpret this remarkably stable incorporation of four different confessions within a single polity has been the subject of a good deal of attention. A consensus has emerged that in the face of both Ottoman and Habsburg threats to the Transylvanian state, the princes and the diet made a pragmatic calculation that it was imperative to sideline the issue of religious differences within the domestic elite. The toleration of a range of religions came to be seen as vital to the stability and security of the state. In addition, the diet's attempts to accommodate religious divisions within local communities over time fostered a mostly peaceful balance between the loyalty of ordinary people to their own confession on the one hand; and, on the other, widespread social acceptance of religious differences in towns, neighbourhoods and even within families. The religious settlement in Transylvania therefore highlights the importance of the successful operation of state laws in limiting popular religious violence in sixteenth-century Europe, and suggests a greater social capacity to cope with the difficulties of living in multi-confessional communities than is often acknowledged.

catholic recovery and protestant rebellion

Elected princes who ruled the Transylvanian principality accepted that protecting the rights of four religions was a means by which to promote their own power. Elsewhere in Central and Eastern Europe, from the latter decades of the sixteenth century, Catholic monarchs calculated that imposing a single religion within their territory would increase royal power. However, at least until 1620 it was far from clear whether or not this assumption was correct. The scale of the task facing Catholic activists in Central and Eastern Europe was immense. Around 1600, only about 25 per cent of Hungarian parishes were Catholic, and in Bohemia the Catholic church controlled an even smaller percentage of parishes. Catholic rulers worked in concert with the clergy hierarchy to revive Catholic fortunes and to overturn the liberties granted to other churches. Early efforts to advance Catholic interests had, however, only encouraged Protestant nobles to seek formal legal assurances of their religious rights, and seemed to have entrenched the strength of Protestant churches.

Historians have offered a number of explanations for the scale and speed of Catholic recovery and for the decline of Protestant Central Europe. Some have focused attention on crucial weaknesses in the state of Protestant church life. For one thing, many churches had emerged from localised reform initiatives and were often slow to develop effective organisational structures. Noble supporters of Trinitarian Protestant churches in Bohemia and Poland had also encouraged clergy to agree to common platforms, which, although useful in pressing for legal rights, inhibited the development of autonomous church institutions. When pressure was applied by Catholic authorities, responses were therefore often slow, uncoordinated and ineffective. However, even more significant was the heavy reliance of the Evangelical, Reformed and other churches on noble loyalty for their survival. Nobles were both the political guardians of Protestant legal rights and local patrons of parish churches. The core of the Catholic mission effort in Central Europe was therefore directed towards noble families. Monarchs attempted to attract the loyalty of magnates to their courts with offers of positions and titles for those willing to convert.[16] From the 1560s, new Catholic academies and schools were also established to compete with Protestant institutions in offering education to the children of noble families. A more competent and energetic leadership within the clerical hierarchy also encouraged efforts through sermons and printed literature to bring the Catholic message to noble houses as well as to towns and the countryside.

Above all, it was the drift of nobles back to the Catholic fold which ensured the decline of Central European Protestantism. The pace and chronology of this process differed across the region. In Poland, from the latter decades of the sixteenth century as the numbers of Protestant nobles dwindled, so did the numbers of Protestant churches. Some churches which were not under the immediate protection of nobles were early casualties of a revived Catholicism, as at Cracow in 1591. Apart from the German-speaking towns of Royal Prussia, Catholicism had become normative among the Polish elite by the early seventeenth century. In 1638 the anti-Trinitarian church lost its legal rights, and in 1668 the diet prohibited repudiation of the Catholic church. Within the Habsburg monarchy, members of the ruling family came to identify dynastic interests with the Catholic cause ever more closely over time and curtailed Protestant freedoms wherever possible. In Inner Austria, for example, Lutherans had been able to use financial pressure to extract guarantees of religious rights during the 1570s. However, Archduke Karl invited Jesuits to set up a college at Graz which successfully competed against a Lutheran school in the town. This Catholic college was expanded into a university during the 1580s, which then attracted students from around the region. Karl's son Ferdinand (later Ferdinand II) was also determined to promote the Catholic cause in Inner Austria. In 1598 Ferdinand was able to close down the Lutheran school in Graz and expelled Protestant ministers from towns across the province. Elsewhere in the Austrian provinces clergy and urban laity took the initiative in promoting Catholic preaching and new styles of piety. For example, Melchior Khlesl, who had been converted by the Jesuits as a child, promoted reform as chancellor of Vienna university and as bishop of Vienna. In 1599 Khlesl demonstrated his personal commitment to a new lay piety by accompanying a procession of 23,000 pilgrims to a Marian shrine at Mariazell.[17]

There were limits to the degree of pressure which the Habsburg court could apply against its Protestant nobles. The court continued to need the support of the estates in Royal Hungary in particular to ensure the defence of the monarchy against further Ottoman encroachment. Many Protestant nobles also proved willing to resist the enforcement of counter-reform measures through rebellion against their Habsburg rulers. In Hungary, Bohemia and Austria religious disputes sparked major noble revolts between 1600 and 1620. The degree to which this Protestant resistance was coordinated across the Habsburg monarchy has been investigated by historians, who have discovered links between Bohemian, Austrian and Hungarian Protestant magnates. There has also been research into how far these rebellions against royal power and counter-reform measures might have been inspired by ideas about rights of resistance. However, rather than developing monarchomach theories, nobles expressed their right to rebel as a matter of native constitutional tradition and aristocratic liberty.[18]

The first major rebellion came in Royal Hungary. Anti-Habsburg feeling among nobles, already fuelled by the impact of inconclusive fighting against the Ottomans, was further aggravated when Rudolf II's army in eastern Hungary expelled Protestants from some towns and a leading Protestant magnate was put on trial for sedition. In 1604 matters came to a head after a Catholic bishop used imperial troops to evict Evangelical ministers from the royal free town of Kassa (Košice). When the diet met in February 1604 complaints were raised about the situation in Kassa. However, Matthias, representing Rudolf, responded by claiming that anyone who raised questions about religion in the diet would be guilty of treason. The court effectively pushed Protestant nobles into the arms of dissidents under the leadership of the Reformed noble, István Bocskai. Bocskai's forces had occupied Kassa by November 1604, and Bocskai was elected as Transylvania's first Reformed prince in February 1605. The crisis for the Habsburgs deepened in the spring of 1605 as the Hungarian diet elected Bocskai as their prince protector (he was described as 'Moses of the Hungarians'). Bocskai's army reached as far as Pozsony (Bratislava), while some of his forces even pushed into Moravia and Lower Austria. This first Habsburg assault on Protestant liberties therefore completely backfired. By the terms of the 1606 peace treaty, Bocskai was confirmed as Transylvanian prince, and Evangelical and Reformed nobles and royal towns in Hungary gained the legal right to practise freely their religion.[19]

The 1608 crisis with the ruling family between Matthias and Rudolf was then used by Protestants across the monarchy to consolidate their legal rights. In Royal Hungary, the diet offered its support for Matthias but only in return for ratification of the terms of the 1606 treaty. The estates in Upper and Lower Austria and in Moravia also backed Matthias, and all managed to extract declarations of support for their religious rights in return. The Bohemian estates remained loyal to Rudolf, and he agreed to a Letter of Majesty in 1609 which confirmed the religious freedoms of nobles and extended rights to towns. However, the results of this crisis within the Habsburg dynasty obscured the impact of the growing number of noble conversions to the Catholic church across the monarchy. In Hungary, for example, the legal rights granted to Reformed and Evangelical churches were almost immediately threatened and undermined as noble converts handed their parishes over to Catholic priests, denying Protestant communities access to a place of worship.

The next trial of strength between royal Catholicism and noble Protestantism in the Habsburg monarchy came to a very different conclusion. The destruction of two Protestant churches on crown lands in Bohemia sparked the deposition of Ferdinand in 1618, and rebellion against his rule spread to Moravia, Hungary, and to Upper and Lower Austria. The rebel estates' cause gained the support of two Reformed princes, Frederick of the Palatinate and Gábor Bethlen of Transylvania. However, this confederation of Protestant princes and estates provided a very short-lived vision of a different political and religious future for Central Europe. The defeat of the rebels at the battle of the White Mountain in November 1620 turned the tide decisively in favour of Catholic monarchy in the region. The collapse of the confederation left Gábor Bethlen to return control over most of Royal Hungary to Ferdinand in 1621. Although the legal rights of Protestants remained in place in Hungary, in practice they proved more and more difficult to uphold. Within the rest of the monarchy, Ferdinand acted to crush his apparently disloyal Protestant subjects. The new Bohemian constitution of 1627 revoked the Letter of Majesty and offered rights only to the Catholic church. Leading rebels were executed or forced into exile, and Protestant nobles faced the choice between conforming or losing their estates. This new constitution was extended to Moravia in 1628, and Protestant nobles in the Austrian provinces faced similar pressure to convert while clergy were expelled. The defeat of the Protestant confederation in 1620 did not cause the death of Central European Protestantism but it did certainly hasten its demise. To the extent to which Protestantism had been simply a badge worn by nobles during the sixteenth century as part of their stand against royal power, then, that badge was largely discarded by aristocrats during the 1620s. However, Protestant preaching and piety had made a much more significant impression upon Central and Eastern European society as was demonstrated by those nobles and clergy who chose exile rather than conformity, and by those who endured Catholic harassment and remained faithful to their Evangelical, Reformed, anti-Trinitarian and other churches into the modern era.

further reading

On the politics of the Reformation period in Central and Eastern Europe the seminal work is Robert Evans, *The Making of the Habsburg Monarchy, 1550–1700* (Oxford, 1979). See also the collection of essays edited by Robert Evans and Trevor Thomas, *Crown, Church and Estates: Central European Politics in the Sixteenth and Seventeenth Centuries* (London, 1991). Among a number of surveys of the early modern history of the region see Robert Kann, *A History of the Habsburg Empire, 1526–1918* (Berkeley, CA, 1977); Jerzy Lukowski and Hubert Zawadzki, *A Concise History of Poland*

(Cambridge, 2001); and Robert Frost, *The Northern Wars: War, State and Society in Northeastern Europe, 1558–1721* (Harlow, 2000).

As yet, no-one has attempted to write a comprehensive history of the Reformation in Central and Eastern Europe. Some impression of the overall impact of reform can be gained by consulting articles in Karin Maag (ed.), *The Reformation in Eastern and Central Europe* (Aldershot, 1997). See also two useful articles: Winfried Eberhard, 'Reformation and Counter-Reformation in East Central Europe', in James Tracey, Thomas Brady, Heiko Oberman (eds), *Handbook of European History, 1400–1600: Late Middle Ages, Renaissance and Reformation*, vol. II (Leiden, 1995); and Robert Evans, 'Calvinism in East Central Europe: Hungary and her Neighbours', in Menna Prestwich (ed.), *International Calvinism, 1541–1715* (Oxford, 1985). In addition, see Maria Crăciun, Ovidiu Ghitta and Graeme Murdock (eds), *Confessional Identity in East-Central Europe* (Aldershot, 2002), especially the introduction.

Recent studies of the Reformation across the Continent have paid a good deal of attention to affairs in Central and Eastern Europe including Diarmaid MacCulloch, *Reformation: Europe's House Divided, 1490–1700* (London, 2003), and Philip Benedict, *Christ's Churches Purely Reformed. A Social History of Calvinism* (New Haven and London, 2002). In addition there are contributions on Central Europe in Andrew Pettegree (ed.), *The Reformation World* (London, 2000), and Ronald Po-chia Hsia (ed.), *A Companion to the Reformation World* (Oxford, 2004). The perspectives of some of the leading historians from Central Europe are also available in a number of translated articles. See articles by Janusz Tazbir on Poland, by František Kavka on Bohemia, and by Katalin Péter on Hungary in Robert Scribner, Roy Porter and Mikuláš Teich (eds), *The Reformation in National Context* (Cambridge, 1994). On questions about moderation and toleration in central Europe see contributions by Jaroslav Pánek and Katalin Péter to Ole Peter Grell and Robert Scribner (eds), *Tolerance and intolerance in the European Reformation* (Cambridge, 1996), and contributions on Central Europe in Howard Louthan and Randall Zachman (eds), *Conciliation and Confession: the Struggle for Unity in the Age of Reform, 1415–1648* (Notre Dame, 2004).

There are a growing number of detailed monographs on particular movements of reform available in English. On the lands of the Bohemian crown, see Thomas Fudge, *The Magnificent Ride: The First Reformation in Hussite Bohemia* (Aldershot, 1998) and Jarold Zeman, *The Anabaptists and the Czech Brethren in Moravia, 1526–1628* (The Hague, 1969). Still useful for Hungary and Transylvania is Earl Wilbur, *A History of Unitarianism: Socinianism and its Antecedents* (Cambridge, MA, 1946). For more recent work on anti-Trinitarianism see Róbert Dán and Antal Pirnát (eds), *Antitrinitarianism in the Second Half of the Sixteenth Century* (Budapest, 1982), and Mihály Balázs and Gizella Keserũ (eds), *György Enyedi and Central European Unitarianism in the Sixteenth and Seventeenth Centuries* (Budapest, 2000). A translation of Imre Révész, *A History of the Hungarian Reformed Church* (Washington, 1956) provides a survey, and for the development of the Reformed movement see Graeme Murdock, *Calvinism on the Frontier, 1600–1660: International Calvinism and the Reformed Church in Hungary and Transylvania* (Oxford, 2000). Affairs in Poland have to date received less attention from authors in Western languages, but there is Janusz Tazbir, *A State Without Stakes: Polish Religious Toleration in the Sixteenth and Seventeenth Centuries* (New York, 1973) and Antanas Musteikis, *The Reformation in Lithuania: Religious Fluctuations in the Sixteenth Century* (New York, 1988). Recent work has substantially moved forward debates on Austrian Protestantism and Catholic recovery in Austria. Key works include Regina Pörtner, *The Counter-Reformation in Central Europe: Styria, 1580–1630* (Oxford, 2001); Howard Louthan, *The Quest for Compromise: Peacemakers in Counter-Reformation Vienna* (Cambridge, 1997); Rona Johnston Gordon, 'The Bishopric of Passau and the Counter-Reformation in Lower Austria, 1580–1636' (DPhil thesis, University of Oxford, 1997); and Elaine Fulton, 'Catholic Belief and Survival in Late Sixteenth-Century Vienna: the Case of Georg Eder (1523–87)' (PhD thesis, University of St Andrews, 2002).

notes

1. For example, see the work of Jenõ Zoványi: *Puritánus mozgalmak a magyar református egyházban* (Budapest, 1911); *A reformáczió Magyarországon 1565-ig* (Budapest, 1921); *A magyarországi protestántizmus 1565-tõl* (Budapest, 1977).
2. Boris Paternu (ed.), *Protestantism and the Emergence of Slovene Literature: Slovene Studies 6/1–2* (Munich, 1984); Karl Schwarz and Peter Svorc (eds), *Die Reformation und ihre Wirkungsgeschichte in der Slowakei* (Vienna, 1996). See also contributions to Robert Scribner, Roy Porter and Mikuláš Teich (eds), *The Reformation in National Context* (Cambridge, 1994).
3. For example László Makkai, *A magyar puritánusok harca a feudálizmus ellen* (Budapest, 1952) and idem, 'The Hungarian Puritans and the English Revolution', *Acta Historica*, 5 (1958), 13–45.
4. Larry Wolff, *Inventing Eastern Europe: the Map of Civilization on the Mind of the Enlightenment* (Stanford, 1994); Vesna Goldsworthy, *Imagining the Balkans* (Oxford, 1997).
5. For Hussitism as 'the region's main "Western" type of reaction to the first crisis of feudalism' see Jenõ Szücs, 'Three Historical Regions of Europe: An Outline' in John Keane (ed.), *Civil Society and the State* (London, 1988), p. 326.

6. Thomas Fudge, 'Luther and the "Hussite" Catechism of 1522', in Maria Crăciun, Ovidiu Ghitta and Graeme Murdock (eds), *Confessional Identity in East-Central Europe* (Aldershot, 2002).
7. G. Szabó, *Geschichte des Ungarischen Coetus an der Universität Wittenberg, 1555–1613* (Halle, 1941); H. R. Guggisberg, *Basel in the Sixteenth Century: Aspects of the City Republic Before, During, and After the Reformation* (St Louis, MO, 1982), pp. 44, 48–9.
8. For example, Christine Peters, 'Mural Paintings, Ethnicity and Religious Identity in Transylvania: the Context for Reformation', in Karin Maag (ed.), *The Reformation in Eastern and Central Europe* (Aldershot, 1997).
9. Robert Evans, 'Calvinism in East Central Europe: Hungary and her Neighbours', in Menna Prestwich (ed.), *International Calvinism, 1541–1715* (Oxford, 1985); David Daniel, 'Hungary', in Andrew Pettegree (ed.), *The Early Reformation in Europe* (Cambridge, 1992); David Daniel, 'Calvinism in Hungary: the Theological and Ecclesiastical Transition to the Reformed Faith', in Alastair Duke, Gillian Lewis and Andrew Pettegree (eds), *Calvinism in Europe, 1540–1620* (Cambridge, 1994); M. Kosman, 'Programme of the Reformation in the Grand Duchy of Lithuania and How it was Carried Through (c. 1550– c. 1650)', *Acta Poloniae Historica*, 35 (1977), 21–50.
10. E. Wilbur, *A History of Unitarianism: Socinianism and its Antecedents* (Cambridge, MA, 1946).
11. B. Hall, *John à Lasco, 1499–1560: a Pole in Reformation London* (London, 1971); Michael Springer, 'Protestant Church Building and the *Forma ac Ratio*: the Influence of John a Lasco's Ordinance in Sixteenth-Century Europe' (PhD thesis, University of St. Andrews, 2004).
12. Krista Zach, *Orthodoxe Kirche und rumänisches Volksbewusstsein im 15. bis 18. Jahrhundert* (Wiesbaden, 1977); Maria Crăciun, 'Orthodox Piety and the Rejection of Protestant Ideas in Sixteenth-Century Moldavia', in Maria Crăciun and Ovidiu Ghitta (eds), *Ethnicity and Religion in Central and Eastern Europe* (Cluj, 1995).
13. For a print depicting this see Philip Benedict, *Christ's Churches Purely Reformed. A Social History of Calvinism* (New Haven and London, 2002), p. 356.
14. On the powers and role of the diet see Graeme Murdock, '"Freely Elected in Fear": Princely Elections and Political Power in Early Modern Transylvania', *Journal of Early Modern History*, 7 (2003), 213–44.
15. The diet records are published as Sándor Szilágyi (ed.), *Monumenta Comitialia Regni Transylvaniae. Erdélyi Országgyűlési Emlékek. Magyar Történelmi Emlékek Harmadik Osztály*, 21 vols (Budapest, 1875–1898). For the 1568 resolution see vol. II (1877), p. 343.
16. P. G. Schimert, 'Péter Pázmány and the Reconstitution of the Catholic Aristocracy in Habsburg Hungary, 1600–1650' (PhD thesis, University of North Carolina, 1990).
17. Regina Pörtner, *The Counter-Reformation in Central Europe: Styria, 1580–1630* (Oxford, 2001), p. 241.
18. K. MacHardy, 'The Rise of Absolutism and Noble Rebellion in Early Modern Habsburg Austria, 1570–1620', *Comparative Studies in Society and History*, 34 (1992), 407–38; Joachim Bahlcke, 'Calvinism and Estate Liberation Movements

in Bohemia and Hungary (1570–1620)', in Maag (ed.), *Reformation in Eastern and Central Europe*.

19. David Daniel, 'The Fifteen Years' War and the Protestant Responses to Habsburg Absolutism in Hungary', *East Central Europe*, 8 (1981), 38–51; Kálmán Benda, 'Habsburg Absolutism and the Resistance of the Hungarian Estates in the Sixteenth and Seventeenth Centuries', in Robert Evans and Trevor Thomas (eds), *Crown, Church and Estates: Central European Politics in the Sixteenth and Seventeenth Centuries* (London, 1991), pp. 123–8.

3
the swiss reformation

bruce gordon

The historiography of the Swiss Reformation is something akin to the medieval fascination with the fabulous Christian ruler Prester John. There are reports of its existence, but few have actually seen it.[1] The reason for this lies in the fact that although it is now usual to speak of the 'Swiss Reformation', the term remains quite foreign to the Swiss, who prefer to speak of the Zurich, Bernese, or Basle reformations, or, at a stretch, the Zwinglian Reformation, though even that term is not widely used for events outside Zurich. For the most part, the term 'Swiss Reformation' is of recent vintage and used almost exclusively by historians from the English-speaking world, beginning most significantly with the Swiss-born American historian Philip Schaff.[2] Even today, the category finds little resonance among historians from the non-English-speaking world, who, for the most part, are not inclined to accept that the events of the Reformation in the Swiss Confederation formed a coherent narrative.

This inclination has been augmented by the tradition of dividing the German- and French-speaking lands of the Swiss Confederation into distinct entities with few connections: the former shaped by Zwingli and Zurich and the latter by Calvin and Geneva. Indeed the very relationship between Geneva and the Swiss is highly problematic. Despite the fact that the city is now central to our very image of Switzerland, and most English histories place it in the Swiss Reformation, it did not officially join the country until the nineteenth century, and Calvin certainly did not think of himself as Swiss. Connections with the Swiss tended to be more acrimonious than fraternal.

If we are to understand the fragmented nature of the subject we must begin in the sixteenth century. From the earliest days of the Reformation there was no tradition of integrating histories of the various parts of the Swiss Confederation, as it was known till the founding of modern

Switzerland in 1848. Zwingli had appealed to Swiss patriotism, but the ill-disguised, and perhaps legitimate, suspicions harboured by the cities of Basle and Berne meant that the reformer of Zurich would never be a Swiss hero. Zwingli had played the dangerous game of coupling the reform movement to the hegemonic interests of the Zurich ruling elite. Whilst this may have made the Reformation possible, it won him few friends outside the city. Thus, the story of the Swiss churches was from the start fractured. There were few common experiences or shared moments. The narrative structure of the German reformation provided by the Ninety-Five Theses, the Diet of Worms, the Protest at Speyer, and so forth, was entirely absent in the Swiss context. For the Swiss, the Reformation could only be spoken of in terms of local reformations, each with its own cast of characters.

The first expansive historical treatment of the Reformation emerged from the hand of Zwingli's successor Heinrich Bullinger, whose *Reformation History*, written at the end of his life, took cognizance of Berne and Basle, but really as parallel phenomena.[3] Bullinger, however, introduced a further difficulty that was to influence the writing of Swiss Reformation history till well into the twentieth century. He ended his account in 1531 with the death of Zwingli, and thereby more or less wrote himself out of the script, despite the pivotal role he played in the almost five decades he led the movement. This paradigm proved highly influential, and successive overviews of the Swiss Reformation in all languages have tended to follow a well-trodden path: discussing Zwingli's rise to prominence in Zurich, the civic Reformation, subsequent events in Berne and Basle, with some attention to the genesis of the Anabaptist movement. With the death of Zwingli, however, the story then briskly shifts to Calvin and Geneva, largely abandoning the German-speaking lands, with the exception of the *Consensus Tigurinus* of 1549. More attention has been given to Bullinger's international influence, in particular in England – an honour recognised in the nineteenth century with the publication of his works in English by the Parker Society.[4]

As Robert Kolb has recently demonstrated, the image of Martin Luther as a hero of the German people that emerged in the early 1520s remained a fixed part of German Protestantism till the twentieth century.[5] Even Communist East Germany had to reconcile itself to its native son. Huldrych Zwingli, however, could never occupy such a place for the Swiss. Although in pious church history accounts he was frequently portrayed as the 'Third Man' of the Reformation, struggling with Luther and setting the scene for Calvin, Zwingli could not have been more different. Not only was his theological orientation significantly other, but he built the Reformation around a very conscious linking of religious reform to the hegemonic interests of the Zurich civic rulers. This was not lost on his contemporaries, even those who espoused similar ideas. Zwingli's death on a battlefield naturally fell into the early modern category of a 'bad death' and his supporters struggled in vain to turn him into a martyr. Although his theology was deeply influential, Zwingli was inextricably linked to the city of Zurich and his machinations were profoundly mistrusted by the other Swiss, who saw his work as part of a long tradition of attempts by Zurich to control the Confederation. Thus, whereas Luther successfully united the movement in the Empire, Zwingli was an extremely divisive figure and his influence tended to be downplayed by non-Zurich supporters of the Reformation. The Eucharistic debate between Luther and Zwingli also cast a pall over the formation of the Reformation narrative. Luther had successfully tagged Zwingli with the label of heretic and the latter's sudden death fatally interrupted the debate in favour of the Germans. The Reformation was cloven in two and there could be no reconciliation. Thus the force of writing from Zurich, both theological and historical, was directed towards the defence of Zwingli's orthodoxy.

This focus on Zwingli remained while Heinrich Bullinger lived, but after his death and the decline of Swiss influence on the European stage there was a desire to forget the Zurich reformer. In the seventeenth century, the theological legacy of the Swiss reformers was increasingly portrayed as a slightly weak opening act for Calvin, and on the all-consuming question of predestination the Zwinglians were thought to have only partially beheld the light. Bullinger's unwillingness to accept double predestination, for example, was understood as an indication that the Swiss had been misguided and had required correction by Geneva. John Calvin himself had contributed to this perception in significant ways. Although deeply dependent upon the Swiss churches from the 1530s till at least 1555, he largely refused to discuss Zwingli. When he did it was hardly flattering.[6] If there were to be some sort of reconciliation of wider Protestantism, as Calvin had hoped, the Swiss legacy would need to be pushed to the side on account of the undying enmity of the Lutherans.

This hostility towards the Swiss remained a fundamental part of German church politics and scholarship for four centuries after the Reformation. In the seventeenth century German Lutherans rejected overtures towards Protestant union from the Scot John Dury on the grounds that the Swiss were heretics.[7] The rise of Pietism and Enlightenment philosophy brought much greater influence from German lands on the Swiss, whose own Reformation theology receded into the background. Following the catastrophe of the Thirty Years War, Reformed thought lost its perch in German intellectual life and was largely marginalised. Calvinism replaced any strong sense of Swiss theology as the foundation of Reformed life and thought. In German church histories of the nineteenth and twentieth centuries the Zwinglian Reformation was distinguished by two characteristics: its departure from Luther on the Eucharist and its close association with radicalism. Nineteenth-century Lutheran scholars such as Karl Müller and Karl Holl continued to portray the Swiss in unflattering terms.[8]

In our survey of scholarship of the Swiss Reformation it makes sense to begin in the early eighteenth century. In 1719 the bicentenary of the Swiss Reformation and the centenary of the Synod of Dordt gave new impetus to efforts at the unification of Protestant confessions. It was in this spirit that the first author of note appears: Johann Jakob Hottinger (1652–1735) wrote a series of irenic works, including a four-volume church history of the Swiss Confederation, in support of broader Protestant unity against Rome. Hottinger believed that the Bible could be the foundation of all teaching, and that differences of opinion have to be accepted by the denominations in all other questions. In writing his history Hottinger was able to draw upon the prodigious efforts of his father, the noted orientalist Johann Heinrich Hottinger (1620–1667). Johann Heinrich had produced the *Thesaurus Hottingerianus* (fifty-two volumes), which remains one of the most important collections of manuscripts and printed works of the Swiss sixteenth-century writers. This enormous collection was augmented in the eighteenth century by another scion of a prominent Zurich family, Johann Jakob Simler (1716–1788), whose collection of sources ran to an incredible 196 volumes.

The first author to examine both the French- and German-speaking lands was the influential Swiss historian Abraham Ruchat in his six-volume *Histoire de la reformation de la Suisse* (1727–28).[9] Although Ruchat leaned heavily on Hottinger for his account of the German-speaking Reformation, for his native Pays de Vaud he turned to the sources. His treatment of the Reformations among the two principal linguistic groups had a higher purpose. He believed that the French- and German-speaking Swiss had little understanding of one another and that their greater Swiss identity was threatened by France, whose role in the Confederation had continually brought about moral calamity. Ruchat found the locus of the Swiss Reformation in the moral cleansing of the people. To this day Ruchat's work remains an exquisite expression of Enlightenment history writing.

The lofty heights of Enlightenment idealism gave way to the Romantic 'Sturm and Drang' in the work of Johannes von Müller (1752–1809), who was deeply influenced by the German philosophy of Johann Gottfried von Herder with its interplay of God, the mind, and history.[10] This is reflected in Müller's 1778 *History of the Swiss*, which would play a key role in the formation of Swiss historical mythology. Müller was interested above all in the emergence of liberty, not only in the state but also in the human condition. A Protestant from Schaffhausen, Müller believed passionately in Scripture as the foundation of human society, particularly the New Testament. Although he was sceptical of absolute forms of religion and was not entirely persuaded by the Reformation, in his historical writing he clearly harkened back to some clear lines of thought from Swiss Reformers. Müller believed the Swiss were in great peril as the waves of revolution from France washed over the Confederation. In the face of such tumult he advised his fellow Swiss to fall back on religion, moral purity, and the old forms of the Confederation. This nostalgic and mythological view of the Swiss past, derived from chronicles, profoundly shaped the Swiss mentality of the nineteenth and twentieth centuries and clearly placed the Reformation in that narrative. Part of Müller's legacy was also the imperative that the sources for Swiss history be located and used in order that the Swiss character might be properly understood.[11]

For Reformation historical writing of the mid-nineteenth century, we need to look at yet another Hottinger, Johann Jakob (1783–1860); and at Louis Vulliemin (1797–1879), both of whom were committed to the writing of Swiss history in the spirit of Müller. This is seen in their intertwining of ideas of historical evolution and a deep reverence for the sources. Both viewed the Middle Ages as the darkness before the dawn and contrasted the clericalism of that period with the humanity of the Enlightenment, although, it must be said, neither was entirely persuaded by Enlightenment rationalism. For both men their Protestantism deeply informed their historical accounts: Hottinger produced a biography of Zwingli, *Huldreich Zwingli und seine Zeit, dem Volke dargestellt* (1842), and an edition of Heinrich Bullinger's *Reformation History* (1838–40). He portrayed the Reformation as a light shining upon the benighted Middle Ages, although his writings on Zwingli rather eschew theology, which Hottinger did not regard as a suitable subject for the profane historian. Above all, the Reformation was the liberation of the conscience and the freeing of the intellect. Hottinger dwelt in particular upon the educational advances brought about by the Reformation, which he saw as opening the way to the Enlightenment. It was an optimistic and forward-looking perspective, embracing nineteenth-century notions of progress. The Reformation was the start of the modern world.

Hottinger's close friend and colleague Louis Vulliemin from the Pays de Vaud inherited the daunting task of re-editing Ruchat's monumental *Histoire*, which had appeared a hundred years earlier. Vulliemin was, however, less inclined to see the Reformation as crucial to the evolution of the modern state. Following Müller, he viewed the Swiss in terms of a cultural identity, not as a monolithic state. The Reformation was a key element in the formation of this identity through its embracing of both linguistic groups. Vulliemin emphasised concord over disputes in the Reformation, though he did not shrink from identifying those whom he regarded as the enemies of the Swiss and reform. Principally vilified were the French, whose money had led to the moral disintegration of the Confederation, and the aristocracy, whose avarice had shamed the Swiss. In reflecting upon why the Confederation had not collapsed between 1550 and the end of the eighteenth century as a result of the French and the treacherous aristocracy, Vulliemin's explanation reflected his core belief: it was the hand of God that had preserved the Swiss.

By the second half of the nineteenth century the Swiss Reformed churches were under sustained pressure from the combined forces of liberalism, democracy, and secularism. In Basle, Zurich, St Gall, Berne, Neuchâtel and Geneva, powerful movements undertook to undo the church–state relations set up in the Reformation and reaffirmed by the constitution of 1848. In Basle virtually every aspect of the church was debated in a series of confrontations between Pietists, Liberals, and orthodox Reformed.[12] The primary issues included the role of the Apostles' Creed and the Reformation confessions, the place of religion in the schools, the election of ministers, and the new hymnbook. Reformation scholarship played an important role in these struggles, and of particular significance was the work of the Basle native Karl Rudolf Hagenbach (1801–1874), who along with his close friend Georg Diethelm Finsler of Zurich, led the moderate party in the church debates. Hagenbach argued for a balance between the veneration of tradition and the openness of the church to new currents of thought. His interest in recovering the work of the reformers lay in his profound belief in their relevance as models of piety for the contemporary church. This attitude was reflected in his examination of the historical background to the *Basle Confession* of 1827 and his book on Johannes Oecolampadius and Oswald Myconius (1859). During the middle decades of the century he edited a series of biographies of the principal protagonists of the Swiss Reformation; for many of these men, these nineteenth-century works remain our best sources. The series was entitled *Leben und ausgewählte Schriften der Väter und Begründer der reformirten Kirche* and ran to more than a dozen volumes. The works were scholarly in nature and made extensive use of the sources, though frustratingly often without referencing them. One of the most important volumes was the great pedagogue Carl Pestalozzi's biography of Heinrich Bullinger (1858), which is remarkable for its use of Bullinger's writings, many of which he included in translation for the reader. Had it prevailed, this argument that Bullinger, not Zwingli, was the model of Christian piety suitable for contemporary Christians would have changed the face of research on the Swiss Reformation. Pestalozzi wrote a similar biography of Zwingli's companion Leo Jud.[13]

In Zurich, Reformation scholarship was likewise wed to the fracturing world of church politics.[14] From the 1840s there was a growing movement within the Zurich church towards liberalisation, focused once again on an enforced church confession. By 1865 a growing chorus of voices supported the separation of church and state, following the model of Basle. The defence of the state–church arrangement was led by the zealous and industrious Georg Finsler, who was primarily responsible for the resurrection of Zwingli's reputation in the nineteenth century.[15] As a weapon against liberal and secular tendencies Finsler adopted Zwingli as hero of the Zurich church and model reformer. The Reformation was restored as a central part of Zurich's history and Zwingli presented once more as the liberator from papal tyranny. This was conducted against the background of Vatican I and, for the Protestants, the alarming realisation that the Catholic populations of the larger, Protestant cities were growing exponentially. Zwingli thus became a central symbol in the war against the established church and Finsler both held numerous lectures on Zwingli and commissioned printed works. The emphasis in these works was upon the greatness of Zwingli's spirit and his construction of the perfect polity, the unity of church and state.

The decisive moment in the emergence of Zwingli as the symbol of the Reformed Church was the 1884 celebration of the four hundredth anniversary of his birth. The occasion had spurred Finsler to raise money by public subscription for a monument to the reformer. This was followed by the commencement of work on a critical edition of Zwingli's works, the founding of the *Zwingliverein* in Zurich in 1897 and of its journal *Zwingliana* a few years later, and the appearance of Rudolf Staehelin's masterful two-volume biography *Huldreich Zwingli. Sein Leben und sein Wirken nach den Quellen. dargestellt* (1895–97). This soon became the standard work on the reformer and represented the epitome of the text-critical historical school of the late nineteenth century. The search was for the 'true' or 'authentic' Zwingli through scrupulous use of the texts. Staehelin's study is a wonderfully rich narrative that focuses on Zwingli's 'Weltanschauung'.

The author, as promised, makes extensive use of the sources, drawing heavily upon two notable collections of documents that appeared in 1870s, Emil Egli's *Aktensammlung zur Geschichte der Zürcher Reformation* (1879) and Johannes Strickler's *Aktensammlung* (1878). These volumes for Zurich presaged a series of document collections for the Protestant Reformed cities: Moriz Stürler's *Urkunden der Bernischen Kirchenreform* (1862) and *Valerius Anshelm: Berner Chronik* (1884–86) and Rudolf Steck and Gustav Tobler's *Aktensammlung zur Geschichte der Berner-Reformation 1521–1532* (1918–23).

A Swiss scholar and churchman who left his homeland and emigrated to America wrote the most influential work on the Swiss Reformation for the English-speaking world. Philip Schaff (1819–1893) was born in Chur and studied at Tübingen under August Neander, who had famously sought to link piety and historical scholarship.[16] Schaff's work is essential for several reasons. As this was the first extensively document-based account of the Swiss Reformation to appear in the English language, Schaff played a key role in establishing the essential narrative that would define the Anglo-Saxon understanding of the movement. Schaff devoted the eighth volume of his *History of Christianity* to the Reformation in the German-speaking lands of the Confederation, while the French-speaking Reformation was treated along with Calvin in volume nine. Drawing heavily on the work of Hottinger, Ruchat, and Hagenbach, as well as the newly published volumes of Egli, Strickler, and Stürler, Schaff's account was intended to make sense of the Reformation for his American audience. He laid out his purpose in his preface:

> Between Switzerland and the United States there has always been a natural sympathy and friendship. Both aim to realize the idea of a government of freedom without license, and of authority without despotism; a government of law and order without a standing army; a government of the people, by the people, and for the people, under the sole headship of Almighty God. At the time of the Reformation, Switzerland numbered as many Cantons (13) as our country originally numbered States, and the Swiss Diet was then a loose confederation representing only the Cantons and not the people, just as was our Continental Congress. But by the revision of the Constitution in 1848 and 1874, the Swiss Republic, following the example of our Constitution, was consolidated from a loose, aristocratic Confederacy

> of independent Cantons into a centralized federal State, a popular as well as a cantonal representation. In one respect the modern Swiss Constitution is even more democratic than that of the United States; for, by the Initiative and the Referendum, it gives to the people the right of proposing or rejecting national legislation. But there is a still stronger bond of union between the two countries than that which rests on the affinity of political institutions. Zwingli and Calvin directed and determined the westward movement of the Reformation to France, Holland, England, and Scotland, and exerted, indirectly, a moulding influence upon the leading Evangelical Churches of America. George Bancroft, the American historian, who himself was not a Calvinist, derives the republican institutions of the United States from Calvinism through the medium of English Puritanism. A more recent writer, Douglas Campbell, of Scotch descent, derives them from Holland, which was still more under the influence of the Geneva Reformer than England. Calvinism breeds manly, independent, and earnest characters who fear God and nothing else, and favours political and religious freedom. The earliest and most influential settlers of the United States – the Puritans of England, the Presbyterians of Scotland and Ireland, the Huguenots of France, the Reformed from Holland and the Palatinate – were Calvinists, and brought with them the Bible and the Reformed Confessions of Faith. Calvinism was the ruling theology of New England during the whole Colonial Period, and it still rules in great measure the theology of the Presbyterian, Congregational, and Baptist Churches.[17]

The key themes in Schaff's work are the historicity of the church, spirituality, religious freedom, consensus rather than discord among the reformers, and institutional reform. He painted a dismal picture of the late-medieval world, but saw beacons of hope that had preserved the true faith. Schaff forged a narrative of the Reformation that ran as follows: Zwingli was the decisive reformer of Zurich and his influence led to the reformations in St Gall (Vadian), Berne (Haller), and Basle (Oecolampadius). With his death the Swiss Reformation essentially came to an end and leadership passed to Calvin in Geneva, who more or less followed Zwingli's ideas, though would not admit it. Bullinger and Beza were the internationalists of the movement and their work consisted entirely in the consolidation of the Reformation and in exercising influence over nascent reform movements throughout Europe. The whole of the Swiss Reformation created the Reformed culture that would shape the world of the Puritans who would create America.

Schaff's work is essentially about Zwingli and Calvin and the emergence of Reformed thought as faithful to the Apostolic tradition and the proper religious foundation of the Christian republic, though he argued that the freedom of the Reformed faith was the basis of the separation of church and state expressed in the American Revolution. It is suffused with Swiss patriotism combined with a great love for his adopted country, and he went to great pains to explain the relationship between the events of the Reformation and their implications for the changing world of the nineteenth century:

> In the sixteenth and seventeenth centuries Church and State, professing the same religion, had common interests, and worked in essential harmony; but in modern times the mixed character, the religious indifferentism, the hostility and the despotism of the State, have loosened the connection, and provoked the organization of free churches in several cantons (Geneva, Vaud, Neuchatel), on the basis of self-support and self-government. The State must first and last be just, and either support all the religions of its citizens alike, or none. It owes the protection of law to all, within the limits of order and peace. But the Church has the right of self-government, and ought to be free of the control of politicians.[18]

The nineteenth century also took a great deal of interest in John Calvin, though in a much more polarised manner. The plethora of histories of Geneva and biographies of Calvin embraced the spirit of returning to the sources, but virtually all were dominated by the question of whether he was a hero or villain. The most important history of Geneva to emerge in the early part of the century was Jacques and Augustus Galiffe's *Matériaux pour l'histoire de Genève* (1829–30). The Galiffes were a father and son team who belonged to an old Genevan family and this deeply influenced their perspective. They lamented Calvin's rise to prominence, which they saw as the destruction of the true Protestantism that had already taken hold in the city before Calvin's arrival. In common with much of the literature that would follow on Calvin they focused on his treatment of opponents such as the Perins, Bolsec, Castellio, and Servetus. Calvin was portrayed as the opponent of religious toleration (in a nineteenth-century sense) and his political dealings are detailed in a most unflattering light. This debate over Calvin and religious toleration would have a long life and dominate important lines of work on Servetus, Bolsec and especially Castellio.[19] A less polemical account came half a century later with the posthumous appearance of Amédée Roget's *Histoire du peuple de Genève depuis la réforme jusqu'à l'escalade* (1870–03). Although a professor in Geneva, Roget did not share the Galiffes' view and offered a fairly neutral assessment of Calvin that was dedicated to the use of sources made available during the nineteenth century. Roget's impartiality was much favoured by Schaff, who regarded his as the best history of Reformation Geneva.

These sources included editions of Calvin's writings and letters. The most important development was the production of a critical edition of Calvin's works by the Strasburg scholars G. Baum, E. Cunitz, and E. Reuss: *Johannes Calvini: Opera quae supersunt omnia*, for the *Corpus Reformatorum*. This series, which ultimately ran to more than fifty volumes, began appearing in 1863 and before its completion all three of the editors were dead. It includes Calvin's letters in ten volumes, which the Strasburg editors had tracked down with meticulous care. Alongside these volumes came the publication of Calvin's letters, together with those of other Swiss reformers, by the Swiss scholar A. S. Herminjard, who likewise did not live to see his work completed. His *Correspondance des réformateurs dans les pays de langue française, etc.* (1866–88) remains to this day not only an essential source on the French-speaking Reformation but the foundation for future work on the relationship between the German and French Swiss reformations.

The biographers of Calvin were also extremely busy. The first account of Calvin's life had been written by Beza (1564) and modified by his friend Nicolas Colladon. These accounts continued to be reprinted and enjoy wide circulation in the nineteenth century and they were included in the critical edition of Calvin's works. One of the most remarkable books to appear was Vincent Audin, *Histoire de la vie, des ouvrages et des doctrines de Calvin* (1841), whose account, according to Schaff, read like a piece of fiction. Audin's Catholicism is evident from the opening page and he sided resolutely with Calvin's numerous opponents, most prominently Jerome Bolsec. Calvin, Audin declared unequivocally, was a disaster for human civilisation. Interestingly, Audin's perspective, whilst shared by many, was not the universal Catholic judgement of Calvin. F. W. Kampschulte, who ultimately left the church to become an Old Catholic, was motivated by a dual devotion to sources and a spirit of ecumenism, and he found in Calvin an admirable portrait of spiritual renewal.[20]

There was a long string of eulogies of Calvin in French, German, and English, mostly written by Protestant clergy and highly derivative in character. The most significant works are those of Paul Henry, Ernst Staehelin, Abel Lefranc, and Emil Doumergue. Paul Henry was a minister of a French Reformed Church in Berlin, and in 1835–44 he produced his three-volume *Das Leben Johann Calvins des grossen Reformators*. The biography was explicitly written against Audin and the Galiffes and attempted to use the sources to present a much more enthusiastic view of Calvin. Liberal Protestantism had very much involved itself in the question of the relationship between the Reformation and the modern world. This endeavour is clearly seen in the work of Emil Doumergue who was without doubt the most influential of the biographers. His massive seven-volume study, entitled *Jean Calvin. Les hommes et les choses de son temps* (1899–1917), provided an exhaustive but by no means impartial account of the reformer's life. The work continued to be extremely influential in Reformed circles and to this day is often cited, particularly in reference to the Servetus affair, from which Doumergue fully exonerated Calvin. He portrayed Calvin as attempting to save the heretic and argued that the order for execution came not from the reformer but from a civic council that contained many opposed to Calvin. According to Doumergue Calvin was not only the hero of the Reformation but also the 'founder of the modern world'. This position was in accord with the Lutheran scholarship of Rietschl, who had made the same claim for Luther. This dimension of liberal thought was, of course, to receive its most influential treatment at the hands of Max Weber, whose *The Protestant Ethic and the Spirit of Capitalism* argued that Calvin's notion of vocation formed the basis of modern capitalism. Weber famously remarked about the reformers:

> They were not founders of societies for ethical culture nor the proponents of humanitarian projects for social reform or cultural ideals. The salvation of the soul alone was the centre of their life and work. Their ethical ideals and the practical results of their doctrines were all based on that alone, as were the consequences of purely religious motives. We shall thus have to admit that the cultural consequences of the Reformation were to a great extent, ... unforeseen and even unwished-for results of the labours of their reformers. They were often far removed from or even in contradiction to all that they themselves thought to attain.[21]

The carnage of the First World War took the wind out of the liberal sails and the concept of the Reformation as the foundation of the modern world lost momentum. The central figure in the refutation of the liberal theological / historical vision was the Swiss theologian Karl Barth, whose dialectical theology stressed the gulf between God and humanity and thereby any possible link between Protestantism and modernity. Barth had a life-long engagement with the classical works of the Protestant reformers with a particular interest in the writing of Jean Calvin, on whom he delivered a series of lectures at Gottingen University in 1922.[22] Although Barth rejected the nineteenth-century view of Calvin, his book on the Genevan reformer was an attempt to make Calvin comprehensible to the modern world.

In the period following the First World War, other areas of the French-speaking Reformation were also given attention and two works above all merit mention. The first is the monumental and still authoritative study by Henri Vuilleumier, *Histoire de l'Église réformée du Pays de Vaud sous le régime bernois*.[23] The second remains our essential study of Calvin's colleague and friend and the man who really linked the German and French Swiss reformations, Guillaume Farel: a collaborative effort published in 1930 entitled *Guillaume Farel, 1489–1565: biographie nouvelle: écrite d'apres les documents originaux par un groupe d'historiens, professeurs et pasteurs de Suisse, de France et d'Italie.*

Till the 1960s, twentieth-century work on the Reformation in Switzerland generally fell into a number of porous categories: cantonal histories, theology of the principal reformers, local histories, and Anabaptist studies. In the histories of the major cities which appeared during the twentieth century the Reformation naturally played a crucial role. Three of the most significant were K. Dändliker's *History of Zurich* (1908–10), Richard Wackernagel's *History of Basle* (1916) and Richard Feller's *Geschichte Berns* (1946–60). These impressive works form in themselves a map of Swiss historical thought in the twentieth century and reflect very closely the ongoing debate about the roles of historians and history in the Republic during the turbulent decades surrounding the two World Wars.[24] While it is unfair to elide them into one category, these cantonal histories reflected the growing sense of Swiss patriotism defined principally by its neutrality. The Reformation is here treated not so much as a religious event, though the ideas and actions of the reformers are discussed, but rather as a decisive moment in the formation of the modern Cantons. The term 'Fatherland' is frequently used and there was a clear effort to establish the distinct local and wider Swiss character which was generally thought to reside in a commitment to freedom, republican liberties, and the refusal to become entangled in foreign wars and affairs. Traditional Swiss values such as 'Heimat', institutional order, and the link between religion and freedom were stressed. Naturally, the stories of the major Protestant cantons were written from a distinctly confessional perspective, but religion was for the most part treated in institutional forms. The separation of so-called profane and religious history was strictly adhered to and these historians did not interest themselves in 'lived religion'.

Closely related to these larger narratives was the plethora of local histories that appeared from the end of the nineteenth century and continued in force during the twentieth. These works were generally drawn from archive sources and remain a rich mine of information on parish communities, though they were intended, naturally, to serve the great cause of local pride. Not infrequently they were doctoral dissertations written in local universities; in the case of the cities they could take the form of studies of particular institutions, such as marriage courts. The wealth of local history owed much to the growth in historical associations (*Vereine*) that began at the start of the century and to the journals produced by these bodies. *Tagebücher*, typical for virtually all of the Swiss cantons, became a focal point for short studies of communities, particular individuals, and significant events. Many of the authors were clergymen who served in these parishes. Yet the atomisation of this work is still striking. Bernese, Basle, Zurich, and Schaffhausen history continued to be published in cantonal journals and there was a very low level of cross-pollination.

A broader perspective is found in more theological works, as this field drew the attention of scholars from Germany, France, and the United States. A few examples must suffice here to draw attention to some highly significant developments. In the field of Zwingli studies the major landmark remains the meticulous re-creation of the Marburg Colloquy by the Heidelberg church historian Walter Koehler, *Zwingli und Luther* (1924). Koehler was a liberal theologian who emerged from the school of Adolf Harnack and he brought to his a task a formidable skill in textual criticism that profoundly reshaped the view of Zwingli. He followed this work with a short and more popular biography of Zwingli that remains one of the most lucid treatments of the reformer. During the twentieth century, scholarship on Zwingli was much influenced by the ongoing production of a critical edition that led professors of church history in Zurich to continue to engage closely with the reformer's work, though to the virtual abandonment of the other reformers in the city. While the critical edition was distinguished by scholarship of the highest quality, the biographies of Zwingli continued to fight old battles. Illustrative of this was the four-volume work by Oskar Farner, professor in Zurich, published 1943–60. Farner's work is rich in its use of the sources and provides a passionate account of the reformer's life, a strong corrective to the tendency to see him as a one-dimensional humanist. His biography is also distinguished by his intense interest in and sensitivity to spirituality, in which he picked up on the traditions of Pestalozzi and Staehelin. Zwingli appears as a man of spiritual renewal, a biblical scholar and preacher. The biography remains in many ways the best portrait of Zwingli the man, but it is by no means free of hagiography. Farner's own intense religiosity is reflected in both this study and in his other works on men such as Leo Jud.

Far and away the most significant scholar to emerge in the middle of the century was the half-Dutch Bernese theologian Gottfried Locher. Over a long and distinguished career Locher devoted himself to the study of Zwingli's theology in the broadest context, producing *Die Theologie Huldrych Zwinglis im Lichte seiner Christologie* in three parts: *Die Gotteslehre, Christologie und Abendmahl* and *Das christliche Leben*. Shortly after he retired he produced his opus magnum, *Die Zwinglische Reformation im Rahmen der europäischen Kirchengeschichte* (1979). Locher was by far the greatest Zwingli scholar of the twentieth century, though his interests were by no means parochial. As the title of his 1979 book suggests, he sought to place Zwingli in the context of the wider European Reformation. His collected essays, which have been translated into English, are dominated by the attempt to elucidate the points of contact and difference between Zwingli, Luther, and Calvin. Locher's interests also extended to the historical situation of the Reformation and he played a key role in the edition of the records for the 1528 Berne Synod. He was the first major Swiss scholar to link theology and historical research. His *Die Zwinglische Reformation* is a veritable catalogue of thinkers associated with Swiss Reformed thought and was a groundbreaking achievement. Locher demonstrated the wide circulation of 'Zwinglian' ideas throughout Europe, though it must be said that he rather overemphasises Zurich's influence in many cases (notably England and Scotland). Locher was also a theologian for whom the Reformation was relevant to the contemporary world and his emphasis upon social ethics, community, and existential relations between God and humanity very much shape his reading of Zwingli.

Without doubt the most emotive area of research has been work on the provenance of Anabaptism in the Swiss Reformation. Largely due to the confessional importance of the subject to scholars of the Mennonite tradition in North America, questions such as the nature of Swiss Anabaptism, infant baptism, the role of the magistrates, and the connections between Anabaptism and the Peasants' War remain fiercely contested. During the first part of the century the religious sociology of Ernst Troeltsch played a decisive role in the formation of North American, particularly Mennonite studies of the rise of Anabaptism. In the 1960s, as part of a wider change in Reformation studies, research on the radical reformers turned to more social/political models. Two classic works that had a profound influence on the field were Fritz Blanke's study of the Anabaptist community in Zollikon, *Brüder in Christo Die Geschichte der ältesten Täufergemeinde* (1955) and George Huston Williams's *The Radical Reformation* (1962). Blanke's work was informed by his own deeply pietistic beliefs and he regarded the emergence of the radicals as a tragic consequence of misunderstanding. Williams's magisterial work, which continues to be reprinted, placed Swiss Anabaptism as part of a broader rejection of Catholic theology and ecclesiology in the Reformation. In the late 1950s a circle of scholars around Harold Bender challenged the standard interpretations of Anabaptism as a doctrinal misfire. In the work of Harold Bender, Robert Friedman and J. C. Wenger there was an attempt to formulate a coherent theological and ecclesiological view of Anabaptism as a voluntary church that required believers' baptism for adults and a separation from political authority. This revisionist movement led to the rehabilitation of the principal radical leaders, and the subsequent work of Peter Clasen and James Stayer revealed that Anabaptism was much more than a one-dimensional revolt of the lower classes. Recent work on Anabaptism is also discussed elsewhere in this volume.

On Calvin the most important scholar through the middle decades of the century was Wilhelm Niesel, whose work dates back to the early part of the century with his five-volume *Opera Selecta*, which he co-edited with Peter Barth. Among Niesel's numerous studies are *Calvins Lehre vom Abendmahl* (1930) and his *Reformed Symbolics* (trans. David Lewis, 1962). Two other works that were of the first rank in their importance for Calvin scholarship were Edward Dowey's *The Knowledge of God in Calvin* and Francois Wendel's *Calvin: Origins and Development of His Religious Thought* (1963). There has been no let-up in work on Calvin himself. Of particular note are the work of Max Engammare on Calvin's sermons, the efforts to produce new editions of Calvin's *Opera Exegetica*, and an ambitious Dutch-led plan to produce a new edition of Calvin's letters in approximately forty volumes. To this we must add the various endeavours to write a biography of Calvin. Most controversial was William Bouwsma's psychological study (*John Calvin: a Sixteenth-Century Portrait*, 1988), which spoke of Calvin's fear of chaos and the abyss and his response to this fear in creating order. This interesting perspective has been fiercely attacked by Richard Muller, who has persuasively shown that Bouwsma relied too heavily on modern ideas and not enough on the sixteenth-century context of Calvin's language.[25] Two recent studies on Calvin in French must also be noted: Denis Crouzet's *Jean Calvin, Vies parallèles* (2000) and Bernard Cottret's *Calvin: a Biography* (2000). Both authors have read extensively Calvin's vast body of work and seek to place the reformer in his historical context, but it is Crouzet who manages to provide a more engaging and multidimensional study.

Bernd Moeller's classic study of the impact of the Reformation on urban cultures was the catalyst for a new generation of studies in the 1960s that examined more closely the institutional character of the Reformation and the relationship between religious reform and social and economic conditions. For Geneva two American scholars, Robert Kingdon and William Monter, had already established a tradition of historical study of religious reform.[26] Kingdon also played a key role in the editing and publication of key document collections, such as those for the Company of Pastors.[27] This work is now supplemented by the massive new critical edition of the *Registres du Consistoire de Genève au temps de Calvin* edited under Kingdon's guidance by Tom Lambert and Jeff Watt, both of whom have written extensively on religious and social life in the city.

Our understanding of the relationship between Calvin and Geneva was radically revised in 1996 with the publication of William G. Naphy's *Calvin and the Consolidation of the Genevan Reformation*. Naphy persuasively demonstrated how Calvin was able to work effectively within the political world of Geneva, often by acting quite mendaciously against his opponents. Calvin's role was further explored by Kingdon's use of the consistory records in his *Adultery and Divorce in Calvin's Geneva* (1995). Kingdon's students have continued to produce studies of various dimensions of religious and social life in Geneva.

Contemporary research on the German-speaking Reformation flourished in Zurich under the guidance of Leonhard von Muralt and Fritz Büsser, who founded the monograph series *Zürcher Beiträge zur Reformationsgeschichte* which makes available the work (mostly doctoral dissertations) of a new generation of historians examining issues such as the political leadership in Zurich and the wider scope of the Kappel Wars. Muralt himself had in 1930 made a major contribution to the study of the Swiss Reformation by his pioneering comparative examination of Zurich, Basle, and Berne.[28] By the middle of the 1970s research on Zurich had far outstripped the other major areas of the Swiss Reformation, with the inevitable result that the subject was skewed towards Zwingli's city. Out of this period came two major works on Zwingli that have remained essential reading: Robert C. Walton's *Zwingli's Theocracy* (1967), and the beautifully written and meticulously researched biography of the reformer by George Potter, which has now established itself as the standard work in the field, though curiously it has not received wide attention in the German-speaking world.[29] Walton engaged with the emerging school of work on Anabaptism and offered the most comprehensive study of institutional reform in Zurich. Potter's shrewd and non-confessional depiction of Zwingli was remarkable for its attention to the wider dimensions of events in Zurich. He drew heavily upon the earlier work of Koehler as well as the social history movement that had begun in the 1960s. Zwingli comes across as a complex individual whose work and thought must be seen in the context of the rapid events of the 1520s. Potter was not a theologian and the principal weakness of the study lies in the rather mechanical manner in which he reviewed Zwingli's ideas. As a study of Zwingli's relations and the political and social background to the reform movement in Zurich, however, the work has not been surpassed in any language.

In 1975 the four hundredth anniversary of Bullinger's death was marked by a conference in Zurich under the leadership of Fritz Büsser, initiating a new stage in scholarship that shifted the focus to the period beyond 1531.[30] An ambitious plan was unveiled to edit Bullinger's works and a start was made to producing editions of his writings, but with limited success. More successful was a critical edition of Bullinger's correspondence, at over twelve thousand letters the largest surviving collection of any sixteenth-century reformer. This project continues to this day, and has already made available a wealth of material for a new generation of scholars of the Swiss Reformation and its influence abroad. In the wake of the 1975 conference a series of monograph studies also opened up the post-1531 world. The most significant was Hans Ulrich Bächtold's *Heinrich Bullinger Before the Council* (1982).[31] As the first in-depth study of the nature of post-Zwingli church, Bächtold's work demonstrated how Bullinger operated behind closed doors to hold together the extremely fragile alliance between the church and the civic magistrates. The book revolutionised the field by undercutting previous assumptions about the nature of the Zwinglian ecclesiology. It remains a seminal study and deeply influenced the more recent scholarship of Pamela Biel, Bruce Gordon, and Andreas Mühling.[32]

The late 1970s also saw the field profoundly changed by work done by scholars outside Switzerland. Particularly significant contributions came from Bob Scribner, Tom Brady, and Hans-Jürgen Goertz. Scribner's study of local religion and popular piety made frequent use of Swiss material and initiated the study of religion from a non-theological perspective. The most radical and controversial reinterpretation of the field came from the German scholar Peter Blickle, who became professor of history in Berne. Blickle had begun his work on the Peasants' War and had cultivated strong contacts with Marxist historians of East Germany, notably Gunther Franz. Blickle's work on the Swiss Reformation very much complemented

Scribner's, in that both looked to lines of continuity in religious practice between the Middle Ages and the early modern period. In Blickle's seminal *The Communal Reformation*, he argued that the situation in the Alpine regions was unique in that a strong communal movement had existed from the late medieval period and that this had profoundly shaped the reception of the Reformation.[33] Blickle demonstrated that this communal movement had taken the form of a desire to control the sacred through election of ministers and the appropriation of church goods and finances. This meant that the rural communities were particularly receptive to the communal message of Zwingli and that they saw in his preaching an affirmation of what they had already begun. Blickle contrasted Zwingli with Luther's more authoritarian approach and sought to demonstrate the particular attractiveness of the Zurich reformer in both Swiss and southern German lands. He argued that this period of communal reformation came to an end with the Peasants' War of 1525, and was replaced by the magisterial or princely Reformation. In Goertz's work the early Reformation is examined in terms of late medieval anticlericalism. Goertz made extensive use of Swiss material in formulating his important argument on the emergence of the reform movement.[34]

Blickle's thesis was highly controversial and was much attacked by scholars such as Heinz Schilling, but its influence was without doubt. His numerous students undertook detailed research on localities throughout the Swiss lands in the 1520s opening up a whole new era of scholarship. Three of the most significant scholars to emerge from this school were Heinrich Richard Schmidt, Francizka Conrad, and Andreas Holenstein.[35] Schmidt's 1986 dissertation on the reform in the southern imperial cities was a major advance in the urban studies begun by Moeller. Schmidt demonstrated that the nervous ruling magistrates had been deeply affected by the preaching and the response to the religious ideas among the guilds. He argued that there was no easy dovetailing of the reformers' ideas with urban reform and that the process involved a complex interplay of religious, political, and economic factors. It was at this time that Hans Guggisberg, author of *Basel in the Sixteenth Century: Aspects of the City Republic Before, During, and After the Reformation* (1982), gave a series of lectures on the Basle Reformation making a similar case, marking a major revision of the subject.

Although Schmidt's work was unfortunately never translated into English, it formed part of a series of important studies of urban reformations and of the relations between urban and rural territories to emerge in the mid-1980s. All of these took part in the debate launched by Blickle. Two of the most notable were Tom Brady's *Turning Swiss: Cities and Empire, 1450–1550* (1985) and Tom Scott's *Freiburg and the Breisgau: Town-Country Relations in the Age of Reformation and Peasants' War* (1986). Brady argued that there was a time in the first two decades of the sixteenth century when the southern German cities were strongly attracted by the communal traditions of the Swiss. Scott was more explicitly critical of Blickle, but he made major advances in the field by demonstrating the importance of trade and communication systems in the propagation of religious ideas. The most recent major study to follow in this line is Randolf Head's *Early Modern Democracy in the Grisons: Social Order and Political Language in a Swiss Mountain Canton, 1470–1620* (1995), a seminal treatment of the Reformation in the canton of Graubunden.

Recent scholarship on the Swiss Reformation has built on the work of Blickle, Brady and others who have brought the subject into a more comparative, international context. Heinrich Richard Schmidt has made a major contribution to the discussions of confessionalisation and moral disciplining in early modern Europe with his massive study of the moral courts in Berne, *Dorf und Religion. Reformierte Sittenzucht in Berner Landgemeinden der Frühen Neuzeit.* This examination of the interrelationships between religion and social history have also informed the more recent study of Franciska Loetz on swearing and blasphemy in Reformation Zurich, *Mit Gott handeln: Von den Zürcher Gotteslästerern der Frühen Neuzeit zu einer Kulturgeschichte des Religiösen* (2002). For Basle, one of the most exciting developments in recent years has been the exploitation of the extremely rich world of Basle printing and the city's diverse religious culture in the sixteenth century. Following from the research of Hans Guggisberg on Castellio, Carlos Gilly produced in 1985 his *Spanien und der Basler Buchdruck bis 1600: ein Querschnitt durch die spanische Geistesgeschichte aus der Sicht einer europäischen Buchdruckerstadt,* which explored the international context of Basle printing and the backgrounds of some of the most prominent printers in the city. This work needs to be set alongside Peter Bietenholz's groundbreaking study, *Basle and France in the Sixteenth Century* (1971).

The most recent work has concentrated on textual editions and reassessments of the primary reformers. In addition to the Genevan consistory records, and the above-mentioned letters of Calvin and Bullinger, critical editions of Beza's letters continue to be made available.[36] In 2005 there will be a major international congress on the work of Beza, and under the leadership of Irena Backus and Max Engamarre Geneva remains a centre of research into Erasmus and Calvin. In Zurich, there has been a concentration on the work of Heinrich Bullinger. In 2004 several new publications appeared, including a collection of essays on the reformer and a major study of his *Decades*.[37] In addition, critical editions of Bullinger's *Decades* and of his *Reformation History* are in preparation. In Berne, Andreas Holenstein is examining rural and urban history in the late medieval and early modern periods with an emphasis on the Bernese context, and he will soon be producing a major study entitled *Berns mächtige Zeit. Das 16. und 17. Jahrhundert neu entdeckt,* which will completely revise our understanding of the city and its territory in the Reformation period. In Basle, Kaspar von Greyerz is leading a project on self-portrayal in the early modern period that is making use of the rich seam of material presented by autobiographical literature. In all these areas theology and history are fully in conversation and important current themes such as communication networks, identity, gender, and social structures are treated.

further reading

The only comprehensive overview of the subject is Bruce Gordon, *The Swiss Reformation* (Manchester, 2002). Essential reading are Peter Blickle, *The Communal Reformation* (New Jersey, 1992); Tom Brady, *Turning Swiss: Cities and Empire 1450–1550* (Cambridge, 1985); Randolf Head, *Early Modern Democracy in the Grisons: Social Order and Political Language in a Swiss Mountain Canton, 1470–1620* (Cambridge, 1995); Glen Ehrstine, *Theater, Culture, and Community in Reformation Bern, 1523–1555* (Leiden, 2002), and Philip Benedict, *Christ's Churches Purely Reformed: A Social History of Calvinism* (New Haven, 2002).

On Zurich see G. R. Potter, *Zwingli* (Cambridge, 1976); E. J. Furcha and H. Wayne Pipkin (eds), *Prophet, Pastor, Protestant: The Work of Huldrych Zwingli after Five Hundred Years* (Allison Park, 1984); Ulrich Gabler, *Huldrych Zwingli: His Life and Work* (Philadelphia, 1986); Lee Palmer Wandel, *Always Among Us: Images of the Poor in Zwingli's Zurich* (Cambridge, 1990); Lee Palmer Wandel, *Voracious Idols and Violent Hands: Iconoclasm in Reformation Zurich, Strasbourg, and Basel* (Cambridge, 1995); Iren L. Snavely, 'Zwingli, Froschauer, and the Word of God in Print', *Journal of Religious and Theological Information* 3/2 (2000), 65–87; Pamela Biel, *Doorkeepers in the House of Righteousness: Heinrich Bullinger and the Zurich Clergy 1535–1575* (Berne, 1991); Bruce Gordon, *Clerical Discipline and the Rural Reformation: the Synod in Zurich, 1532–1580* (Berne, 1992).

On Basle, see Hans Guggisberg, *Basel in the Sixteenth Century: Aspects of the City Republic Before, During, and After the Reformation* (St Louis, 1982). On Anabaptism, Leland Harder (ed.), *The Sources of Swiss Anabaptism* (Scottdale, 1985). On religious radicals, Mark Taplin, *The Italian Reformers and the Zurich Church, c.1540–1620* (Aldershot, 2004).

On Calvin and Geneva, Bernard Cottret, *Calvin: a Biography* (Grand Rapids, 2000); Alister E. McGrath, *A Life of John Calvin: a Study in the Shaping of Western Culture* (Oxford, 1990); Richard Muller, *The Unaccommodated Calvin: Studies in the Foundation of a Theological Tradition* (Oxford, 2000); William Naphy, *Calvin and the Consolidation of the Genevan Reformation* (Manchester, 1994); Jeff R. Watt, *Choosing Death: Suicide and Calvinism in Early Modern Geneva* (Kirksville, 2001), and Karen E. Spierling, *Infant Baptism in Reformation Geneva: the Shaping of a Community, 1536–1564* (Aldershot, 2005).

notes

1. For a recent review of scholarship in the field see the 'Selected Bibliography' in my *Swiss Reformation* (Manchester, 2002). A useful annual review of current literature is found in the Swiss journal *Zwingliana*.
2. On Schaff, see Gary K. Pranger, *Philip Schaff (1819–1893): Portrait of an Immigrant Theologian* (New York, 1997).
3. Heinrich Bullinger, *Reformationsgeschichte*, ed. J. J. Hottinger and H. H. Voegeli, 3 vols (Frauenfeld, 1838–40).
4. Heinrich Bullinger, *The Decades*, ed. T. Harding (Cambridge, 1849–51), from the London edition of 1577.
5. Robert Kolb, *Martin Luther as Hero, Teacher, Prophet: Images of the Reformer 1520–1620* (Grand Rapids, 2000).
6. Bruce Gordon, 'Calvin and the Swiss Reformed Churches', in Andrew Pettegree, Alistair Duke, and Gillian Lewis (eds), *Calvinism in Europe, 1540–1620* (Manchester, 1996).
7. Bruce Gordon, '"The second Bucer": John Dury's Mission to the Swiss Reformed Churches in 1654–55 and the Search for Confessional Unity', in John M. Headley, Hans J. Hillerbrand and Anthony J. Papalas (eds), *Confessionalization in Europe, 1555–1700* (Aldershot, 2004).
8. This view is laid out in Karl Holl, *Gesammelte Aufsätze zur Kirchengeschichte*, 3 vols (Tübingen, 1921–28).
9. Richard Feller and Edgar Bonjour, *Geschichtsschreibung der Schweiz*, vol. II (Basle and Stuttgart, 1962), pp. 614–18.
10. On Swiss historical writing in the nineteenth century see Oliver Zimmer, *A Contested Nation: History, Memory and Nationalism in Switzerland 1761–1891* (Cambridge, 2003).
11. Richard Feller and Edgar Bonjour, *Geschichtsschreibung der Schweiz*, 2 vols (Basle, 1962, 1979).
12. The best treatment of this in English is Lionel Gossman, *Basel in the Age of Burckhardt: a Study of Unseasonable Ideas* (Chicago, 2000).
13. Carl Pestalozzi, *Leo Judä: Nach handschriftlichen und gleichzeitigen Quellen* (Elberfeld, 1860).
14. The best study of this period in English is Gordon A. Craig, *The Triumph of Liberalism: Zurich in the Golden Age, 1839–1869* (New York, 1988).
15. Georg Finsler and Rudolf Finsler, *Diethelm Georg Finsler, der letzte Antistes der zürcherischen Kirche* (Zurich, 1916–17).

16. See Klaus Penzel, *The German Education of Christian Scholar Philip Schaff: the Formative Years, 1819–1844* (Toronto, 2004).
17. Philip Schaff, *History of the Christian Church*, 3rd edn, vol. VIII (New York, 1892).
18. *Ibid.*
19. Consult the historiography of this debate in the preface to Hans R. Guggisberg, *Sebastian Castellio, 1515–1563: Humanist and Defender of Religious Toleration* (Aldershot, 2003).
20. F. W. Kampschulte, *Johannes Calvin, seine Kirche und sein Staat in Genf* (Leipzig, 1869).
21. Max Weber, *The Protestant Ethic and the Spirit of Capitalism*, trans. T. Parsons (London, 1930), p. 90.
22. Available in English as Karl Barth, *The Theology of John Calvin*, trans G. Bromiley (Grand Rapids, 1995).
23. Four vols (Lausanne, 1927).
24. Ulrich Im Hof, *Mythos Schweiz. Identität – Nation – Geschichte. 1291–1991* (Zurich, 1991).
25. Richard A. Muller, *The Unaccommodated Calvin: Studies in the Foundation of a Theological Tradition* (Oxford, 2000). Muller also provides a useful examination of Calvin historiography.
26. Robert M. Kingdon, *Geneva and the Coming of the Wars of Religion in France, 1555–1563* (Geneva, 1956); idem, *Geneva and the Consolidation of the French Protestant Movement, 1564–1572* (Geneva, 1967); William Monter, *Calvin's Geneva* (New York, 1967).
27. Robert M. Kingdon and Jean-François Bergier (eds), *Registres de la Compagnie des Pasteurs de Genève au temps de Calvin*, 2 vols (Geneva, 1962–64).
28. Leonhard von Muralt, 'Stadtgemeinde und Reformation in der Schweiz', *Zeitschrift für schweizerische Geschichte*, 10 (1930), 349–84.
29. G. R. Potter, *Zwingli* (Cambridge, 1976).
30. The proceedings of this influential conference were published in Ulrich Gäbler and Endre Zsindely (eds), *Bullinger-Tagung 1975. Vorträge, gehalten aus Anlaß von Heinrich Bullingers 400. Todestag* (Zurich, 1977).
31. Hans Ulrich Bächtold, *Heinrich Bullinger vor dem Rat. Zur Gestaltung und Verwaltung des Zürcher Staatswesens in den Jahren 1531 bis 1575* (Berne, 1982).
32. Pamela Biel, *Doorkeeper at the House of Righteousness: Heinrich Bullinger and the Zurich Clergy 1535–1575* (Berne, 1991); Bruce Gordon, *Clerical Discipline and the Rural Reformation* (Berne, 1990); Andreas Mühling, *Heinrich Bullingers europäische Kirchenpolitik* (Berne, 2001).
33. Peter Blickle, *Gemeindereformation: die Menschen des 16. Jahrhunderts auf dem Weg zum Heil* (Munich, 1985).
34. Hans-Jürgen Goertz, *Pfaffenhass and gross Geschrei* (Munich, 1987).
35. Heinrich Richard Schmidt, *Reichsstädte, Reich und Reformation. Korporative Religionspolitik 1521–1529/30* (Stuttgart, 1986).
36. Theodore Beza, *Correspondance de Théodore de Bèze*, ed. Hippolyte Aubert (Geneva, 1960–). The series has now reached volume 26.
37. Bruce Gordon and Emidio Campi (eds), *Architect of Reformation: an Introduction to Heinrich Bullinger, 1504–1575* (Grand Rapids, MO, 2004); Peter Opitz, *Heinrich Bullinger als Theologe. Eine Studie zu den Dekaden* (Zurich, 2004). The latter is the first full examination of Bullinger's theological method.

4
the low countries

judith pollmann

an eclectic reformation

The Low Countries have something of a special place in Reformation history, both because of the great diversity of religious groups that emerged there, and because the Habsburg Netherlands were a place where the Reformation undoubtedly and genuinely emerged 'from below', against all the odds of a government repression that was fiercer than anywhere else in Europe. Yet, however popular, the Netherlandish Reformation is also a very slippery beast. First, its historiography has been bedevilled by the political outcome of the reformation process. The Dutch Revolt split the sixteenth-century Habsburg Netherlands into a Dutch Republic, the predecessor of the current kingdom of the Netherlands, and an area known as the 'Southern' Netherlands, that continued to be ruled by various branches of the Habsburg family for another two centuries, and that developed into the kingdom of Belgium. Not only is this confusing for students (the terms 'Netherlands' and 'Netherlandish' in this chapter refer to all of the Low Countries, whilst the term 'Dutch' refers only to the new Dutch Republic and its successor); scholars have also often projected these distinctions back onto the pre-Revolt era. Thus, it has taken long to uproot the idea that there had *always* been different religious cultures in North and South; while the main cultural divide was arguably that between the urbanised, literate Western provinces of Holland, Zeeland, Brabant and Flanders and the less developed areas in the East.

Yet there are other reasons, too, why the Netherlandish Reformation has often seemed intractable. Although some Netherlandish thinkers, notably Cornelis Hoen, contributed directly to early Reformation debates, the theological and practical links between Netherlandish dissidents and the great reformers were often complex and fraught.[1] From the start there emerged a very wide spectrum of dissenting religious views, only some of which can be related meaningfully to emerging new churches, or even sects. Besides Lutherans and Calvinists, there were not only the Anabaptists, who developed into many shades of Mennonite; but also the smaller sects like the David Jorists, the Family of Love, and those known as 'spiritualists', 'libertines' and 'sacramentarians', whose views did not always amount to more than a rejection of some aspects of traditional religion. Throughout the later nineteenth and early twentieth century, historians spilled much ink in trying to assign a clear religious identity to early dissenters and to define 'phases' in the Netherlandish Reformation – an effort that was closely related to establishing the weight of each group.

From the 1940s, historians began to question the usefulness of this approach, and became more aware of the diffuse character of the early Reformation especially. This struck an important chord with political historians who were pleading for more attention for the 'middle groups' in the sixteenth-century political landscape.[2] J. J. Woltjer, in *Friesland in Hervormingstijd* (1962) and much later work, highlighted that Netherlandish society before and during the Revolt could not be divided between clear 'Protestant' and 'Catholic' camps, but that there were many shades of religious allegiance. Johan Decavele's *De dageraad van de Reformatie in Vlaanderen* (1975), and the seminal essays that Alastair Duke began to publish about the early Reformation, confirmed that the emphasis should be on diversity. Even if the authorities (and later scholars) had confidently assigned labels to dissenters ('lutheran', 'anabaptist', 'sacramentarian'), most evidence suggests that dissenting believers were highly eclectic in their adoption of new ideas.[3] As long as there were no separate churches, and books were rare, many people acquired their dissenting views piecemeal, from conversations, the odd daring sermon, ephemeral publications, ballads, and the refrains and plays that were performed by the *rederijkerskamers*, or chambers of rhetoric – the poetry societies whose membership included members of the elite as well as artisans.[4]

The Low Countries were a highly urbanised and very literate society. While modern scholarship has found that Netherlandish priests were by no means as 'corrupt' and poorly educated as scholars used to believe, the 'knowledge gap' between the priesthood and part of the (male) laity had undoubtedly been shrinking.[5] Moreover, although Erasmus' teachers from the *Devotio Moderna* were not the proto-Protestants scholars once imagined, they had taught laypeople that they could not just leave religion to the professionals. Even in the fifteenth century, the *rederijkerskamers* had habituated their members to 'scriptural' thinking, and to imagining themselves in the role of popular preachers.[6] In the early decades of the sixteenth century there was an ever-growing number of vernacular Bibles for sale, and the first vernacular Psalter, the *Souterliedekens* (1539), became hugely popular. Yet scripture reading did not automatically turn people into Protestants. In the land of Erasmus opponents of Lutheranism were just as proud of their 'scripturality' as their opponents.[7] Neither was anticlericalism confined to heretics; both the fiscal privileges of the religious orders and the changing intellectual climate had, for example, created an endemic anti-monasticism.[8]

Since Luther disapproved of clandestine conventicles, dissenters in the Netherlands did not form evangelical churches along the German model. Yet they continued to read the works of Luther and other evangelical thinkers, and many developed their own dissenting views, *sola scriptura*. While in Germany the priesthood of all believers had been abandoned after the Peasants' War, in the Netherlands the reformation of the common man continued unfettered by clerical interventions. Charismatic lay leaders like Hendrik Niclaes and David Joris, or the elusive Loists, constantly developed new strands of reformist thought.[9]

The recognition that only few of the early Netherlandish dissenters were, or aspired to be, card-carrying heretics, has had profound implications for the understanding of the early Reformation. As long as few people experienced their dissent as a form of 'conversion', and made little effort to organise themselves along doctrinal lines, it remained hard to tell them apart from the many critical, scripture-reading and anticlerical Catholics. The Habsburg authorities ignored this problem and handled heresy as a form of sedition rather than as a 'thoughtcrime'. They criminalised discussions about doctrine by non-theologians, but also the 'aiding and abetting' of heretics and the possession of heretical books, for instance. For local authorities, however, it was precisely the hybrid and fluid way in which heresy manifested itself that made it undesirable to punish it as severely as the crown wanted. Too many magistrates seem to have recognised at least some of their own views in the errors of those whom the government deemed to be 'heretics'.[10]

Yet if we want to argue that dissent could spread in the Netherlands precisely because it was so eclectic and unconfessional, we also need to explain how and why two groups of early dissenters, Anabaptists and Calvinists, went further and set up separate communities. Until 1950 the historiography of early Anabaptism in the Netherlands was shaped by one overriding concern – to explain, and preferably explain away, the relationship between the later communities and the radical Anabaptists who in 1534–35 in Münster had tried to forge the New Jerusalem. This concern was as old as the defeat of the Münsterites itself – and had been particularly urgent for those Anabaptists who had regrouped into strictly pacifist communities, who in the Netherlands became known as *doopsgezinden*. Thus the conversion narrative of the *doopsgezinde* leader Menno Simons, whose name lives on today in the Mennonite churches, was profoundly shaped by Menno's concern to distance himself from those Anabaptists who had accepted a baker as their king, run around naked in preparation for the second coming, practised polygamy, instituted community of goods, and tried to overthrow the government of Amsterdam as well as that of the city of Münster in Westphalia.[11]

Although there was a growing interest among Marxist historians in the radical Reformation since at least 1900, most *doopsgezinde* historians in the Low Countries were very resistant to attempts to see early Anabaptists as proto-revolutionaries. Although not all of his conclusions have stood up to subsequent analysis, the 1954 study by A. F. Mellink was, then, a real breach with the historiographical tradition. Mellink argued that the movement had especially appealed to the poor, and to those hit by the economic crisis that struck in the province of Holland in the 1530s.[12] Since then, much more work has been done to contextualise the movement, the events in Münster have been charted with more precision, and the diversity of the Anabaptist response both during and after Münster has been highlighted. The transformation in Netherlandish Anabaptist studies was crowned by the appearance of Samme Zijlstra's authoritative new book on early modern Anabaptism in the Netherlands in 2000.

This modern scholarship cites a combination of reasons why a sudden following emerged in the 1530s Netherlands for the millenarian prophet Melchior Hoffmann. By offering adult baptism to Netherlandish believers, Hoffmann created a sacramental distinction between dissenting and traditional believers – a distinction that became all the more significant when the Netherlandish Anabaptists came to believe that only 144,000 rebaptised people would be saved on the day of judgment whose imminence Hoffmann preached. Because the Anabaptists developed Bible-reading conventicles into a very simple form of organisation, in which small communities could worship without needing any clerical input, the movement could spread easily. Moreover, the movement built on and developed a strong current of apocalyptic expectation – further fired by economic hardship. This is why so many people from the Northern Netherlands, especially, heeded the call to come to Münster to await the end of time there.[13]

After the defeat and gruesome punishment of the Münsterites in 1535, Netherlandish Anabaptists regrouped through a highly effective system of conventicles. These were guided by Menno Simons as well as lesser known leaders like Leenaert Bouwens and David Joris, but precisely because their gatherings were led by lay elders, because worship happened in private homes, and because they were pacifist and rejected links with the secular powers, these groups could remain relatively invisible. Joris even recommended that believers should continue to attend Mass so as to avoid suspicion (so-called 'Nicodemism'). This made it difficult for the authorities to suppress them. Verheyden, Decavele and Marnef have shown that by the 1560s, there was a considerable Anabaptist presence in the Southern provinces,[14] whilst in the later sixteenth century there were probably more Mennonites than Calvinists in the Dutch Republic.[15]

Even so, Calvinists had become ever more visible and more dominant in the religious landscape. It was only in 1555, and probably under the influence of Calvin's rejection of Nicodemism, that the first of his Netherlandish followers made a clean break with the Catholic church, and formed separate churches 'under the cross'. Like the Anabaptist gatherings, Reformed churches in the Netherlands (as in France) were organised from the bottom-up, and did not depend on a hierarchical structure that would have made them more vulnerable to persecution. Yet more than the Anabaptists, Calvinists in the Netherlands benefited from an international network, and from upper- and middle-class support. Although fewer nobles were attracted to Calvinism in the Netherlands than in France, the Reformed recruited adherents among lawyers and merchants as well as artisans.[16]

Calvinism reached the Netherlands from different sources. Unsurprisingly, we find some of the earliest Calvinist activity among the French-speaking Walloon communities on the French border; by the early 1560s, Tournai and Valenciennes were heaving with Reformed activity. The first confession the Netherlandish Reformed churches adopted in 1561, written by Guy de Brès, also was the spitting image of that of the French churches.[17] Yet it used to puzzle scholars that there is so little evidence that Calvin himself coordinated, or even encouraged, the spread of Calvinism in the Netherlands. Recent research has solved that riddle, by showing that many Netherlandish Calvinists learned of the Reformed faith not in the Low Countries, but whilst they were in exile. Dissident believers who had been found in breach of the Habsburg heresy placards and had been banished, or who had fled, had few possible refuges to choose from: principally London in England; Emden, Wesel, Frankfurt and Strasburg in the Holy Roman Empire; and also Geneva, Zurich and Basle. Here many of them first encountered a coherent version of Reformed thought. Through the work of Decavele and Pettegree, especially, we have come to realise that people who had left the Netherlands as eclectic dissenters became Calvinists while in exile, and then exported their new faith back to their native lands. Particularly important for the development of the Netherlandish Reformed churches was Jan Łaski, or John a Lasco, a Polish theologian who played a key role in the stranger churches of London and Emden. Emden came to function as the 'mother church' for the churches under the cross; and from 1558, when Elizabethan England readmitted Protestant refugees, the contact between exiles in England and the Southern Netherlands, especially, intensified. It was members of the London stranger churches who first brought Calvinism to the West Flanders countryside where, in August 1566, the great iconoclasm was to begin.[18]

repression and revolt

Until quite recently, scholars relied on martyrologies to establish the numbers of victims of the heresy legislation in the Netherlands. Recent scholarship has established that martyrs' books, however interesting in their own right, are inappropriate sources for quantitative research, not least because martyrologists tended to omit anyone whose orthodoxy was questionable.[19] Using court records and other sources, scholars have now arrived at much better estimates of the scale of repression. These have confirmed that the repression of heresy in the Low Countries was fiercer than anywhere else in Europe, but have also shown that (with the important exception of Flanders, which had a very active inquisitor) by the late 1550s the executions in the Low Countries had ground to a virtual halt.[20] Even many Catholics had by now developed a thorough distaste for the way in which the crown handled heresy. In some cases this was for commercial reasons; Antwerp did not want to lose the German Lutheran merchant community, for instance. Yet mostly it was based on a disapproval of the legal tools which the crown used. Having turned heresy into a form of treason (violation of divine majesty), the crown argued that local legal privileges no longer applied in heresy cases. Crucially, as Duke has shown, this affected local rights, like the limit on confiscations and the right to be tried in one's own city, that were central to the benefits of citizenship.[21]

Moreover, there was a growing fear that worse infringements on the subjects' rights were round the corner. The plan to reorganise the archaic episcopal structure of the Netherlands that was unveiled in 1561 was sensible, but it was explicitly connected to Philip II's desire to counter heresy, and that was one of the many reasons why it met with widespread opposition. Much of this fear came to be expressed in the rumour that the crown sought to introduce the 'Spanish inquisition' in the Netherlands.[22] As generations of Catholic historians have noted, the crown had no such intention. A bemused King Philip II pointed out that the existing placards were, in fact, more draconian than rules by which the Spanish inquisition operated (which allowed mercy for those who were prepared to renounce their views).[23] Yet as scholars have come to realise, to his Netherlandish subjects that was not the point – to fear the 'Spanish' inquisition meant that one feared arbitrary justice.[24]

The Netherlandish nobles had many reasons to resent crown policies, but both among the grandees and lesser nobles there was, as Henk van Nierop has argued, also a genuine and growing conviction that the stringency of the crown about heresy was a political mistake. In their view the situation in the Low Countries had become a European anomaly. What was possible in the Holy Roman Empire, what was now being tried in France, surely was the road forward. William of Orange, the mightiest nobleman in the Netherlands, had already granted freedom of worship to Protestants in his principality of Orange. Several other nobles reformed their own estates.[25]

On 5 April 1566, three hundred armed noblemen, soon nicknamed *gueux* or 'beggars', marched into the palace of the Habsburg regent, Margaret of Parma, in Brussels with a petition demanding an end to the heresy laws. The regent referred the matter to Madrid, but in the meantime suspended the placards. Soon the country was awash with Reformed activity. Returned exiles were preaching open-air sermons (*hagepreken*, or 'hedgepreachings') just outside the cities that attracted hundreds, and sometimes thousands of people. As summer progressed, and activists became ever bolder, thoughts were turning to the acquisition of buildings in which the preachings might continue in the winter months. Existing churches could be used, but they would need 'cleansing'. On 10 August 1566 the first images were toppled in Steenvoorde and in the ensuing weeks, most cities in the Netherlands witnessed outbreaks of iconoclasm in one or more churches. These were very violent in some places; in others the secular authorities pre-empted violence by stripping one or more churches of imagery themselves.[26]

Unsurprisingly, the iconoclasm has been one of the most hotly contested issues in Netherlandish Reformation history. Reformed historians, who disapproved of the violence as much as anyone else, were traditionally inclined to blame the violence on 'senseless mobs'. Others pointed to the evidence that preachers had encouraged, and elders had taken part in, the image breaking. Kuttner's attempt to relate the events of 1566 to the misery of the proletariat in the West Flanders countryside and rising grain prices briefly seemed to strengthen the former interpretation of events, yet his theory has not, in fact, proven tenable. The work of Phyllis Mack Crew and Andrew Pettegree has confirmed that there can be little doubt of the Calvinist leadership's involvement in encouraging the violence – even if they had not always been able to control it.[27] David Freedberg and Alastair Duke have also pointed to the enormous cultural and religious significance of image-breaking for the Reformed. Many image breakers exposed images to ritual humiliation or 'dared' images to fight back; iconoclasm was in some ways a test for the validity of the new faith itself.[28]

A more problematic issue remains of what we are to make of the passive response of Netherlandish Catholics to the violence; not just in 1566 but also during the later episodes in the Dutch Revolt when churches and clerics were attacked, the saints mocked, crosses broken, and Catholics impeded from worshiping. In the light of the new historiography on the French wars of religion, that has charted the violent response of French Catholics to the Calvinist threat, the lack of Catholic resistance to the Calvinist onslaught in the Netherlands seems quite remarkable. The discussion about this issue is ongoing, and has, perhaps, brought to light most of all how little is known about Catholic laypeople in the sixteenth-century Netherlands, but it seems likely that any explanation will have to refer not just to the different political context, but also to the attitude of the Netherlandish clergy – who for a variety of reasons refrained from mobilising the laity against the heretics.[29]

By 1567 those clergy could have been forgiven for thinking that the worst was behind them. Philip II had dispatched an army to the Netherlands, under the command of the Duke of Alba. Alba's Council of Troubles, arrogating supreme judicial power, set out to exact retribution for the image-breaking, not just from those who had committed it, but also from those who had negotiated and compromised with the Calvinists. Nine thousand of those condemned had already fled abroad, but more than a thousand people were executed, among them a number of high-ranking noblemen. From abroad, resistance continued under the leadership of the exiled William of Orange, who now found himself forced into an (often uneasy) alliance with the Reformed.[30] Together they sought support from the French Huguenots, themselves beleaguered; help from England was, for the moment, limited to the hospitality it offered to refugees as well as to the very violent Calvinist privateers whom Orange had licensed. It was Elizabeth's decision in March 1572 at last to expel these 'seabeggars' that precipitated their attack on the city of Den Briel on 1 April, and so started a second Revolt that the Habsburgs would never be able to fully suppress.

How Calvinist was that Revolt? There was a time when historians believed that without Calvinist resistance theory the Revolt and the rejection of royal authority would have been unthinkable. That no longer seems likely. Calvinist political theory was not as innovative as once thought. In practice it was formulated to justify rebellions once they had already happened. The rebels, moreover, could and did also deploy other ideological resources, like defence of their liberties and privileges, and (newly invented) concepts of 'Netherlandish' patriotism and antihispanism.[31] That is not to deny that the Revolt relied heavily on Calvinist input; moreover, the need to guarantee religious rights for Calvinists ruled out a compromise of the type that the Habsburgs could and did reach with other rebellious subjects. Yet Calvinists provided one extreme end of the political spectrum. Their numbers were small, and although they were armed, any town council that could rely upon its civic militia could defend itself against them. It was the 'middle groups' in society, Juliaan Woltjer has been arguing, who ultimately determined how much room for manoeuvre the Calvinist minorities were given – and could retain.[32]

This is also borne out by the fate of the Calvinist Republics that emerged in Flanders and Brabant between 1577 and 1585. When all the other Netherlandish provinces joined Holland and Zeeland in their Revolt in 1576, this was mainly out of despair over the behaviour of the underpaid, mutinying Spanish troops. In the so-called Pacification of Ghent that they signed in that year, the provinces agreed that no-one was to be persecuted for heresy but also resolved that everywhere, except in Holland and Zeeland, Catholicism was to be retained as the sole religion. Soon, however, Calvinist refugees from Flanders and Brabant emerged in their hometowns, demanding freedom of worship; attempts to negotiate forms of 'religious peace' did not put a stop to Calvinist activism. In 1577 Calvinists in Ghent seized power, and in the following months imposed a Calvinist regime on the other Flemish cities, whilst in Brussels and Antwerp, Calvinists also effectively took over the town councils. Catholics and Protestant moderates were sidelined, harassed and in many cases had to flee their towns. Yet there was not enough support for a Calvinist theocracy. Disgust over the scenes in Flanders, especially, moved the Walloon provinces to abandon the Revolt and make their peace with the king, leaving William of Orange to deal with what was now effectively a revolt within a revolt. Building on these rifts among the rebels, the new Habsburg governor-general Alessandro Farnese, Duke of Parma, accelerated his campaign of reconquest. One after the other, the Calvinist cities were retaken; after the fall of Antwerp in 1585, the Revolt in the South was over.

What happened in these Calvinist Republics had a crucial impact on the Revolt, but has not been studied as thoroughly as one might expect. We know that the Ghent Calvinist leadership had a complex political agenda, that partly aimed to restore Ghent to its former economic importance and to reverse the policies that Charles V had imposed when punishing the city for its last revolt in 1539. Yet the religious, cultural and social dimensions of this extraordinary episode in Ghent history are still only partly understood, whilst of other cities we know even less. Fortunately there is now work in progress to remedy this.[33]

the formation of confessional cultures

The fall of Antwerp in 1585 not only marked a watershed in the Revolt, but also in the history of Netherlandish Catholicism. Parma's victory is commonly treated by scholars as the beginning of the Counter-Reformation in the Southern Netherlands. Since Alba's departure, the Spanish strategy for dealing with dissent had changed dramatically. Parma offered an amnesty, under which subjects might either 'reconcile' themselves with the Crown (which implied accepting Catholicism) or take advantage of a period of grace in which they could realise their assets, and depart for other lands. Tens of thousands of people left the Southern Netherlands in the later 1580s, on top of the enormous flow of refugees (both Catholic and Protestants) who had already fled the conflict. Many but not all of these migrants were Protestants – some of them simply departed to flee the disastrous economic downturn that the war had brought in the South. Most of them initially went to Germany and England, before moving to the increasingly prosperous Dutch Republic, where they were to make a profound impact on society – both culturally and religiously.[34]

If the departure of large numbers of dissenters from the Southern Netherlands was a precondition for successful recatholicisation of the southern Low Countries, it was not the whole answer. Many of those who stayed behind had at some point or other flirted with Protestantism, even more had relatives among the refugees, and everybody still hoped and expected that the Netherlands would, at some point, be reunited. Little research has been done into the 'reconciliation' process and we have very little idea of the number of Protestants who opted for reconciliation rather than to leave.[35] What is quite clear is that Parma and his successors did not start a campaign to flush every last Protestant out of hiding.[36] Although there were many witch-trials in the Habsburg Netherlands, repression of heresy was now largely replaced by indoctrination: through processions and sermons, plays and pamphlets, songs and confraternities, Sunday schools and cheap print like the tens of thousands of prayer cards that were distributed among believers.[37]

An important study of the Jesuit contribution to this process and its political dimensions was published in the 1950s by J. A. Andriessen. Much less used to be known, however, about the role of the bishops and other secular priests. Using decanal visitation records, especially, the work that Michel Cloet and his students have done since 1968 has highlighted the slow process by which the Netherlandish bishops increased their grip, especially on the countryside, and tried to transform popular culture. Craig Harline and Eddy Put's excellent and entertaining case study on the work of bishop Matthias Hovius gives a fascinating insight in the many hurdles, within and outside the church, that bishops faced in imposing their will, and in changing popular religious perceptions.[38]

One of Harline and Put's achievements is to highlight the importance of seeing churchmen's activities in a social and political context. Although the complementary studies by Thijs and Marinus of the Counter-Reformation process in Antwerp and the work of Lottin on Lille have yet to be followed up by other local studies, in recent years there has also appeared some other exciting new work on the political use of, and input into, Catholic reform. An important exhibition in 1998 showed how the regime of the popular 'Archdukes' Albert and Isabella (1598–1621) seamlessly combined politics and devotion, dynastic interest and Counter-Reformation zeal. Duerloo has continued to develop this theme, most recently in a study on the shrine of the Virgin of Scherpenheuvel, on the frontier with the rebels in the North, that developed into a symbolic bastion for the church militant.[39]

The success of the Counter-Reformation in the Southern Netherlands clearly owed much to the support of the state, but less is known about its reception on the ground. Lay believers have generally been studied as the objects of indoctrination rather than as the agents of change. Yet there was more to Counter-Reformation success among the laity than opportunism among the elite and the passive acquiescence of the majority. The people in those Southern Netherlandish areas that were reconquered by the Dutch Republic in the 1620s, for instance, had become so committed to Catholicism that they proved impervious to all subsequent attempts to Protestantise them.[40] There are some interesting leads on how this lay commitment may have come about; good studies of education and catechism teaching now exist, Harline has offered stimulating ideas about the role miracles played in this process, and there is also a promising initiative to research the emergence of 'networks' of influential Catholics through the school system. Yet more work will need to be done on people of flesh and blood before we can integrate the lay Catholics of the Southern Netherlands into the story of the Counter-Reformation.[41]

This remains equally true, however, for the Catholics of the Dutch Republic, who were living as a religious minority in what had from 1580 become in many (though not all) respects a Protestant state. The Union of Utrecht of 1579, through which the rebellious provinces committed themselves to mutual defence, maintained that no-one would be persecuted for religious reasons, but otherwise stated that each province could make its own religious legislation. Holland and Zeeland had already outlawed Catholic worship in 1573, but they had done so on security grounds, and had not infringed 'freedom of conscience' – the freedom to believe what one wanted. This distinction, later buttressed by theorists like Simon Stevin and Justus Lipsius, was to remain a cornerstone of the Republic's religious policies.

As we have seen, when the other provinces first joined the Revolt, they initially all chose to remain Catholic, but that commitment was eroded by Calvinist agitation and political developments. In 1580 Catholic worship was banned throughout the northern provinces, a ban that lasted until the late eighteenth century. Everywhere, the Reformed church now became the 'most favoured' church. Its ministers were paid out of the sequestered funds of the old church, church buildings were granted for Reformed use only and the Reformed churches made efforts to control charity and education. They were, mostly, committed to the church order that had been designed at the Synod of Emden in 1571, and, mostly, abided by the Dutch confession and the Heidelberg catechism. Ministers and lay elders together admitted members to communion and administered discipline, deacons distributed charity, whilst delegates from these consistories gathered in regional *classes* (singular *classis*) to discuss policy issues and complex disciplinary cases.

But how Calvinist was the Dutch Republic? Unlike the Calvinist radicals in Flanders and Brabant, the authorities in the Northern provinces proved unwilling to grant too much power to the churches; many among the elites did not want to exchange the Spanish for a 'Genevan' inquisition. Until at least 1618, membership of the Reformed church was no precondition for officeholding. And while most towns initially denied even Lutherans and Mennonites the right to worship in public, in practice it emerged that both they and Catholics were left plenty of scope to organise and practise their faith in private homes. Although Catholic priests, especially, ran the risk of arrest, expulsions and fines, such harassment was mostly incidental. A system emerged whereby dissenting communities paid fixed bribes, called 'recognition money', to local law enforcers, and were otherwise left in peace. Catholics were a discriminated minority, which, as the work of Willem Frijhoff has shown, always longed for better days. Yet their existence was less precarious than was once thought; one measure for the security of the Catholic community, especially, was its considerable art patronage. Its vitality was not just sustained by the 'mission' priests, often trained in the Southern Netherlands, who worked in many cities, but also by large numbers of lay sisters, or *kloppen*, who taught, catechised and did community work.[42]

It is not easy to explain why some groups and communities in the Dutch Republic proved less susceptible to Protestantism than others. Just after the end of the Second World War, the Catholic historian L. J. Rogier produced an innovative study of early modern Dutch Catholicism. Developing an idea of Pieter Geyl, Rogier argued that the 'Protestantisation' of Catholics in the Northern Netherlands had usually not preceded, but followed on from the Revolt. 'Protestantisation' had only succeeded, he said, where three conditions were met: where believers' commitment to the faith had been weakened by exposure to clerical 'corruption' of some sort or other, where secular authorities supported Protestantism, and where Catholics had to make do without clergy for too long. The attraction of this thesis was that it seemed to account for the enormous regional variations in the acceptance of Protestantism; in many parts of the Netherlands villages with Catholic majorities could be found in close proximity to Protestant communities. Yet both Rogier's assumption that Catholicism was the 'default' position of the Dutch, and his view of 'Protestantisation' as a top-down process were critically received.[43]

In fact, subsequent research has shown that we cannot simply divide the Dutch population into Catholics and Protestants. In 1974 A. Th. van Deursen published a study that revolutionised our understanding of the role of the Reformed church in the Dutch Republic. Van Deursen argued that we should clearly distinguish between the two different roles that the Reformed church played. On the one hand it acted as an inclusive 'public church' – which preached for all, baptised and married all comers, and which, on days of public prayer and penance, besought the Almighty in the name of all the commonwealth. Yet in its other role, as a 'communion', the Reformed church was an exclusive, 'voluntary' church, that reserved access to Lord's Supper to its membership. Membership did not come by virtue of citizenship; adults over eighteen had to make a confession of faith, and were only admitted to communion if they had a minimum knowledge of the Reformed faith, and lived an 'honest' life. To keep the communion table pure, those who wanted access to communion also had to submit themselves to church discipline, which was exercised over members only, and did not have any civic repercussions.[44]

Traditionally, scholarly estimates of the Reformed support-base had relied on the number of baptisms in the churches, yet van Deursen showed that the 'public' church also baptised children whose parents belonged to other confessions. As a consequence he argued that the only reliable indicator of the support for Calvinism was church membership. Examining a sample of Holland villages and small towns, he demonstrated that, at least until 1620, the membership of many Reformed communities had actually remained very small indeed. In most churches the members were outnumbered by what were sometimes known as *liefhebbers* or 'sympathisers', people who came to church but who did apparently not care to get access to communion. His findings thus alerted scholars to the importance of a neglected category of believers, those who did not formally belong to any church at all.[45]

Although van Deursen saw *liefhebbers* as 'members in the making', he nevertheless highlighted that the acceptance of the Reformation had been a very slow process. As a series of local studies has subsequently confirmed, the pace, flavour and implementation of the Reformation differed from town to town, from village to village and depended heavily on local conditions. In rural areas, especially, the Reformation often arrived very belatedly.[46] Towns were better served, with better clergy, but in several cities there were profound disagreements about the nature of the local Reformation. Benjamin Kaplan, for instance, has shown how in Utrecht the 'libertine' Reformed tradition, instituted by Hubert Duifhuis, clashed with the Calvinist 'consistorial' party who wanted Utrecht's church to conform more closely to Holland models. In Leiden there were fierce conflicts about the appropriate level of government intervention in the church. But even where there were no major conflicts the Reformation presented towns governments with formidable problems.[47] Building both on van Deursen's work, and on studies of German cities, Jo Spaans' study of Haarlem between 1577 and 1620 explored the interaction between the various religious groups and the authorities. This led to two important conclusions. First, whilst it was possible to assign a definite confessional colour to about half the city's population, it appeared that as late as

1620 about half of Haarlem's adults did not take communion in any church, legal or illegal. Secondly, the Haarlem authorities consciously tried to develop policies that, whilst emphasising general Christian values, underplayed confessional differences. Guilds, civic militia companies, neighbourhoods, and the authorities, were encouraged to retain a non-confessional sense of Christian community. Kaplan's book on Utrecht developed this thesis further, and showed that it was not just the authorities but many believers themselves who retained what he called a commitment to 'metaconfessional' values.[48]

By the late 1990s it was therefore clear that we should not treat Dutch religious culture as the sum of distinct confessional groups. Church membership was a matter of choice, and this choice was often exercised individually. It seems that more women than men joined churches, and that many believers only joined late in life, but it would be desirable to have more and better data on patterns of church membership after 1620. Quite why people opted to join a church is an even more intractable matter. There is some evidence to suggest that for single women, especially, church membership could be a marker of respectability, and that for men, the decision to join might coincide with the taking up of other responsibilities. One case study has argued that it was not just private considerations, but also a strong sense of the importance of the collective bond with God that could motivate people to join a church.[49]

The realisation that all church members were *de facto* members of a religious minority has also prompted scholars to ask how Dutch toleration really worked. Until at least 1650, confessional segregation was impossible. Many civic institutions retained a mixed membership; many families were religiously divided. Willem Frijhoff has coined the term *omgangsoecumene* for the 'everyday ecumenism' that was practised between Dutch believers of different faiths. Confessional identity certainly mattered to believers, many of whom were very intolerant about people of other faiths; yet it seems that in day-to-day contacts they managed to suspend most of that hostility by concentrating on the behaviour and tangible morality of others, rather than their doctrines.[50]

The voluntary character of church membership also affected the extent to which churches could aspire to influence their members. On the one hand, it meant that they could expect a high level of compliance with confessional norms; on the other, they had to rely completely on social control to enforce these. Thus the Amsterdam Reformed church could suspend Rembrandt's mistress Hendrikje Stoffels from the Lord's Supper, but since Rembrandt himself was not a member and the state did not back up Reformed disapproval, there were no other consequences. Because church charity in most cases was not restricted to the membership of Reformed churches, even the poor were not exposed to forced confessionalisation.[51]

In the seventeenth century, the Republic was to become ever more confessionally diverse. A rift over the doctrine of double predestination that broke out in 1610 almost tore the Reformed church apart and brought the Republic to the brink of civil war. After the Synod of Dordt of 1618–19 tightened the definition of orthodoxy in the Reformed church, the expelled 'Remonstrants' or 'Arminians' set up their own community. Whilst the Arminians had started from a position, both on predestination and discipline, that was really very close to Calvin's own views, their move to clandestinity was accompanied by a rapid intellectual radicalisation through which their community would become the cradle for many new forms of religious dissent. Jews had also settled in parts of the Republic, as did religious refugees from elsewhere in Europe. It is the presence of these groups that lent the Republic its tolerant reputation. Yet Jonathan Israel and others have reminded us that such toleration hinged on connivance only. When the authorities felt the need, books were banned, careers were broken, and lives disrupted.[52]

Looking at recent research on the Netherlands from a European perspective, two features stand out. One of the striking aspects of the story of the Netherlandish Reformations is the eagerness with which a wide range of believers seized the chance to claim some sort of religious agency: from the late medieval poets who liked to act as popular preachers to the sacramentarians who could not believe in a 'bread God'; from the blind Mennonite woman who dictated her best-selling religious songtexts to a group of friends to the *kloppen* who catechised village children. These men and women were confident about claiming a religious voice for themselves.

Secondly, the Netherlandish case suggests that once early modern believers had persuaded themselves that salvation was a matter between man and God, there was nothing self-evident about 'confessionalisation'. Given the choice, as the Dutch were in the seventeenth century, only half of adults thought they needed access to communion. Many of these same people were happily attending sermons and reading devotional texts, and quite a few of them undoubtedly thought that church membership was good for other people. But however Christian they felt, they did not equate religiosity and church membership.

So does this make the Netherlands the odd culture out in the 'confessional age'? Scholars used to devote a great deal of energy to showing how uniquely Netherlandish the Reformations of the Low

Countries were. That seems unhelpful; what is much more likely is that in the Netherlands, highly urbanised, highly literate, highly particularist, we see a high concentration of what elsewhere in Europe was either much more diluted or stymied by much more effective coalitions between church and state. This lay pride, the ability and eagerness of literate laypeople to shape their own religious life, appeared in many shapes and guises. But without a way to accommodate it and to find a stakeholding role for these believers, no church, however well supported by the state, could hope to dominate early modern societies.

further reading

Essential reading for any student of the Netherlandish Reformations are A. C. Duke, *Reformation and Revolt in the Low Countries* (London, 1990) and his essay 'The Netherlands' in Andrew Pettegree (ed.), *The Early Reformation in Europe* (Cambridge, 1992). Illuminating on the spread and eclectic nature of Reformation ideas is also Gary K. Waite, *Reformers on Stage: Popular Drama and Religious Propaganda in the Low Countries of Charles V* (Toronto, 2000), while the role of print is discussed by A. G. Johnston, 'Printing and the Reformation in the Low Countries, 1520–c. 1555', in J. F. Gilmont (ed.), *The Reformation and the Book* (Aldershot, 1998). Important critical notes on the existence of an 'Erasmian' tradition in the Netherlands (and Europe) can be found in M. E. H. N. Mout et al. (eds), *Erasmianism: Idea and Reality* (Amsterdam, 1997). There is no modern overview in English of the history of the Anabaptist and Mennonite tradition in the Netherlands, but a helful insight into the world of the sects is offered by G. K. Waite, *David Joris and Dutch Anabaptism, 1524–1543* (Waterloo, Ontario, 1990). The spread of the Reformation in the most important urban centre in the Netherlands before and during the first phase of the Revolt is analysed by G. Marnef, *Antwerp in the Age of Reformation: Underground Protestantism in a Commercial Metropolis, 1550–1577*, trans. J. C. Grayson (Baltimore and London, 1996).

The role of exiles and the importance of Emden for the Dutch Reformed church are highlighted in A. Pettegree, *Emden and the Dutch Revolt: Exile and the Development of Reformed Protestantism* (Oxford, 1992). On the iconoclasm of 1566, see P. Mack Crew, *Calvinist Preaching and Iconoclasm in the Netherlands, 1544–69* (Cambridge, 1978). The classic overview of the Dutch Revolt remains G. Parker, *The Dutch Revolt* (London, 1977), but a useful introduction to more recent scholarship is G. Darby (ed.), *The Origins and Development of the Dutch Revolt* (London, 2001). On the Counter-Reformation in Flanders, see C. E. Harline and E. Put, *A Bishop's*

Tale: Mathias Hovius Among his Flock in Seventeenth-Century Flanders (New Haven, 2001); some useful essays on Catholics in the Dutch Republic can be found in W. Frijhoff, *Embodied Belief: Ten Essays on Religious Culture in Dutch History* (Hilversum, 2002). An important study of a local reformation and the debates within the Reformed church in the Republic is B. J. Kaplan, *Calvinists and Libertines: Confession and Community in Utrecht, 1578–1620* (Oxford, 1995). On the Dutch Republic and the problem of toleration see the essays in R. Po-chia Hsia and H. F. K. van Nierop (eds), *Calvinism and Religious Toleration in the Dutch Golden Age* (Cambridge, 2002).

notes

1. A. C. Duke, 'The Netherlands', in Andrew Pettegree (ed.), *The Early Reformation in Europe* (Cambridge, 1992); idem, *Reformation and Revolt in the Low Countries* (London, 1990). On Hoen, see B. J. Spruyt, *Cornelius Henrici Hoen (Honius) and his Epistle on the Eucharist (1525)* (Houten, 1996).
2. A. J. Roelink's chapter XI in J. A. van Houtte (ed.), *Algemene geschiedenis der Nederlanden*, 12 vols (Utrecht, 1949), vol. IV; J. Smit, 'The Present Position of Studies Regarding the Revolt of the Netherlands', in J. S. Bromley and E. H. Kossmann (eds), *Britain and the Netherlands* (London, 1960).
3. J. Decavele, *De dageraad van de Reformatie in Vlaanderen (1520–1565)* (Brussels, 1975); Duke, *Reformation and Revolt*; J. J. Woltjer, *Friesland in Hervormingstijd* (Leiden, 1962); H. Trapman, 'Le Rôle des "sacramentaires" des origines de la Réforme jusqu'en 1530 aux Pays-Bas', *Nederlands archief voor kerkgeschiedenis*, 63 (1983), 1–24.
4. Gary K. Waite, *Reformers on Stage: Popular Drama and Religious Propaganda in the Low Countries of Charles V* (Toronto, 2000); Anne-Laure Van Bruaene, '*In principio erat verbum:* Drama, Devotion, Reformation and Urban Association in the Low Countries', in C. Black and P. Gravestock (eds), *Early Modern Confraternities in Europe and the Americas: International and Interdisciplinary Perspectives* (Aldershot, forthcoming 2005). On ephemeral print see A. Duke, 'Posters, Pamphlets and Prints: the Ways and Means of Disseminating Dissident Opinions on the Eve of the Dutch Revolt', *Dutch Crossing*, 27 (2003), 24–44.
5. A. J. A. Bijsterveld, *Laverend tussen Kerk en wereld. De pastoors in Noord-Brabant 1400–1570* (Amsterdam, 1993); R. R. Post, *Kerkelijke verhoudingen in Nederland vóór de Reformatie van c. 1500 tot 1580* (Utrecht, 1954); Duke, *Reformation and Revolt*, pp. 80–5.
6. J. Trapman, 'Erasmianism in the early Reformation in the Netherlands', in M. E. H. N. Mout et al. (eds), *Erasmianism: Idea and Reality* (Amsterdam and Oxford, 1997); Duke, *Reformation and Revolt*, pp. 1–28; Van Bruaene, 'In principio'.
7. Duke, *Reformation and Revolt*, pp. 54–6.
8. Ibid., pp. 77–85; J. D. Tracy, 'Elements of Anticlerical Sentiment in the Province of Holland under Charles V', in P. A. Dykema and H. A. Oberman (eds), *Anticlericalism in Late Medieval and Early Modern Europe* (Leiden, 1994), pp. 257–69.

9. A. G. Johnston, 'Printing and the Reformation in the Low Countries, 1520–c. 1555', in J.-F. Gilmont (ed.), *The Reformation and the Book* (Aldershot, 1998); Duke, 'The Netherlands'; Duke, *Reformation and Revolt*; A. Hamilton, *The Family of Love* (Cambridge, 1981); G. K. Waite, *David Joris and Dutch Anabaptism, 1524–1543* (Waterloo, Ontario, 1990); E. Braekman, 'Eloi Pruystinck', in A. Séguenny (ed.), *Bibliotheca Dissidentium. Répertoire des non-conformistes religieux des seizième et dix-septième siècles. Biographie et bibliographie des oeuvres dissidentes et état de la recherche sur eux* (Baden-Baden, 1989).
10. J. J. Woltjer, *Tussen vrijheidsstrijd en burgeroorlog. Over de Nederlandse opstand, 1555–1580* (Amsterdam, 1994), pp. 9–30.
11. S. Zijlstra, *Om de ware gemeente en de oude gronden. Geschiedenis van de dopersen in de Nederlanden, 1531–1675* (Hilversum, 2000), pp. 1–26; W. Bergsma and S. Voolstra (eds), 'Uyt Babel gevloden, in Jeruzalem getogen. Mennon Simons' verlichting, bekering en beroeping' , *Doperse Stemmen*, 6 (1986).
12. W. J. Kühler, *Geschiedenis der Nederlandsche Doopsgezinden in de zestiende eeuw* (Haarlem, 1932); A. F. Mellink, *De wederdopers in de Noordelijke Nederlanden 1531–1544* (Groningen, 1954).
13. Duke, 'The Netherlands'; Duke, *Reformation and Revolt*; Zijlstra, *Om de ware gemeente*; A. Jelsma, *Frontiers of the Reformation: Dissidence and Orthodoxy in Sixteenth-Century Europe* (Aldershot, 1998), pp. 52–74.
14. A. L. E. Verheyden, *Anabaptism in Flanders, 1530–1650: a Century of Struggle* (Scottdale, PA, 1961); Decavele, *De dageraad van de reformatie*; G. Marnef, *Antwerp in the Age of Reformation: Underground Protestantism in a Commercial Metropolis, 1550–1577*, trans. J. C. Grayson (Baltimore and London, 1996).
15. Zijlstra, *Om de ware gemeente*; Waite, *David Joris*.
16. Henk van Nierop, 'The Nobility and the Revolt of the Netherlands: Between Church and King, and Protestantism and Privileges', in Philip Benedict et al. (eds), *Reformation, Revolt and Civil War in France and the Netherlands 1555–1585* (Amsterdam, 1999). Belgian scholars, especially, have worked on Calvinism as a bourgeois religion: see J. Decavele, 'Historiografie van het zestiende eeuwse protestantisme in België', *Nederlands Archief voor Kerkgeschiedenis*, 62 (1982), 1–27; cf. R. S. DuPlessis, *Lille and the Dutch Revolt: Urban Stability in an Era of Revolution, 1500–1582* (Cambridge, 1991).
17. G. Moreau, *Histoire du protestantisme à Tournai jusqu'à la veille de la révolution des Pays Bas* (Paris, 1962).
18. Decavele, *De dageraad van de reformatie*; A. Pettegree, *Emden and the Dutch Revolt: Exile and the Development of Reformed Protestantism* (Oxford, 1992); H. Schilling, *Niederländische Exulanten im 16. Jahrhundert: ihre Stellung im Sozialgefüge und im religiösen Leben deutscher und englischer Städte* (Gütersloh, 1972).
19. A. L. E. Verheyden, 'De Martyrologia in de optiek van de hedendaagse Martelaarslijsten' and J.-F.Gilmont, 'Un instrument de propagande religieuse: les martyrologes du XVIe siècle', both in *Sources de l'Histoire Religieuse de la Belgique* (Louvain, 1968); A. Pettegree, 'Adriaan van Haemstede: the Heretic as Historian', in B. Gordon (ed.), *Protestant History and Identity in Sixteenth-Century Europe*, 2 vols (Aldershot, 1996), vol. II.
20. Duke, *Reformation and Revolt*; A. L. E. Verheyden, *Le martyrologe Protestant des Pays-Bas du Sud au XVIme siècle* (Brussels, 1960), p. 71; A. Goosens, *Les inquisitions modernes dans les Pays-Bas méridionaux, 1520–1633* (Brussels, 1997); W. Monter, 'Heresy Executions in Reformation Europe, 1520–1565', in O.

P. Grell and R. W. Scribner (eds), *Tolerance and Intolerance in the European Reformation* (Cambridge, 1996); P. Benedict, *Christ's Churches Purely Reformed: a Social History of Calvinism* (New Haven and London, 2002), p. 176.

21. Duke, *Reformation and Revolt*, pp. 152–74; J. D. Tracy, 'Heresy Laws and Centralisation under Mary of Hungary', *Archiv für Reformationsgeschichte*, 73 (1982), 284–308.
22. M. Dierickx, *De oprichting der nieuwe bisdommen onder Filips II, 1559–1570* (Utrecht, 1950); Duke, *Reformation and Revolt*, pp. 152–74.
23. M. Dierickx, 'La politique religieuse de Philippe II dans les Pays-Bas', *Hispania: Revista española de historia*, 16 (1956), 130–43.
24. Duke, *Reformation and Revolt*, pp. 152–74; K. W. Swart, 'The Black Legend during the Eighty Years War', in J. S. Bromley and E. H. Kossmann (eds), *Britain and the Netherlands: Papers Delivered to the Anglo-Dutch Historical Conference* (The Hague, 1975).
25. H. F. K. van Nierop, 'Alva's Throne: Making Sense of the Revolt of the Netherlands', in G. Darby (ed.), *The Origins and Development of the Dutch Revolt* (London, 2001); van Nierop, 'The Nobility and the Revolt'.
26. P. Mack Crew, *Calvinist Preaching and Iconoclasm in the Netherlands, 1544–69* (Cambridge, 1978); S. Deyon et al., *Les Casseurs de l'été 1566. L'Iconoclasme dans le Nord* (Paris, 1981); Duke, *Reformation and Revolt*, pp. 125–51.
27. Crew, *Calvinist Preaching*; E. Kuttner, *Het hongerjaar 1566*, trans. J. M. Romein (Amsterdam, 1949); Pettegree, *Emden and the Dutch Revolt*; J. Scheerder, *De Beeldenstorm* (Bussum, 1974).
28. A. Duke, 'De Calvinisten en de "paapse beeldendienst". De denkwereld van de beeldenstormers in 1566', in M. Bruggeman et al. (eds), *Mensen van de Nieuwe Tijd* (Amsterdam, 1996); D. Freedberg, *Iconoclasm and Painting in the Netherlands, 1566–1609* (New York, 1988).
29. J. J. Woltjer, 'Violence during the Wars of Religion in France and the Netherlands: a Comparison', *Nederlands Archief voor Kerkgeschiedenis*, 76 (1996), 26–45; Henk van Nierop, 'Similar Problems, Different Outcomes: the Revolt of the Netherlands and the Wars of Religion in France', in K. Davids and J. Lucassen (eds), *A Miracle Mirrored: the Dutch Republic in European Perspective* (Cambridge, 1995); Benedict et al., *Reformation, Revolt and Civil War*, pp. 149–63; J. Pollmann, 'Countering the Refomation in France and the Netherlands: Clerical Leadership and Catholic Violence, 1560–1585', *Past and Present*, forthcoming 2006.
30. K. W. Swart, *William of Orange and the Revolt of the Netherlands, 1572–84*, ed. A. C. Duke et al. (Aldershot, 2003).
31. Q. Skinner, *The Foundations of Modern Political Thought*, 2 vols (Cambridge, 1978), II, pp. 189–358; M. van Gelderen, *The Political Thought of the Dutch Revolt, 1555–1590* (Cambridge, 1992); M. E. H. N. Mout, 'Van arm vaderland tot eendrachtige Republiek', *Bijdragen en Mededelingen voor de Geschiedenis der Nederlanden*, 101 (1986), 345–65; J. J. Woltjer, 'Dutch Privileges, Real and Imaginary', in Bromley and Kossmann (eds), *Britain and the Netherlands*.
32. Woltjer, *Tussen vrijheidsstrijd en burgeroorlog*.
33. The only book-length studies of the Republics are H. A. Enno van Gelder, *Revolutionnaire Reformatie. De vestiging van de Gereformeerde kerk in de Nederlandse gewesten gedurende de eerste jaren van de Opstand tegen Filips II, 1575–1585* (Amsterdam, 1943) and, from a Marxist perspective, T. Wittman et al., *Les*

Gueux dans les 'bonnes villes' de Flandre (1577–1584) (Budapest, 1969). See also J. Decavele et al. (eds), *Het eind van een rebelse droom. Opstellen over het calvinistisch bewind te Gent (1577–1584) en de terugkeer van de stad onder de gehoorzaamheid van de koning van Spanje (17 september 1584)* (Ghent, 1984); G. Marnef, *Het Calvinistisch Bewind te Mechelen, 1580–85* (Heule, 1987).

34. R. Esser, *Niederländische Exulanten im England des 16. und frühen 17. Jahrhunderts* (Berlin, 1996); Schilling, *Niederländische Exulanten*; J. Briels, *Zuidnederlanders in de Republiek, 1572–1630. Een demografische en cultuurhistorische studie* (Sint Niklaas, 1985); and cf. G. Marnef, 'Protestanten in "Noord" en "Zuid". Kerkhistorische beschouwingen n.a.v. een recente studie', *Bijdragen tot de geschiedenis*, 70 (1987), 139–45.
35. M. J. Hendrickx, 'Enkele cijfers in verband met de bekering van protestanten te Antwerpen in 1585–89', *Ons Geestelijk Erf*, 41 (1967), 302–9.
36. Goosens, *Inquisitions modernes*; M. J. Marinus, 'De protestanten te Antwerpen, 1585–1700', *Trajecta*, 2 (1993), 372–91.
37. A. K. L. Thijs, *Van Geuzenstad tot katholiek bolwerk: Antwerpen en de contrareformatie* (n.p., 1990); E. Put, *De cleijne schoolen. Het volksonderwijs in het hertogdom Brabant tussen Katholieke Reformatie en Verlichting (eind 16e eeuw – 1795)* (Leuven, 1990); M.-S. Dupont-Bouchat et al., *Prophètes et sorciers dans les Pays-Bas, XVIe-XVIIIe siècle* (Paris, 1978); F. Vanhemelryck, *Heksenprocessen in de Nederlanden* (Leuven, 1982).
38. J. Andriessen, *De Jezuieten en het saamhorigheidsbesef der Nederlanden, 1585–1648* (Antwerp, 1957); J. Tracy, 'With and Without the Counter-Reformation: the Catholic Church in the Spanish Netherlands and the Dutch Republic, 1580–1650', *Catholic Historical Review*, 71 (1985), 547–75; M. Cloet, 'Algemeen verslag over de kerkgeschiedenis betreffende de nieuwe tijd sinds 1970', in F. Daelemans and M. Cloet (eds), *Godsdienst, mentaliteit en dagelijks leven. Religieuze geschiedenis in België sinds 1970* (Brussels, 1988); C. E. Harline and E. Put, *A Bishop's Tale: Mathias Hovius Among his Flock in Seventeenth-Century Flanders* (New Haven, 2001).
39. W. Thomas and L. Duerloo (eds), *Albert and Isabella, 1598–1621: Essays* (Leuven, 1998); A. Lottin, *Lille, citadelle de la Contre-Réforme? (1598–1668)* (Dunkerque, 1984); M. J. Marinus, *De contrareformatie te Antwerpen (1585–1676). Kerkelijk leven in een grootstad*, (Brussels, 1995); Thijs, *Van Geuzenbolwerk*; L. Duerloo and M. Wingens, *Scherpenheuvel. Het Jeruzalem van de Lage Landen* (Leuven, 2002).
40. G. Rooijakkers, *Rituele repertoires: Volkscultuur in oostelijk Noord-Brabant, 1559–1853* (Nijmegen, 1994); M. Wingens, *Over de grens. De bedevaart van katholieke Nederlanders in de zeventiende en achttiende eeuw* (Nijmegen, 1994).
41. C. E. Harline, *Miracles at the Jesus Oak: Histories of the Supernatural in Reformation Europe* (New York, 2003); Tracy, 'With and Without'; Put, *De cleijne schoolen*.
42. W. Frijhoff, *Embodied Belief: Ten Essays on Religious Culture in Dutch History* (Hilversum, 2002); X. van Eck, *Kunst, twist en devotie. Goudse katholieke schuilkerken 1572–1795* (n.p., 1994); M. Monteiro, *Geestelijke maagden. Leven tussen klooster en wereld in Noord-Nederland gedurende de zeventiende eeuw* (Nijmegen, 1996).
43. L. J. Rogier, *Geschiedenis van het katholicisme in Noord-Nederland in de 16e en 17e eeuw*, 3 vols (Amsterdam, 1945–47); H. A. Enno van Gelder, 'Nederland geprotestantiseerd?', *Tijdschrift voor geschiedenis*, 81 (1968), 445–64; P. Geyl,

'De protestantisering van Noord-Nederland', in idem, *Noord en Zuid. Eenheid en tweeheid in de Lage Landen* (Utrecht, 1960).

44. A. Th. van Deursen, *Bavianen en Slijkgeuzen. Kerk en kerkvolk ten tijde van Maurits en Oldenbarnevelt*, (Assen, 1974); Duke, *Reformation and Revolt*, pp. 269–93.
45. Deursen, *Bavianen en Slijkgeuzen.*
46. Duke, *Reformation and Revolt*, pp. 227–68.
47. B. J. Kaplan, *Calvinists and Libertines: Confession and Community in Utrecht, 1578–1620* (Oxford, 1995); C. Kooi, *Liberty and Religion: Church and State in Leiden's Reformation, 1572–1620* (Leiden, 2000), pp. 204–15.
48. Kaplan, *Calvinists and Libertines*; J. W. Spaans, *Haarlem na de Reformatie. Stedelijke cultuur en kerkelijk leven, 1577–1620* (The Hague, 1989).
49. W. Bergsma, *Tussen Gideonsbende en publieke kerk. Een studie over het gereformeerd protestantisme in Friesland, 1580–1650* (Hilversum, 1999); J. Pollmann, *Religious Choice in the Dutch Republic: the Reformation of Arnoldus Buchelius (1565–1641)* (Manchester, 1999); idem, 'Women and Religion in the Dutch Golden Age', *Dutch Crossing*, 24 (2000), 162–82.
50. Frijhoff, *Embodied Belief*; R. Po-chia Hsia and H. F. K. van Nierop (eds), *Calvinism and Religious Toleration in the Dutch Golden Age* (Cambridge, 2002).
51. C. Parker, *The Reformation of Community: Social Welfare and Calvinist Charity in Holland, 1572–1620* (Cambridge, 1998); H. Roodenburg, *Onder censuur. De kerkelijke tucht in de gereformeerde gemeente van Amsterdam, 1578–1700* (Hilversum, 1990); H. Schilling, *Civic Calvinism in Northwestern Germany and the Netherlands* (Kirksville, MO, 1991); J. Pollmann, 'Off the Record: Problems in the Quantification of Calvinist Church Discipline', *Sixteenth-Century Journal*, 33 (2002), 423–38.
52. Deursen, *Bavianen en Slijkgeuzen*; C. Berkvens-Stevelinck et al. (eds), *The Emergence of Tolerance in the Dutch Republic* (Leiden, 1997); Hsia and van Nierop, *Calvinism and Religious Toleration.*

5
france

penny roberts

The French Reformation is often presented as a *revolution manquée*: at once a powerful force that threatened to sweep all before it and a cause that was bound to stall in the face of insuperable hurdles. The dichotomy between its near success but predictable failure marks the traditional as well as more recent historiography of the Reformation's course in France. It also reflects the French position somewhere between the states which embraced Protestantism and those which rejected it. In this regard, it has most in common with the experience of the Netherlands where Reformed Protestants formed a substantial and influential minority, and which were eventually divided into two confessionally and politically distinct territories. Likewise, in France, there were fears that religious division would lead to a fragmentation of the French kingdom. That it did not do so is due to various factors, amongst which it might be assumed was the political integrity of France. Yet its unity was only recently acquired and, consequently, rather fragile. Studies of the French Reformation characteristically concentrate on the contribution of various social groups, individuals and institutions to the confessional ferment and the highlighting of what are seen as key watershed moments: the Affair of the Placards (1534), the Saint Bartholomew's Day massacre (1572), the promulgation of the Edict of Nantes (1598). A further, ongoing focus of discussion is the relative importance of a distinctively French movement and the eventual dominance of Calvinism. The French Reformers, or Huguenots as they are commonly known, are in turn presented as a dynamic and potent force and as a suffering and beleaguered minority. Indeed, there is something tragic and heroic in the traditional accounts of the French Reformation's failure as viewed by sympathetic contemporaries, by later Revolutionaries who saw Protestantism as a force for liberation and progress in opposition to the *ancien régime*, as well as by more recent historians.[1] Characteristically, this 'tragedy' is divided into three acts – pre-wars (or at least up to 1560 and the movement's perceived zenith), the religious wars (1562–98), and post-wars (from the Edict of Nantes of 1598 to its revocation in 1685) – all of which have proved significant in shaping past and current interpretations.

Despite harbouring the most notorious medieval heretical sect, the Cathars, and the continued existence of pockets of Waldensianism, by 1500 the French kingdom boasted an enviable reputation as a bastion of Catholic orthodoxy. The monarchy coveted its title of 'most Christian king' and its claim to be the eldest son of the Church. Its enthusiasm in upholding the primacy of the French Church was more than matched by that of the *parlement* of Paris (the chief sovereign law court) and the theological faculty of the University of Paris. Moreover, these institutions maintained a Gallican position, supporting a Catholic Church independent from Rome, which they felt had been compromised by the crown's recognition of papal authority in the 1516 Concordat of Bologna. In 1517, when Erasmus famously declared France to be 'the purest ... part of Christendom', there was no indication of the Lutheran tremor which was about to convulse Europe and destabilise this certainty.[2] Nevertheless, historians have long identified a strain of so-called French 'pre-reform', with efforts to shake up certain religious orders, and to encourage education and preaching amongst the clergy.[3] Combined with elite enthusiasm for Christian humanism, this is believed to have fed into, and prepared the way for, acceptance of the new ideas in France. The contribution of sixteenth-century French scholars, such as Jacques Lefèvre d'Etaples, to biblical scholarship in particular, have been highlighted to suggest that home-grown as well as foreign influences should be taken into account when considering the emergence of the French Reform.[4] Nevertheless, it is widely accepted both that prior to Calvin no comparable figure emerged to direct the French Reformation, and that Luther's stance jolted some evangelical thinkers into adopting a more radical position. The potential threat posed by the spread of Lutheran ideas in France was recognised early, but both the development of French Protestantism and a concerted official response would take time. This delayed and piecemeal development, compared with the more rapid progress of the Reformation elsewhere in Europe, requires explanation.

Although organised Protestant churches did not develop in France until the 1540s, and much more so the 1550s, the early decades of French religious dissent are presented as a vibrant and exciting time. In Lucien Febvre's memorable phrase, they formed 'a long period of magnificent religious anarchy', during which heterodox ideas of many different kinds were able to flourish, despite official attempts to root them out.[5] The anarchic aspects of the movement were also emphasised by Pierre Imbart de la Tour in his classic four-volume work on its origins.[6] Yet, despite the radical pronouncements of a few eccentric individuals, this so-called 'anarchy' was largely confined to mainstream Protestant ideas. Anabaptist sects would never attain even a foothold in France. The apparent disorder was due more to the movement's amorphous nature and, therefore, the authorities' difficulties in locating and identifying it. Indeed, for the early decades, it is misleading to talk of a movement at all. Local groups came together on an *ad hoc* and informal basis, and there was a bewildering range of sources for religious ideas. Itinerant preachers, banned literature (much of it published abroad and imported), and debates over the workbench, the kitchen table or the tavern trestle, made for a varied evangelical shopping basket. The lack of major vernacular texts exacerbated this lack of uniformity: the first fully translated Bible only appeared in 1535, and the French edition of Calvin's *Institution* in 1542. What literature was available was dependent on trade links with, or proximity to, publishing or evangelically-active centres, chance encounters with pedlars, or the initiative of local booksellers. Short pamphlet-type works such as the Lutheran 'Book of true and perfect prayer' (1528), or the more obviously provocative, 'The antithesis of the deeds of Jesus Christ and the pope', had considerable appeal whether read individually or in groups.[7] The prohibition of such delicacies, like the notoriety of individual preachers, only made them more enticing to the curious. In addition, the dependence on local authorities to respond to unorthodoxy allowed it to develop almost unhindered.[8] Attempts at enforcing censorship enjoyed only limited success, so much so that the late 1530s have been termed the 'heroic period of the reformed book'.[9] The authorities were aware of the problem of distinguishing deliberate heresy from error, yet its separation from blasphemy was only clarified in the early 1540s.[10] There is general consensus among historians that by the time the judicial machinery had ground into action, the problem was too widespread and diffuse for the courts to deal with effectively.

Accounts of this period generally highlight particular individuals and incidents which reveal the coexistence of conservative and radical strands of dissent. Most well-known is the 'Circle of Meaux', the 'anarchic diocese where the religious war was lit' just to the east of Paris in the 1520s.[11] Here key figures of the early French evangelical movement came together: Guillaume Briçonnet, the orthodox bishop who instituted a moderate reform programme based on preaching; Marguerite d'Angoulême, the king's sister and evangelical author, who corresponded with and gave patronage to the Meaux group; Lefèvre d'Etaples, the 'French Erasmus', already influential for his 1523 translations from the New Testament; as well as open adherents of the Reform, notably Guillaume Farel. Granted protection in the highest places, Janus-like, the Meaux group combined sincere Catholic reform with more radical pronouncements and a popular following of the town's clothworkers concerned about rising grain prices.[12] This galvanised an unprecedented alliance between intellectuals and people, an explosive combination which was bound to attract adverse attention. The suppression of the Meaux group in 1524–25 led to the first of many periods of exile for important figures in the movement and some of the first martyrs of the French Reformation. But this group is also an isolated example of organised dissent in these early decades, quickly checked and dispersed by the authorities. Might the official suppression of Protestant tendencies have been more effective, therefore, than traditionally thought? Or does the fact that the Meaux group was only suppressed during the king's absence underline the importance of the crown's position?

Another commonly-cited pivotal event of the Reformation's early years, particularly in its impact on royal policy, was the 1534 Affair of the Placards. Broadsheets were posted in various Loire towns (penned by the exile Antoine Marcourt), whose savage attack on the Mass is said to have shocked the crown out of its complacency about the new religious ideas.[13] The incident was followed by the king's leading an ostentatious procession of the Host in Paris, and a concentrated burst of executions. But more recently historians have argued that the Affair's importance should not be overstated.[14] Nevertheless, the monarchy did have good reason to be concerned about the growth of religious dissent. Heresy threatened Christian orthodoxy, but could also be interpreted as a challenge to royal authority. In particular, it undermined the sacral powers of the French monarchy; for they depended on the validity of the real presence in the Eucharist, which the reformers attacked.[15] Furthermore, the coronation oath included a commitment to the protection of the Church and the extirpation of heresy. The salvation of the people was in the king's care. The harm which heterodoxy could do to both the body social and the body politic was emphasised by its opponents, who commonly compared its impact to the spread of venom, plague, or vermin.[16] These more sinister aspects appeared to confirm, too, the association of evangelical ideas with political sedition.

One notable result of the Affair was that several reformers chose 'to scuttle to safety' in exile: among them Jean Calvin, whose exile would be lifelong.[17] It was at this time, in Basle, that Calvin began his most famous work, the *Institution of the Christian Religion* (1536). Despite its dedication to Francis I, and its emphasis on obedience even to an impious king, Calvin stressed that the loyalty of the faithful was to God first, which could be construed as justifying resistance to a monarch of a different faith.[18] This potent mix of apparent conservatism and potential radicalism would characterise the position of the French Protestant elite throughout the sixteenth century. Plotting and agitating for change involved pastors and nobles as well as popular militancy.

One of the primary factors blamed both for the initial spread of the new ideas, and the faltering official response, is the inaction and ambivalence of the monarchy. In the final analysis, the success of the French Reformation would be dependent on the support of the king. While the parameters of orthodoxy remained largely undefined, Francis I (1515–47) was reluctant to act against Christian humanists with whom he was sympathetic. This led him into a direct clash with the theological faculty and the *parlement*. He and his successor Henry II (1547–1559) were also concerned to curry favour with the German Lutheran princes, as potential allies against the Emperor Charles V. French rulers were pragmatic in their approach to religious policy, and curious both about the religious ferment and about what it might mean for their authority over the Church; but this does not mean that they were insincere in upholding Catholicism. Alain Tallon has recently suggested that Catholicism was fundamental both to France's Gallican identity and, as we have seen, to the monarchy's sacred status. On this view, France was destined to remain Catholic, even under the (initially) Protestant King Henry IV (1589–1610).[19] However, this conviction was certainly not shared by contemporaries, Protestant or Catholic, who hoped or feared that the crown might yet be won round. After all, such a move would have reinforced the French crown's opposition to the Spanish Habsburgs, its alliance with Scotland and the French Church's independence from Rome. The monarchy's approval sometimes appeared tantalisingly close. It was no more predictable that England would have instituted a state-led Reformation than France.[20] Protestant optimism was buoyed by the apparently fortuitous deaths of Henry II and Francis II (1559–60) and the accession of the more conciliatory regime of Charles IX (1560–74) under the regency of Catherine de Medici. The 1561 colloquy at Poissy which attempted to reconcile the churches' theological differences; similar overtures to the Council of Trent via the French delegation; and a recognition of the Huguenot right to worship, first in the Edict of January 1562 and in subsequent edicts of pacification, underscored this apparent royal favour. France's confessional destiny, therefore, was by no means assured. This undoubtedly contributed to its drawn-out religious wars.

Nevertheless, William Monter has argued that, despite royal reticence, the *parlements*' effort against 'heretics' in the 1520s and 1530s largely succeeded, 'driving the most committed Protestants into exile'.[21] The records of heresy trials suggest that, ironically, it was in the late 1540s, just at the point when the new king, Henry II, was introducing more stringent anti-heresy measures, that the battle began to be lost. Contrary to the usual impression of a weak official response, Monter asserts, legal repression was one of the factors in delaying the rapid spread of Protestantism in France until the 1550s, when pressure of numbers swept away the judicial barriers to the formation of a French Reformed Church. Other arguments have also been put forward to explain how Protestantism was initially kept in check in France. The effectiveness of the crackdown on heresy in a given locality was dependent on the vigilance of the higher clergy; a reforming bishop might contain the situation for a time.[22]

Conversely, a lack of unity and mutual suspicion between different social groups is blamed for inhibiting the early growth of French Protestantism. The Reformation in France is commonly portrayed as comprising two discrete developments: broadly, elite and popular. This is comparable to the magisterial experience of Lutheran Germany, with intellectuals, nobility and the professions on the one hand, and artisans and journeymen on the other providing an unpredictable and unruly element. The story of these disparate groups is brought together in the official history of the French Reformed Church, the *Histoire ecclésiastique*, but even here they are shown to be pursuing parallel tracks rather than enjoying harmonised convergence.[23] This dichotomy was first identified by Henri Hauser in a short but influential article of 1899, in which he argued that the French Reformation was primarily a popular (and therefore home-grown) movement before being hijacked by the nobility and urban elites.[24] In particular, Hauser declared that this was a social as well as a religious revolution, seeing these as separate rather than complementary movements and giving primacy to the former. More recently, his mantle has been taken up energetically by Henry Heller. In a series of local studies, Heller emphasises the 'gulf of social hostility' which ensured that prior to the 1550s these groups remained separate until they were brought together by the organised churches introduced by Calvin.[25] Like Hauser, he identifies urban artisans as the earliest adherents of the Reformation, due to socio-economic dislocation which helped to promote the growth of heresy.[26] However, Heller also blames the socio-economic instability of the 1540s and 1550s for the retardation of the French Reformation.

Whilst the elites conformed in the hope of Reformation from above, popular elements publicly defied of the authorities, generating the very disorder which their co-religionists condemned.

Social historians such as Philip Benedict, Natalie Davis and David Nicholls have further explored the 'popular' aspects of Protestantism, but unlike Heller, they have restored the primacy of religion, integrating the social and religious more firmly together.[27] For Mack Holt, this is encapsulated in the notion of 'a body of believers rather than ... a body of beliefs', and the clash between the faiths reveals 'the social rather than the theological' impact of Reformation.[28] In addition, there has been an ongoing interest in identifying which craft groups within the population adhered to the new faith and why (with diverse and sometimes rather inconclusive results), concerning trade solidarity or novelty, family and neighbourhood ties, but also demonstrating the impact of local events and the decisions made by local officials.[29] The role of the elites in the French Reformation has then, in no sense, been marginalised, and continues to attract major studies.

This focus on the elites can be subdivided into studies of the respective roles of the sword and robe nobilities, as well as the identification of the universities and schools as hotbeds of dissent.[30] Throughout its existence, French Protestantism was dependent on noble protection for its survival whenever it emerged into the public gaze. High-ranking nobles, in particular, were able to provide for worship on their estates, a position officially sanctioned in the edicts of pacification from 1563. They were also important conduits for the representation of Huguenot interests to the crown. Unsurprisingly, the Genevan Church under Calvin made a point of targeting certain great nobles as ripe for conversion: principally the Bourbon princes of the blood (Navarre and Condé) and the Châtillon brothers, a branch of the powerful Montmorency family. Calvin's successor Theodore Beza nurtured a close relationship with Henry of Navarre before and after his accession to the French throne in 1589.[31]

The nobility's contribution has also been viewed more negatively, as diverting their co-religionists into a series of destructive civil wars, allowing factional division to override more pressing concerns. The idea that the Reformation was subsumed into the infighting of the great noble families who prolonged the religious wars for their own interests is longstanding. Yet the oft-cited calculation that half of the nobility had embraced the Reform by 1560 underlines the substantial support for the movement. The figure is not easily verifiable, but it conveniently fits the perceived clientage split caused by hostility between the Houses of

Guise and Bourbon/Montmorency – although patterns of confessional allegiance were never so clear-cut. Nevertheless, it is undeniable that the Protestants were able to rally enough of the sword nobles to their side, alongside their other resources, to present a serious challenge to the royal army.[32] Without their support, the Huguenots would have appeared a less challenging and a more easily suppressed threat.

For several decades, historians have also examined the important role of the *parlements* and the legal profession to the Reform movement and the religious wars.[33] This has partly been done through studies of the Huguenot political assemblies, the extraordinary chambers established to examine cases involving Protestants, and the judicial commissions sent out to enforce the peace edicts.[34] As well as its contribution to the ideology of the Reform movement, the office-holding robe nobility has also been credited with a more tolerant and pacific approach to the troubles dividing France and an influential role in the formation of royal policy to address these issues.[35] Whilst, on the one hand, its position is seen as restrained and conservative, the more radical assertions of popular sovereignty and of the right to resist an ungodly monarch also came from this quarter. At a time of increasing royal assertion of the power of the law exercised by a highly educated judiciary, it is perhaps not surprising that its members should be in the vanguard of the most dynamic political manifestations of both faiths: the Huguenot party and the Catholic League. In a tense but mutually beneficial alliance with the traditional power brokers, the nobility of the sword, they posed a formidable threat to the French monarchy. Yet they were also, as much as the monarchy, wedded to the maintenance of law and order, and just as unsettled by the prospect of popular insurgency. Keeping the people in check was a shared concern of crown, robe and sword, and shaped the elite's response to the more destabilising aspects of Reformation.

Indeed, from the outset, it was public dissent rather than private beliefs which determined the official response.[36] Accompanied by uncertainty as to just how great the threat was, and how many people had been influenced by the new ideas, the authorities' main fear was popular militancy and open disobedience. This seemed to be confirmed by the Affair of the Placards. Although this incident helped to define the boundaries of orthodoxy, the persecution it inspired was short-lived. Nevertheless, the further discovery of clandestine meetings and increasing acts of iconoclasm reinforced fears that this was only the tip of a much more radical iceberg.[37] The nocturnal nature of such activities fuelled official and popular anxiety. Even Huguenot elites were swift to condemn them, fearful of the impact on official recognition of their churches. The anonymity of such acts also allowed the Huguenots to accuse Catholics of staging the attacks on images, thus fomenting confessional tension. Other challenges were more public. Psalm-singing, popularised through Clément Marot and Beza's metrical psalter, became a badge of Protestant identity and, consequently, a source of provocation.[38] Likewise, the sight and reports of Protestant martyrs going joyfully to their place of execution functioned as defiant propaganda for their faith. The authorities' growing realisation that these executions were counterproductive led to a shift, around 1560, to hanging as traitors those who they would previously have burned as heretics.[39] The French Reformation had its fair share of martyrs, but this was of necessity a clandestine movement, encouraged into the open only when conditions appeared more auspicious for its acceptance.

As their congregations burgeoned from the mid- to late 1550s, the Huguenots became more openly defiant, meeting to hold their services, or *prêches*, in public with an armed guard. Such overt challenges to the law and to Catholic sensibilities unsurprisingly provoked flashpoints of confessional conflict: notoriously in Paris in 1557 (in the rue Saint Jacques) and 1561 (in the parish of Saint-Médard), but also in cities and towns across France. The sudden surge in Huguenot numbers on the eve of the religious wars fuelled both their growing confidence and their Catholic neighbours' fears. Triumphalism bred militancy; Huguenots were increasingly prepared to flout the law, convinced of the righteousness of their cause.[40] The years 1559–61 have been labelled the 'wonder years' of the Huguenot movement, a period that witnessed a remarkable coalescence of major landmark events: the first national synod in Paris (May 1559) which produced a decidedly French confession of faith and discipline for the whole church; notable royal concessions including the interconfessional Colloquy of Poissy (September 1561) and the negotiations which culminated in the Edict of January 1562. Yet therein also lay the seeds of future adversity culminating in the opening shot of the wars, the massacre of a Huguenot congregation in the town of Vassy by the Duke of Guise's men in March 1562.[41] Indeed, Timothy Watson believes that, at least in the case of Lyons, 'the broad, rather fluid coalition of interests which enabled the movement to grow so rapidly was not solid enough to coalesce into a firm party grouping' in the face of Catholic counter-action.[42] Yet, the movement had come a long way since the 'anarchy' of the early decades. As Donald Kelley has put it, between the 1520s and 1550s, 'transient and sporadic dissidence' was transformed into a 'coherent movement and an almost corporate organisation'.[43]

This metamorphosis has principally been attributed to the growing influence of Geneva. However, although the increasing impact of the

Genevan Church on the French Reform movement from the late 1550s is not in doubt, its influence prior to this point is less certain. Some have felt that undue attention is given to the role of Calvin and his colleagues in directing the French churches. This can be traced to the distorted view presented by Calvinist histories at the time, notably the *Histoire ecclésiastique* and the martyrology of Jean Crespin, as well as the desire of subsequent French Protestants to present their Church as a coherent and united force. Crouzet sums up the prevailing sceptical position by arguing that the 'multiple heterodox experiences' ensured that the 'Reformed religion' did not exist as an historical fact until 1559–61.[44] The concentration on Geneva has created a monolithic view of the French Reformation which ignores internal developments, such as the emphasis on congregationalism and clerical equality recently highlighted by Glenn Sunshine.[45] Other centres were just as influential in the movement's early years, principally Strasburg under Martin Bucer in the 1530s and 1540s, a model imitated by Calvin himself, and the structures developed in the francophone pays de Vaud under Pierre Viret.[46]

One way of cutting the Gordian knot between those who argue for the autonomous development of a distinctively French Church and those who uphold the centrality of Geneva is to see them as complementary phenomena. Indeed, although explicable among contemporaries, it is nevertheless strange how Calvinism is viewed as 'foreign' by many historians; its leadership was drawn almost exclusively from France, and the Genevan Church was France's major contribution to the Reformation. The pastors at Geneva were the French Church in exile with all the zeal that implies. Yet rather than imposing their will on the churches in France, at least in the 1550s and 1560s, they were often responding to demands for ministers and guidance. Nor were relations between the French churches and Geneva always conciliatory. The survival strategies adopted by some of those on the ground in the face of persecution were condemned by Calvin as Nicodemism. Undoubtedly, the French Church was more robust and self-willed than it has often been given credit for. The Genevan Church sought to control it, but the French movement also saw the advantage of allying itself with a movement which could provide it with a uniform structure, trained ministers, and communication networks which would strengthen it in the face of uncertain royal policy. It is in any case clear that the predominance of Genevan influence in the French Reformation was assured by the mid- to late 1550s. Nevertheless, the role of Calvinism in either stifling or saving the movement is keenly debated.

On the one hand, David Nicholls laments the loss of the vigorous 'religious anarchy' with the dominance of Calvinism, and argues that, as a positive unifying force, leadership came too late to stave off ultimate decline.[47] On the other, Henry Heller acknowledges that through Geneva's guidance, 'a subterranean and fragmented movement of reformation gradually became a cohesive force'.[48] Most historians concede that Calvinism's more systematic approach (through noble leadership) was needed to negotiate with the crown and to fight a civil war, as well as to secure the edicts of pacification. Nevertheless, recent research has demonstrated how fluid and diverse religious dissidence remained in France well into the 1560s.[49] Geneva could not control French events; Mark Greengrass has argued that the expansion of the home-grown movement and noble organisation were more significant factors.[50] Ironically, it was the adversity of the religious wars which consolidated Geneva's hold. As Nicholls concludes, 'Calvinist discipline was revealed as ideal for the needs of an embattled minority', assuring its triumph in the crucible of the first decade of the wars.[51]

Hauser's lament regarding the concentration of historians on the period after 1560, 'when it [the French Reformation] almost ceases to be a religious revolution and becomes a political party', reflects the watershed marked by the outbreak of armed conflict.[52] The French Reform movement on the eve of war was a militant, dynamic and optimistic force, at its height numerically (estimated at some 10 per cent or 1.8 million of the population, gathered in perhaps 1,200 churches).[53] This was also the only point at which Protestant publications outnumbered Catholic.[54] The brutal sectarian violence and forced conversions of the wars steadily eroded the Huguenots' numbers, confidence and vitality. The French Reformation is presented as having suffered an arrested development, its period of greatest expansion curtailed by the outbreak of war blamed variously on religious militancy or noble greed. Yet the erosion of the Reformed cause was very gradual, and there were some more positive outcomes of the conflict for the Huguenots. The period of the religious wars witnessed an unprecedented and virtually unparalleled experiment in confessional coexistence through the successive edicts of pacification, culminating in that of Nantes in 1598. Royal policy continued to fluctuate, as indeed it always had, allowing the Huguenots some breathing space and sporadically permitting them official strongholds and judicial representation. France was not prepared to countenance a Protestant king, as demonstrated by the opposition to Henry IV's accession in 1589 and by his eventual conversion to Catholicism in 1593. But nor would it accept a foreign prince, no matter how Catholic. Though much grumbled about, the concessions to the religious minority embodied in the edicts were broadly complied with. The contents of, and foundations for, the celebrated settlement at Nantes had already been tried, tested and laid.

Despite this growing appreciation of the tenacity of the Reform movement after 1562, the less attractive aspects of confessional division have also received particular scrutiny. The French religious wars stirred up violent passions on both sides, but it was the brutality of Catholic crowds against their Protestant victims which excited most contemporary and recent comment. It was Natalie Zemon Davis' anthropological approach to this subject which first provided us with a rational explanation for such mob violence, locating it in sacral symbolism.[55] She explained disfiguration and mutilation of Protestant corpses as an attack on those whose actions were believed to be polluting society, and the frequent use of fire and water in these rituals (dumping in rivers a speciality) as a means of purification. Castration and the destruction of unborn children represented an attempt to destroy the next generation of contaminants. Conversely, Protestants targeted those aspects of Catholic worship which they found most offensive: saints' statues and other sites of devotion, as well as the priests who perpetuated such theological errors. Despite a lively debate with Janine Garrisson, who argued for more socio-economic motives to the violence, Davis' paradigm has prevailed, culminating in a magisterial two-volume study by Denis Crouzet.[56] Crouzet has used Catholic apocalyptic fears (which he identifies as lying behind the violence against Protestants) to paint a vivid mental world prevailing in France from the outbreak of the Reformation to the reign of Henry IV, all-encompassing and manifesting itself in myriad ways. However, he dismisses the purificatory aspects highlighted by Davis in favour of the destructive outcome of apocalyptic fervour. The Catholic response, both through physical violence and in print, succeeded in limiting the spread of the French Reformation where the authorities had failed.[57]

Like its Catholic counterpart, popular reformation was radical and disruptive. If just a handful of the social elite had embraced the Reform, the authorities could perhaps have turned a blind eye to nobles worshipping discreetly on their estates (as was later embodied in the edicts of pacification) and even to notables in their homes. But confrontation and rebellion was another matter, especially when the nobility's involvement gave rebellion a political dimension. Incidents such as the Conspiracy of Amboise (1560) and the Surprise of Meaux (1567) – failed Huguenot *coups d'état* – unsurprisingly fuelled the movement's association with subversion and conspiracy. Such an association was in any case inevitable for a distinctive and suspect minority.[58] A complex interaction of militancy paired with conservatism marked the movement. These uneasy bedfellows were the product, in particular, of the movement's relationship with the crown. Most Huguenots continued to place their hopes in securing royal support, or at least royal protection. Their petitions of grievance were couched in obedient and deferential terms, and even when the Huguenots took up arms they claimed to be acting on behalf of the king against his (foreign) enemies.[59] Likewise, the various edicts of pacification which embodied and sanctioned Huguenot rights during the wars reaffirmed royal authority and Huguenot subjection to the monarch's will.[60] Only when this relationship broke down, as following the 1572 Saint Bartholomew's Day massacre, did the Huguenots adopt more radical positions.

Responsibility for the massacre remains an area of great historical contention, but the disputes over who gave the orders have given way to reflections on its impact. There is no question that it sent the French Reform movement reeling and represented a severe body blow from which it would never fully recover. Primarily this was due to the effective decapitation of the movement. Denis Crouzet has reinterpreted the royally sanctioned murder of the Huguenot leadership as an attempt to maintain rather than shatter the confessional harmony which the Huguenot nobles' aggression seemed to threaten, in line with the crown's neoplatonic vision of universal concord presided over by a benevolent monarch.[61] The view that the king saw the murders as a pre-emptive strike against treacherous plotters, rather than an ideological blow against heretics, is supported by Charles IX's striking of a medal in commemoration of the event. It depicts Charles seated in triumph, crushing underneath his throne the five principal conspirators. This emphasis contrasts to Pope Gregory XIII's unequivocally confessional medal celebrating the same event, in which the avenging angel smites the Huguenots. Unsurprisingly, however, Charles failed in his attempt to separate the political from the confessional, and to persuade ordinary worshippers to trust in his protection. Massacres of Huguenots ensued in a number of major towns, which combined with other forms of intimidation such as rebaptism, led for many to a stark choice between abjuration and exile.[62] The outcome for royal authority was significant, too: it was not only challenged by another religion, but its validity was questioned by more militant Catholics, suspicious of its policy of appeasement. In the very different circumstances of 1584, when the Huguenot leader Henry of Navarre became heir apparent, neither side dared fear or hope for what might happen should a Protestant inherit the French throne.

The paradox of a movement seen simultaneously as both radical and conservative cannot be explained simply by identifying its diverse social mix. The conservatism of the Huguenots, enshrined in their repeated declarations of loyalty to the crown and anxiety to act within the law, needs to be set off against the more militant aspects of their cause, for which the potential was present even after the wars.[63] They were easily cast as traitors to the kingdom, on the basis of their (spectacularly unsuccessful) conspiracies to seize power, and other violations of the law such as iconoclasm or going armed to assemblies in forbidden locations. Most notorious of all has become their development of theories of resistance to a tyrannical regime, in a series of tracts produced after the 1572 massacres. Chief among these were Francis Hotman's *Francogallia* (1573), Beza's *Du Droit des magistrats* (1574) and Philippe Duplessis Mornay's *A Defence of Liberty Against Tyrants (Vindiciae contra tyrannos)* (1579). However, the transformation of these theories into actions is far harder to discern. We see it in the municipality of La Rochelle's defiance of the royal governor, leading to the port's siege in 1573; in the alliance between the Reformed leadership in the south with the disaffected Catholic nobles, the king's brother Alençon and the governor of Languedoc, Henri Montmorency-Damville; and in the Huguenots' retreat to their southern strongholds, such as Montauban, Nîmes and Montpellier, though the notion that these constituted a 'state within a state' was long ago discredited. However, once again, the Huguenots were aware that the best chance for their survival was getting the monarchy on side. By 1576, noble leadership of the movement had been restored by Navarre's escape from captivity and repudiation of his forced conversion to Catholicism. The short-lived Edict of Beaulieu of the same year promised the most substantial royal concessions ever made to the Huguenots; and despite a precipitous return to war, peace negotiations continued between the crown and Navarre into the late 1570s and early 1580s. There were dark days following the 1585 Treaty of Nemours, which repealed the edicts of pacification, outlawed Protestantism and reinforced the ascendancy of the Guise-led Catholic League, but this was a reaction to Navarre's having become heir apparent. Thereafter, it was the League which would adopt resistance theories to oppose the accession of an ungodly monarch. The Huguenot movement was undoubtedly undone by the conversion of Navarre (now Henry IV) in 1593, but it is hard to see what other tack he might have taken to reconcile the country to his rule. The Huguenots' alternately militant and conservative stance was, therefore, dependent on and a reaction to political circumstance. These apparent contradictions enabled them to adapt and reposition themselves, but made their relationship with the monarchy fraught even under Henry IV.

For Raymond Mentzer and Andrew Spicer, Huguenots are 'a minority that has constantly sought and pursued strategies for survival in a hostile environment ... [which] has indelibly marked and shaped the community'.[64] But what did it mean to be a Huguenot? The Huguenot experience was determined by many factors: whether it was a period of toleration or persecution, growth or decline; the size of a community and its ability to worship freely; the protection and support offered by kin-group or friends; the social status, gender or age of an individual.[65] One of the most revealing insights into the Huguenot mentality is provided by Mark Greengrass in his study of the survivors' accounts of the Saint Bartholomew's Day massacre, which reveal how the trauma of events they could neither understand nor explain was internalised, and how their confidence in God's protection was shaken.[66] As a minority group, one became and remained a member of the Reformed faith through choice, and only the most committed were able to resist the pressure to renounce their beliefs which accompanied each round of repression. Many historians have emphasised how a persecuted minority mentality made the Huguenots increasingly defensive, conservative, and stoical about their fate. Persecution was interpreted as an indication of election, associating them with ancient Israel and the early Christian martyrs, which in turn was reaffirmed by their bestowing of Old Testament names on their children.[67] Such a position was reinforced by the tales of suffering in the martyrology of Jean Crespin (1554 and thereafter), and in other Huguenot histories. Huguenot attachment to the Psalms, with their themes of endurance and trust in God's deliverance, and their use at times of spiritual need and as an act of defiance, is also explicable in this light.[68] Our understanding of the tribulations endured by Huguenot communities has been enriched by a number of urban studies. In major towns such as Dijon, Lyons, Marseilles, Paris, Rouen and Troyes, they formed a minority substantial enough to provoke confrontation but too small effectively to challenge the established order.[69] However, where Huguenots found themselves in the majority, as in some of the towns of the south and west which became their havens of refuge, we can begin to see how their communities would function in a more secure environment.

The so-called 'Huguenot crescent' extending from La Rochelle in the west to the towns of Dauphiné in the east, and encompassing the major strongholds of Montauban, Montpellier and Nîmes, underlines the urban character of the movement. Only in the Cévennes, parts of Normandy, the Midi and Dauphiné do we find substantial rural support for Protestantism.[70] It was from the towns of France that the demand came for trained ministers, and it was to the towns that several hundred pastors were sent, mostly from Geneva, but also from Neuchâtel and Lausanne. This urban concentration lent the movement much of its dynamism and also determined their Catholic neighbours to distance

Huguenot places of worship from large centres of population. Once established or 'gathered', beginning with those of Poitiers and Paris in the mid-1550s, the churches were able for the first time to administer their own sacraments and organise consistories to ensure the discipline and harmony of the congregation. In time, local colloquies, provincial and national synods, and general assemblies completed the Calvinist structure of the French Reformed Church and eased communication between the churches and with Geneva. The proposed regularity of the meetings of these bodies was disrupted by the ensuing civil wars, as is their surviving documentation. Most churches returned to their earlier, clandestine existence. However, in the southern towns of Montauban, Montpellier, Nîmes and La Rochelle, the Huguenots enjoyed majority status which could lead to a certain complacency in upholding the strict observance of the discipline and other behaviour which made their religion distinctive. Nevertheless, their vulnerability to attack and erosion of their autonomy in a predominantly Catholic country also shaped and reinforced their confessional identity.[71] Recent studies have particularly focused on two aspects of these Reformed regimes, the activities of the consistory and the provision of poor relief: 'a defining aspect of Huguenot culture'.[72]

At the end of the wars, with the relative stability provided by the Edict of Nantes, the geography of the Huguenot crescent was confirmed and churches struggled to be re-established and to survive elsewhere in France. From an estimated 10 per cent of the population in 1560, the Huguenots' numbers had been reduced by half through massacre, abjuration and exile. In the seventeenth century, the faithful would see their numbers eroded further: partly through the defection of much of their vital noble leadership; their political defeat in the 1620s, epitomised by Louis XIII's successful siege of La Rochelle; and natural erosion and acculturation.[73] It had by then been many decades since the 'wonder years' around 1560 which seemed to presage the imminent triumph of the Reformed cause. Likewise, throughout Europe, the Protestant Reformation was retreating in the face of a resurgent Catholicism. No state had successfully integrated its confessional minority, aside from the Netherlands which, as in so much else, was the exception which proved the rule. The ultimate failure of the French Reformation, despite its early promise, has led some historians to speculate as to how things might have been different, whilst others have concluded that it could have gone no other way. The story of the French Reformation combines all the drama of an epic struggle and the pathos of a tragic defeat. Its dual nature is tenacious and does not allow for the recounting of a single plotline. Its complexity is reflected in a multi-faceted historiography which denies us any simple answers, and is littered with stark contradictions, but through which emerges a French Reformation neither revolutionary nor hopelessly disparate.

further reading

The essential starting point is Mark Greengrass, *The French Reformation* (Oxford, 1987); cf. his essay on 'France', in B. Scribner, R. Porter and M. Teich (eds), *The Reformation in National Context* (Cambridge, 1994). David Nicholls has contributed many insights into the progress of the early Reformation: see his 'France', in Andrew Pettegree (ed.), *The Early Reformation in Europe* (Cambridge, 1992); 'Heresy and Protestantism, 1520–1542: Questions of Perception and Communication', *French History*, 10 (1996), 182–205; 'The Theatre of Martyrdom in the French Reformation', *Past and Present*, 121 (1988), 49–73; 'The Social History of the French Reformation: Ideology, Confession and Culture', *Social History*, 9 (1984), 25–43.

The classic essay on socio-economic issues is, Henri Hauser, 'The French Reformation and the French People in the Sixteenth Century', *American Historical Review*, 4 (1899), 217–27; on this theme, see also Henry Heller, *The Conquest of Poverty: the Calvinist Revolt in Sixteenth-Century France* (Leiden, 1986).

Other aspects of the French Reformation are explored in Donald R. Kelley, *The Beginning of Ideology: Consciousness and Society in the French Reformation* (Cambridge, 1981); Christopher Elwood, *The Body Broken: the Calvinist Doctrine of the Eucharist and the Symbolization of Power in Sixteenth-Century France* (New York and Oxford, 1999); William Monter, *Judging the French Reformation: Heresy Trials by Sixteenth-Century Parlements* (Cambridge, MA, and London, 1999).

On the Wars of Religion, see Mack P. Holt, *The French Wars of Religion, 1562–1629* (Cambridge, 1995); and his 'Putting Religion back into the Wars of Religion', *French Historical Studies*, 18 (1993), 524–51; R. J. Knecht, *Catherine de' Medici* (London, 1998); Mark Greengrass, *France in the Age of Henri IV*, 2nd edn (London, 1995); Luc Racaut, *Hatred in Print: Catholic Propaganda and Protestant Identity during the French Wars of Religion* (Aldershot, 2002).

Studies of particular towns during the religious wars have provided important perspectives: Philip Benedict, *Rouen during the Wars of Religion* (Cambridge, 1981); Barbara B. Diefendorf, *Beneath the Cross: Catholics and Huguenots in Sixteenth-Century Paris* (Oxford, 1991); Penny Roberts, *A City in Conflict: Troyes during the French Wars of Religion* (Manchester, 1996).

Important essays on aspects of the wars include Natalie Zemon Davis, 'The Rites of Violence: Religious Riot in Sixteenth-Century France', *Past and Present*, 59 (1973); Philip Benedict, 'The Saint Bartholomew's Massacres in the Provinces', *Historical Journal*, 21 (1978), 205–25; Denis Richet, 'Sociocultural Aspects of Religious Conflicts in Paris during the Second Half of the Sixteenth Century', in R. Forster and O. Ranum (eds), *Ritual, Religion and the Sacred: Selection from the Annales* (1982); Mark Greengrass, 'Hidden Transcripts: Secret Histories and Personal Testimonies of Religious Violence in the French Wars of Religion', in M. Levene and P. Roberts (eds), *The Massacre in History* (1999); Penny Roberts, 'Royal Authority and Justice during the French Religious Wars', *Past and Present*, 184 (2004), 3–32.

A number of valuable essay collections have recently appeared, notably Philip Benedict et al. (eds), *Reformation, Revolt and Civil War in France and the Netherlands, 1555–1585* (Amsterdam, 1999); R. A. Mentzer and A. Spicer (eds), *Society and Culture in the Huguenot World, 1559–1685* (Cambridge, 2002); Philip Benedict, *The Faith and Fortunes of France's Huguenots, 1600–85* (Aldershot, 2001).

notes

1. On the historiography of French Protestantism, see in particular David Nicholls, 'The Social History of the French Reformation: Ideology, Confession and Culture', *Social History*, 9 (1984), esp. 25–35.
2. Margaret Mann, *Erasme et les débuts de la réforme française (1517–36)* (Paris, 1934), p. 22.
3. See notably, A. Renaudet, *Préréforme et humanisme à Paris pendant les premières guerres d'Italie (1494–1517)* (Paris, 1916).
4. But see Lucien Febvre, *Au coeur religieux du XVIe siècle* (Paris, 1957), pp. 24, 68, arguing that any such independent development is debatable.
5. *Ibid.*, p. 66.
6. P. Imbart de la Tour, *Les Origines de la Réforme*, 4 vols (Paris, 1905–35), III, pp. ix, 180, 186.
7. Denis Crouzet, *La Genèse de la Réforme française, 1520–1562* (Paris, 1996), pp. 348–52; David Nicholls, 'Heresy and Protestantism, 1520–1542: Questions of Perception and Communication', *French History*, 10 (1996), 201; Mark Greengrass, *The French Reformation* (Oxford, 1987), p. 13; Francis Higman, 'Luther et la piété de l'Église gallicane; le *Livre de vraye et parfaicte oraison*', in his *Lire et découvrir: La circulation des idées au temps de la Réforme* (Geneva, 1998), pp. 179–200; Philip Benedict, *Christ's Churches Purely Reformed: a Social History of Calvinism* (New Haven and London, 2002), p. 139.
8. David Nicholls, 'The Nature of Popular Heresy in France, 1520–42', *Historical Journal*, 26 (1983), 262–3.
9. Francis Higman, 'Le levain de l'Évangile', and 'Genevan printing and French censorship, 1520–1551', in his *Lire et découvrir*.

10. Nicholls, 'Nature of Popular Heresy', 270–1; Crouzet, *Genèse de la Réforme*, pp. 362–3, 404–6.
11. Imbart de la Tour, *Les Origines de la Réforme*, III, p. 186.
12. Henry Heller, 'Famine, Revolt and Heresy at Meaux, 1521–5', *Archiv für Reformationsgeschichte*, 68 (1977), 133–56.
13. Donald R. Kelley, *The Beginning of Ideology: Consciousness and Society in the French Reformation* (Cambridge, 1981), pp. 15–16.
14. Greengrass, *French Reformation*, p. 26; R. J. Knecht, *Renaissance Warrior and Patron: the Reign of Francis I* (Cambridge, 1994), pp. 313–21.
15. Crouzet, *Genèse de la Réforme*, p. 471; Christopher Elwood, *The Body Broken: the Calvinist Doctrine of the Eucharist and the Symbolization of Power in Sixteenth-Century France* (New York and Oxford, 1999), esp. pp. 21–30.
16. A-M. Brenot, 'La peste soit des huguenots: Etude d'une logique d'exécration au XVIe siècle', *Histoire, Economie et Société*, 11 (1992), 553–70; Mack P. Holt, 'Wine, Community and Reformation in Sixteenth-Century Burgundy', *Past and Present*, 138 (1993), 66, 69.
17. Greengrass, *French Reformation*, p. 27.
18. Crouzet, *Genèse de la Réforme*, pp. 308–12.
19. Alain Tallon, *Conscience nationale et sentiment religieux en France au XVIe siècle* (Paris, 2002).
20. Mark Greengrass, 'France', in Robert Scribner, Roy Porter and Mikuláš Teich (eds), *The Reformation in National Context* (Cambridge, 1994), pp. 58–62.
21. William Monter, *Judging the French Reformation: Heresy Trials by Sixteenth-Century Parlements* (Cambridge, MA, 1999), pp. 82–3.
22. D. Nicholls, 'The Origins of Protestantism in Normandy: a Social Study' (PhD thesis, University of Birmingham, 1977), pp. 125–33; David S. Hempsall, 'Measures to Suppress "la peste luthérienne" in France, 1521–2', *Bulletin of the Institute of Historical Research*, 49 (1976), 298–9; Penny Roberts, *A City in Conflict: Troyes during the French Wars of Religion* (Manchester, 1996), p. 32.
23. *Histoire ecclésiastique des églises réformées au royaume de France*, ed. G. Baum, E Cunitz and R. Reuss, 3 vols (Paris, 1883–89), from the original work published in 1580 under the auspices of Theodore Beza.
24. H. Hauser, 'The French Reformation and the French People in the Sixteenth Century', *American Historical Review*, 4 (1899), 217–27.
25. Henry Heller, *The Conquest of Poverty: the Calvinist Revolt in Sixteenth-Century France* (Leiden, 1986), pp. X–XI.
26. *Ibid.*, pp. 26, 69–70.
27. Philip Benedict, *Rouen during the Wars of Religion* (Cambridge, 1981); Natalie Zemon Davis, *Society and Culture in Early Modern France* (Stanford, 1975); Nicholls, 'Origins of Protestantism'; Mack P. Holt, 'Putting Religion back into the Wars of Religion', *French Historical Studies*, 18 (1993), 524–51.
28. Mack P. Holt, *The French Wars of Religion, 1562–1629* (Cambridge, 1995), p. 2.
29. See useful summary of these in Crouzet, *Genèse de la Réforme*, pp. 477–539. Cf. J. M. Davies, 'Persecution and Protestantism: Toulouse, 1562–75', *Historical Journal*, 22 (1979), 31–51; James R. Farr, 'Popular Religious Solidarity in Sixteenth-Century Dijon', *French Historical Studies*, 14 (1985), 192–214.
30. Crouzet, *Genèse de la Réforme*, pp. 368–74; Kelley, *The Beginning of Ideology*, pp. 131–65; G. Huppert, *Public Schools in Renaissance France* (Urbana, IL, 1984);

G. de Groër, *Réforme et Contre-Réforme en France: Le collège de la Trinité au XVIe siècle à Lyon* (Paris, 1995), esp. pp. 51–67; Heller, *Conquest of Poverty*, pp. 70–110.

31. Scott M. Manetsch, *Theodore Beza and the Quest for Peace in France, 1572–1598* (Leiden, 2000).
32. Mark Greengrass, 'Financing the Cause: Protestant Mobilization and Accountability in France (1562–1598)', in Philip Benedict, Guido Marnef, Henk van Nierop and Marc Venard (eds), *Reformation, Revolt and Civil War in France and the Netherlands 1555–1585* (Amsterdam, 1999); James B. Wood, *The King's Army: Warfare, Soldiers and Society during the Wars of Religion in France, 1562–1576* (Cambridge, 1996).
33. D. Richet, 'Aspects socio-culturels des conflits religieux à Paris dans la seconde moitié du XVIe siècle', *Annales: économies, sociétés, civilisations*, 32 (1977), 764–89 (translated in R. Forster and O. Ranum (eds), *Ritual, Religion and the Sacred: Selection from the Annales* (1982)); N. L. Roelker, *One King, One Faith: the Parlement of Paris and the Religious Reformations of the Sixteenth Century* (Berkeley, CA, 1996); J. Powis, 'Gallican Liberties and the Politics of Later Sixteenth-Century France', *Historical Journal*, 26 (1983).
34. Penny Roberts, 'Royal Authority and Justice during the French Religious Wars', *Past and Present*, 184 (2004), 3–32; Jérémie Foa, 'Making Peace: the Commissions for Enforcing the Pacification Edicts in the Reign of Charles IX', *French History*, 18 (2004), 256–74.
35. Arlette Jouanna, 'Idéologies de la guerre et idéologies de la paix en France dans la seconde moitié du XVIe siècle', in M. Yardeni (ed.), *Idéologie et propagande en France* (Paris, 1987).
36. Nicholls, 'Heresy and Protestantism', 203–4; idem, 'Nature of Popular Heresy', 268–70.
37. D. S. Hempsall, 'The Languedoc, 1520–1540: a Study of Pre-Calvinist Heresy in France', *Archiv für Reformationsgeschichte*, 62 (1971), 236–43. On iconoclasm, see O. Christin, *Une Révolution symbolique: L'iconoclasme huguenot et la reconstruction catholique* (Paris, 1991), questioned in Crouzet, *Genèse de la Réforme*, p. 381, and Nicholls, 'Heresy and Protestantism', 202.
38. Crouzet, *Genèse de la Réforme*, pp. 384–91; Kelley, *Beginning of Ideology*, p. 97; Andrew Pettegree and Matthew Hall, 'The Reformation and the Book: a Reconsideration', *Historical Journal*, 47 (2004), 805–7.
39. D. Nicholls, 'The Theatre of Martyrdom in the French Reformation', *Past and Present*, 121 (1988), 49–73.
40. Philip Benedict, 'The Dynamics of Protestant Militancy: France, 1555–1563', in Benedict et al., *Reformation, Revolt and Civil War*.
41. Greengrass, 'France', p. 59.
42. Timothy Watson, 'Preaching, Printing, Psalm-Singing: the Making and Unmaking of the Reformed Church in Lyon, 1550–1572', in R. A. Mentzer and A. Spicer (eds), *Society and Culture in the Huguenot World, 1559–1685* (Cambridge, 2002), p. 27.
43. Kelley, *Beginning of Ideology*, p. 92.
44. Crouzet, *Genèse de la Réforme*, p. 345.
45. Glenn S. Sunshine, *Reforming French Protestantism: the Development of Huguenot Ecclesiastical Institutions, 1557–1572* (Kirksville, MO, 2003), esp. pp. 1–11.

46. Greengrass, *French Reformation*, pp. 21–4, 29–30; Heiko A. Oberman, 'Calvin and Farel: the Dynamics of Legitimation in Early Calvinism', *Journal of Early Modern History*, 2 (1998), 32–60.
47. Nicholls, 'Nature of Popular Heresy', 261–2, 274–5.
48. Heller, *Conquest of Poverty*, p. 111; cf. Crouzet, *Genèse de la Réforme*, p. 345.
49. Thierry Wanegffelen, *Ni Rome, ni Genève: Des fidèles entre deux chaires en France au XVIe siècle* (Paris, 1997).
50. David Nicholls, 'France', in Andrew Pettegree (ed.), *The Early Reformation in Europe* (Cambridge, 1992), p. 129; Greengrass, *French Reformation*, p. 41.
51. Nicholls, 'France', p. 135.
52. Hauser, 'The French Reformation', p. 220.
53. Greengrass, *French Reformation*, pp. 42–5.
54. Pettegree and Hall, 'The Reformation and the Book', p. 804.
55. Natalie Zemon Davis, 'The Rites of Violence: Religious Riot in Sixteenth-Century France', *Past and Present*, 59 (1973); also in her *Society and Culture*.
56. Natalie Zemon Davis and Janine Garrisson, 'Debate', *Past and Present*, 67 (1975), 127–35; Denis Crouzet, *Les Guerriers de Dieu: la violence au temps des troubles de religion, vers 1525-vers 1610* (Paris, 1990). Cf. Mark Greengrass' review article, 'The Psychology of Religious Violence', *French History*, 5 (1991), 467–74.
57. Luc Racaut, *Hatred in Print: Catholic Propaganda and Protestant Identity during the French Wars of Religion* (Aldershot, 2002).
58. Penny Roberts, 'Huguenot Conspiracies, Real and Imagined, in Sixteenth-Century France', in B. Coward and J. Swann (eds), *Conspiracies and Conspiracy Theory in Early Modern Europe* (Hounslow, 2004).
59. Penny Roberts, 'Huguenot Petitioning during the Wars of Religion', in Mentzer and Spicer (eds), *Society and Culture in the Huguenot World*.
60. Roberts, 'Royal Authority and Justice'.
61. Denis Crouzet, *La Nuit de la Saint-Barthélemy: Un rêve perdu de la Renaissance* (Paris, 1994).
62. Philip Benedict, 'The Saint Bartholomew's Massacres in the Provinces', *Historical Journal*, 21 (1978), 205–25.
63. On this mixture of conservatism and militancy, see Benedict et al. (eds), *Reformation, Revolt and Civil War*, especially articles by Benedict and Denis Crouzet; Mentzer and Spicer (eds), *Society and Culture in the Huguenot World*, especially articles by Penny Roberts and Alan James.
64. R. A. Mentzer and A. Spicer, 'Introduction: *Être protestant*', in their *Society and Culture in the Huguenot World*, p. 2, drawing on Janine Garrisson, *L'Homme protestant* (Brussels, 1986).
65. Mark Greengrass, 'Informal Networks in Sixteenth-Century French Protestantism', in Mentzer and Spicer (eds), *Society and Culture in the Huguenot World*; Barbara B. Diefendorf, *Beneath the Cross: Catholics and Huguenots in Sixteenth-Century Paris* (Oxford, 1991), pp. 127–36; Penny Roberts, 'Huguenotes et bigotes: les femmes et la Réforme vues par Nicolas Pithou', in J. Provence (ed.), *Mémoires et mémorialistes à l'époque des Guerres de Religion* (Paris, forthcoming).
66. Mark Greengrass, 'Hidden Transcripts: Secret Histories and Personal Testimonies of Religious Violence in the French Wars of Religion', in M. Levene

and P. Roberts (eds), *The Massacre in History* (New York and Oxford 1999), pp. 69–88.

67. Charles Parker, 'French Calvinists as the Children of Israel: an Old Testament Self-Consciousness in Jean Crespin's *Histoire des Martyrs* before the Wars of Religion', *Sixteenth Century Journal*, 24 (1993), 227–47; Luc Racaut, 'Religious Polemic and Huguenot Self-Perception and Identity, 1554–1619', in Mentzer and Spicer (eds), *Society and Culture in the Huguenot World*; Philip Conner, *Huguenot Heartland: Montauban and Southern French Calvinism during the Wars of Religion* (Aldershot, 2002), pp. 58–61.
68. Diefendorf, *Beneath the Cross*, pp. 136–41.
69. Farr, 'Popular Religious Solidarity'; Davis, *Society and Culture*; Wolfgang Kaiser, *Marseille au temps des troubles, 1559–1596* (Paris, 1992); Diefendorf, *Beneath the Cross*; Benedict, *Rouen during the Wars*; Roberts, *City in Conflict*.
70. Nicholls, 'France', pp. 129–30; E. Le Roy Ladurie, *Les paysans de Languedoc*, 2 vols (Paris, 1966), I, pp. 341–4; Janine Garrisson, *Protestants du Midi, 1559–1598* (Toulouse, 1980).
71. Conner, *Huguenot Heartland*; Judith P. Meyer, *Reformation in La Rochelle: Tradition and Change in Early Modern Europe, 1500–1568* (Geneva, 1996); Kevin C. Robbins, *City on the Ocean Sea: La Rochelle 1530–1650* (Leiden, 1997).
72. Martin Dinges, 'Huguenot Poor Relief and Health Care in the Sixteenth and Seventeenth Centuries', in Mentzer and Spicer (eds), *Society and Culture in the Huguenot World*, esp. p. 173. Cf. three articles by Raymond A. Mentzer: '*Disciplina nervus ecclesiae*: the Calvinist Reform of Morals at Nîmes', *Sixteenth Century Journal*, 18 (1987), 89–115; 'Organizational Endeavour and Charitable Impulse in Sixteenth-Century France: the Case of Protestant Nîmes', *French History*, 5 (1991), 1–29; 'Marking the Taboo: Excommunication in French Reformed Churches', in his *Sin and the Calvinists* (Kirksville, MO, 1994).
73. Philip Benedict, *The Huguenot Population of France, 1600–1685: the Demographic Fate and Customs of a Religious Minority* (Philadelphia, 1991).

6
britain and ireland

alec ryrie

The Reformations in Britain and Ireland have attracted a wholly disproportionate amount of historical attention. In the nineteenth and twentieth centuries Britain was a major European power and a player on the world stage, and so it seemed natural to lavish historical attention onto it. Moreover, the spread of the English language across the world has ensured a plentiful supply of anglophone historians who find learning languages tiresome and who therefore choose to work on British material. The result has been to make the British Reformations look more important than they really are. Sixteenth-century England was a second-rank European power, in political, commercial and cultural terms. One historian has, mischievously, compared England's importance to that of another island realm on the fringes of Latin Christendom – Cyprus.[1] Scotland, proud as it was of its status as a European kingdom, was even more of a bit player. Ireland was a realm whose nominal English overlordship was a cover for the lack of any settled government over most of the island.

Unsurprisingly, then, the historiography of the British and Irish Reformations is not only a dense thicket, but a spiky one. This is not only because too many historians are jostling for too little space, but also because the debates are, even now, more than merely scholarly. At times they have almost been choked by ulterior motives: and it is through those ulterior motives that the subject is best approached.

confessional debates

The first and most influential historians of the British Reformations were John Foxe and his mentor John Bale. Foxe's colossal church history, the *Acts and Monuments* – usually known as Foxe's *Book of Martyrs*, but it is much more than a book of martyrs – has set the tone for Protestant histories of the English Reformation since 1563. Foxe also explicitly saw his work in a British context and included valuable material on the Scottish Reformation. John Bale was England's first Protestant historian. He was also, briefly, bishop of Ossory in Ireland, and his bitter account of his unhappy time in Ireland has coloured much subsequent Protestant writing about the island and its people. Subsequent histories have found it hard to leave Foxe and Bale's shadows behind. 'Protestant' narratives of the Reformation, in their tradition, persist; so too do 'Catholic' narratives which look to Foxe's earliest opponents. Nowadays, not all 'Protestant' or 'Catholic' historians of the British Reformations have partisan religious convictions of their own; but plenty do, and even their most secular or most ecumenical colleagues can find it hard to resist being drawn in sympathy to one side or the other.[2]

The Protestant narrative of the English Reformation is currently out of fashion, but it refuses entirely to disappear. Classically, it is the story of how England threw off an oppressive Romish yoke. It begins with the Lollards, England's late medieval heretics, some of whose beliefs seemed to anticipate Protestantism. These glimmerings of light turned to dawn in the 1520s, as Protestant ideas began to reach England from a group of converts in Continental exile. When Henry VIII decided to take England into schism in the 1530s, he was therefore pushing at an open door; perhaps he had even been half-persuaded by the Protestants' arguments. The light of the Gospel waxed until the fiery midday of Edward VI's reign (1547–53), when an unambiguously Protestant government swept away all remaining corruption. The young king's unexpected death plunged the realm into a sudden eclipse; the accession of his half-sister Mary, who was determined to return England to papal obedience, led to five years of darkness lit only by the fires on which she burned some three hundred Protestants for their faith. Yet her effort was futile, and at her death in 1558 the long, sunlit afternoon of English Protestantism began, under the (more or less) benevolent eye of Elizabeth I.

Modern versions of this narrative are subtler. Its ablest proponent in recent years was A. G. Dickens, whose survey of *The English Reformation* (1964, revised 1989) began modern study of the English Reformation. Dickens too was a Protestant, albeit of a more softly-spoken kind than Foxe. And he added a significant new element, appropriate for a democratic age: unlike Foxe, he argued that the Protestants made a success of winning large numbers of converts early on. Thus in 1558 Elizabeth became queen of a nation that was already mostly Protestant.

This blithe 'Whig-Protestant' narrative of the English Reformation has long been balanced by a mirror-image Catholic narrative. In the twentieth century it was pursued by Catholic scholars of considerable ability and increasing subtlety, such as Aidan Gasquet, Philip Hughes and, more recently, J. J. Scarisbrick and Eamon Duffy.[3] This tradition strongly asserts the rude health of the pre-Reformation church. It dismisses Lollardy as an irrelevance, and the early Protestant converts as little more. The Reformation emerges primarily as a political event, driven by the lust and greed of Henry VIII, whose desire for a divorce was eventually overtaken by his desire to plunder the church's wealth. Edward VI's reign becomes merely an extension of his father's. Mary's reign becomes the great 'what if', a chance to restore England to the Roman fold which was lost only because of the accident of her death. Under Elizabeth, attention turns to the stubborn survival of a Catholic minority, and the eventual martyrdom of many of their number. More recently, this tradition has also questioned how effective even Elizabeth's regime was in building a Protestant nation. Christopher Haigh – not himself a Catholic, but an eloquent exponent of a variant of the 'Catholic' view – famously asked two questions of the English Reformation.[4] From above, or from below? And fast, or slow? Haigh's own answer is clear: from above and slow. This was a Reformation imposed with difficulty on an unwilling country by a *rex ex machina*. Haigh – unlike Duffy – even doubts whether the imposition ever succeeded.

In the 1980s and 1990s, this view was in the ascendant: this was partly the result of a fashion for self-styled 'revisionism' which had spread from studies of the English Civil War, but it also reflected the power of Scarisbrick, Haigh and Duffy's scholarship. The most powerful part of their analysis has been their rehabilitation of the late medieval church. Duffy's heady portrait of a vibrant Catholic piety has proved compelling. Other, more statistical and less passionate studies of the late medieval church on the ground have confirmed his picture, reinforced by the use of new sources such as wills and churchwardens' accounts.[5] Haigh, in one remarkably effective article, has discredited the idea that there was widespread hatred of and contempt for the clergy in late medieval England: 'anticlericalism' now seems to be something between a literary convention and a Protestant myth.[6]

The study of Catholicism in post-Reformation England has also been rescued from genteel confessional isolation. In 1975, Haigh and John Bossy both published books disagreeing over the nature of post-1560 Catholicism: was it (as Haigh argued) a stubborn survivor of the late medieval church, or (Bossy's view) a fresh, Continental transplant?[7]

Intruguingly, it is now clear that in the 1590s English Catholics were themselves divided over essentially the same issue: should they try to exist quietly in a Protestant state (the 'survivalist' view of the secular priests) or treat England as a mission field to be evangelised by any means necessary (the view of the Jesuit missionaries)? It seems that the survivalist view won out, despite or because of the heroic deaths of so many missionary priests. Bossy and Haigh both saw Catholicism becoming confined to a network of gentry families who did little more than endure. This view of an essentially static Catholic community has been vigorously challenged by Michael Questier, who sees a much more fluid set of Catholic circles which continued to win converts from the Protestant population.[8] Perhaps most importantly, Alexandra Walsham has redirected our attention to the 'church papists', those Catholics who conformed outwardly to the established church, but whose Catholicism was no less real for that.[9] They are a reminder that it is misleading to draw confessional lines too sharply.

The other great success of the Catholic revival in Reformation history has been the rescue of Mary's reign. Duffy, Haigh and particularly Jennifer Loach have argued that Mary's Catholic restoration was both popular and successful. Several recent works demonstrate that her religious policy, once dismissed as an unthinking reaction, was both rich and sophisticated.[10] Duffy famously stated in 1992 that 'a convincing account of the religious history of Mary's reign has yet to be written',[11] and this is still true: some recent attempts to reassess this period have had a very mixed reception.[12] Yet we are manifestly moving in the right direction. At the least, this reign can no longer be dismissed as a speed-bump on England's providential road to Protestantism.

As this reassessment has matured, it has also come to look at the most notorious aspect of Mary's reign, her policy towards Protestantism. Her critics have long pointed to the fact that Protestants out-published Catholics during her reign, and used this to suggest that her regime was little more than a rabbit in the headlights of the Protestant juggernaut. It is now clear that she and her bishops were not simply inactive, but rather chose not to confront the Protestants on their own terms. They preferred to counter the threat of heresy by building a Catholic alternative, rather than by dignifying the heretics' vitriolic attacks with a reply.[13] Whether this was a wise choice is another matter. The Protestant opposition which Mary faced was numerically small, but as Andrew Pettegree's *Marian Protestantism* (1996) emphasises, it was both vigorous and well-organised. The burnings, too, remain controversial. No-one wishes to defend them, but in a forthcoming book Thomas Freeman – the most authoritative scholar of Marian Protestantism – will argue that they were grimly effective, both in decapitating the Protestant movement and in disrupting its underground networks.[14] The consensus is still that the ferocity of Mary's anti-heresy campaign backfired. It is less clear whether, when faced with such an active and determined group of dissidents, Mary had any viable alternatives.

Away from these topics, the 'Catholic' view of the Reformation has not always had the upper hand. For the late medieval period, scholars of more Protestant sympathies have questioned the dismissal of Lollardy. John F. Davis's overblown claims for Lollardy's influence on and survival through the Reformation have recently been subjected to a brutally effective mauling by Richard Rex.[15] Rex may not have the last word, however, partly because Lollardy's textual remains are so rich that medievalists – especially literary scholars – are unable to resist them. The best work on Lollardy has focused on their books, which English Protestantism self-consciously embraced.[16] The attempt to link Lollardy geographically to Protestantism now seems shaky, and Lollardy cannot be seen in any simple way as a 'cause' of the Reformation. Yet neither the Lollards themselves, nor the idea that they played some part in the process, will quite go away.[17]

A broader debate has followed from Haigh's 'slow' Reformation. The English Reformation, once seen as a mid-sixteenth-century affair, has become a 'Long Reformation' extending into the seventeenth and possibly even the eighteenth century.[18] Much of the most exciting work done in the past generation has looked at the process by which the Reformation was 'bedded down' in the reigns of Elizabeth I and James I. Ironically, this is the area where the Catholic paradigm has been challenged most successfully. The field has long been dominated by one exceptionally subtle and sophisticated Protestant historian, Patrick Collinson. Collinson pioneered the modern study of the 'Puritans', those Protestants who agitated for Elizabeth's Reformation to go further than the queen was willing to allow.[19] Through his work, and that of his students and successors, Puritanism has been rescued from its stereotype of dour fanaticism. The Puritans (a term Collinson dislikes, for it was originally a term of abuse; he prefers their revealing self-description, 'the godly') emerge as a restless, energetic and committed minority. They took their role as pastors extremely seriously, although how successful they were at this continues to be disputed. (Haigh has even argued that some congregations, faced with such zealous ministers, managed to 'tame' them.)[20] However, the thrust of recent scholarship has been to emphasise the Protestant achievement in Elizabeth's reign and after. The Puritans' grand hopes may have failed, but they succeeded in altering the religious mood of the country. As Haigh has put it, they may not have made a nation of Protestants, but they did make a Protestant nation.[21] Peter Lake has identified a constituency who can be called 'moderate Puritans', whose beliefs were no less sincere but whose political vision was less confrontational. From their point of view, the accounting of profit and loss seemed more complex.[22] Several recent studies have also emphasised how Protestant ideas seeped out into the wider culture. Tessa Watt's *Cheap Print and Popular Piety* uses printed ephemera – ballads, broadsheets, chapbooks and other throwaway items – to argue that there was a real popular Protestantism, even if the more po-faced preachers might not always have approved of it. Ian Green's massive studies of popular Protestant print have reinforced this point.[23] Alexandra Walsham's *Providence in Early Modern England* uses similar sources to examine the impact of the Protestant doctrines of divine providence. Apparently some aspects of Protestantism could be genuinely popular, especially when they resonated with themes which were already present in popular culture.

The current picture of English religious culture in about 1600, then, is one which is – in Watt's words – 'distinctively "post-Reformation", but not thoroughly "Protestant"'.[24] Much of the most interesting recent work in English Reformation studies has traced the contours of this post-Reformation culture. Notable are Peter Marshall's work on beliefs about death and the dead, and Peter Lake's literary-based study emphasising the religious plurality, instability and anxiety of Elizabethan England.[25] Norman Jones' splendid survey of the English Reformation has taken this slow cultural transformation as its central theme, looking at the impact of the Reformation on (among other things) family structures and public morality.[26]

As well as a longer Reformation, we now think of a more localised one. Since the 1950s, both Protestant- and Catholic-leaning historians of the Reformation have tried to outflank one another by examining the impact of the Reformation in local communities. Although Dickens was an early enthusiast for these studies, they have done Catholic historians more favours than Protestants. Haigh's analysis of one notoriously conservative county, Lancashire, left the county's few Protestants looking like the ineffectual and bullying representatives of a distant and arrogant establishment.[27] However, English local studies have also helped us to move beyond a zero-sum game in which Catholic and Protestant historians each try to count their legions. Several studies, notably Susan Brigden's monumental *London and the Reformation* (1989), have detected widespread religious confusion and indifference alongside the noisier religious partisans. Muriel McClendon's study of England's second city, Norwich, even claimed to find a kind of toleration there, with civic virtues trumping religious disputes – although many other scholars have been unconvinced by this.[28] Another of Dickens' innovations – the systematic study of wills – has also helped to complicate the picture. Numerous local studies have attempted to trace shifting religious opinions using wills. The difficulties involved in using wills in this way are formidable, but wills do seem to indicate widespread religious disillusion and detachment by the 1550s.[29]

This picture of a population who were neither enthused nor horrified by religious change, but rather bewildered by it, seems to be an important part of the answer to a question which Catholic scholarship invites. If the English so disliked the Reformation, why did they accept it? Part of the answer is, of course, that they did not; hence the considerable recent interest in the great rebellions of the Tudor period.[30] The biggest of these, the northern risings of 1536, now seem to have been principally a religious affair. The rebels had other grievances, but there would have been no rebellion had they not been provoked by Henry VIII's religious policies. The south-western rebellion of 1549 is even more unmistakably a Catholic rising. By contrast, the periodic attempt to give the south-eastern disturbances of 1549 the character of a Protestant rebellion has not, yet, been so successful. Ethan Shagan's work has made it clear that some of the 'campers' whose protests disrupted much of the south-east that summer echoed the regime's evangelical jargon, but this does not tell us much about their real motives.[31] Yet even unmistakably Catholic rebellions only broke out in those regions where local power structures had collapsed, or where local magnates actively stirred them up. Most English men and women may have disliked the Reformation, but most of them did not resist it either. Why not?

Christopher Marsh calls this problem the 'compliance conundrum', and his sparky survey of popular religion in the period attempts some solutions. He suggests that the dramatic changes of the Reformation era were accompanied by some equally important continuities. The theologies behind auricular confession and Protestant catechising were very different, but for the lay person on the receiving end, the experiences were similar enough. Perhaps more significantly, Marsh suggests that distaste for change was balanced by other concerns. Good neighbourliness and a wish to preserve social peace may have smoothed the way to accepting change imposed from above – especially in a society where the ethic of obedience was so strong.[32] Shagan makes a similar, more cynical argument: that we should think not of compliance but of 'collaboration'. Haigh pointed out in the 1980s that the piecemeal nature of the English Reformation made it easier to accept: 'the English ate their Reformation as a recalcitrant child is fed its supper, little by little, in well-times spoonsful'. Shagan and I have argued that some of these spoonsful were themselves sweetened – with church property, or the chance to throw off tiresome requirements like Lenten fasting and prayer for the dead.[33] Even those who disliked the Reformation often found themselves becoming implicated in it.

However, this view of religious change remains uncomfortably functional. Perhaps many English people accepted the Reformation for essentially pragmatic reasons, and perhaps widespread religious confusion helped to prevent the opponents of change from rallying, but the Protestants themselves cannot be written out of the story entirely. Since the mid-1990s there has been a revival of interest in early English Protestantism, led by Diarmaid MacCulloch. MacCulloch argues that the reign of Edward VI, in particular, deserves the same renaissance of interest as that of Mary has received, and there are signs that this is happening.[34] The Edwardian Reformation no longer looks like an accidental by-product of foreign policy, as Michael Bush rather implausibly argued in the 1970s.[35] Rather, a Protestant clique was steadily implementing a pre-planned revolution, which was much more than a wrecking movement; it called for a complete regeneration of English religious life, and extended into social and economic reform. I and others have also redirected attention to Protestantism in the reign of Henry VIII: a small but socially and politically influential movement, whose shifting convictions and ambitions have a profound importance for the Reformation process as a whole. In particular, Peter Marshall has analysed the phenomenon of evangelical conversion. The determination of these early converts was daunting, even while their precise beliefs were fluid.[36] Yet more remains to be done on Henry VIII's reign: in particular, we badly need a post-revisionist look at the dissolution of the monasteries.

If the 'Protestant' view of the Reformation is not dead yet, nor is a third 'confessional' view of the English Reformation: the 'Anglican' view of a Reformation characterised neither by Catholic zeal nor by Protestant excess, but by the sweet moderation of a *via media*. The attack on this view in the past thirty or more years has been relentless. MacCulloch dismisses it as 'the myth of the English Reformation', emphasising the theological radicalism of the Tudor reformers, and helping to recast the Anglican hero Richard Hooker as a prophet without honour in his own time.[37] John Neale's view of the 1559 settlement as an overture for the Civil War, in which a moderate queen faced a zealously unrepresentative 'Puritan choir' in Parliament, has likewise been thoroughly discredited. The work of Norman Jones and Winthrop Hudson has made plain that the parliamentary opposition Elizabeth faced was conservative, not radical.[38] Stephen Alford has emphasised the real religious radicalism of the regime itself.[39] It now appears that full-blown Reformed Protestantism was the established orthodoxy of the Church of England from the 1560s until it was challenged by a small and influential clique in the 1620s.[40]

Yet the 'Anglican' view has not entirely died out. Hooker has never lost his fascination for theologians. Elizabeth I's own religion – Protestant, but awkwardly and sometimes gracelessly conservative in its flavour – remains enigmatic.[41] However, the most significant 'Anglican' revival recently has again been in the field of popular religion. Judith Maltby's *Prayer Book and People in Elizabethan and Early Stuart England* (1998) has argued that while few English people experienced a Puritan-style conversion, many of them did find something genuinely appealing about the new religion: its Book of Common Prayer. Maltby argues that this liturgy, which was later so important to Anglicanism, had already established itself in the affections of English people by 1600. The virtue of this argument is that it provides an account of the religion of the silent majority without reducing that religion to a soggy indifference. Maltby dubs her subjects 'Prayer Book Protestants': sincere enough in their Protestantism, but distinct from the Puritans who looked to the Bible, not the Prayer Book. It is a powerful argument, and resonates with Marsh's picture of conformist neighbourliness. If there is a lingering sense of unease, it is because here, as elsewhere, the historiography remains confessionally coloured: Maltby is an Anglican priest.

Similar antagonisms can be found in histories of the Scottish and Irish Reformations. In Scotland, the predominant flavour of Reformation history has long been Protestant triumphalism, in a polemical tradition of drawing on the contemporary works of John Knox and David Calderwood. This heroic, but sour, narrative was paralleled by a weaker Catholic historiography, which lamented both the old church's failures and the new church's destructive fanaticism, and which was coloured by the long romanticisation of Mary Stewart and of the Jacobite cause. Catholic scholarship was and is pursued principally through the pages of the *Innes Review*, and a collection of articles from the *Review* published in 1962 as *Essays on the Scottish Reformation* remains the most valuable and enduring Scottish monument to Catholic Reformation studies.

When the *Essays* were published, the Protestant view was already under attack from Gordon Donaldson's *The Scottish Reformation* (1960). Donaldson questioned the truism that the late medieval Scottish church was steeped in corruption: certainly some of its parts were in better condition than others, but an apocalyptic picture of universal decay was polemical rather than historical. More controversially, Donaldson also questioned the assumption that the Scottish Reformation had been essentially Presbyterian from its earliest days; which meant that Knox, the great Protestant hero who died in 1572, might not have been a Presbyterian. It is a view which seems obvious now, but was thoroughly provocative in its time. Yet there are still traces of Protestant triumphalism in Scottish Reformation studies, the most sophisticated current representative of this tradition being James Kirk.[42]

As in England, Scottish Reformation historiography has attempted to break away from conflicting partisan narratives through local studies, but progress has been mixed. In 1978, Ian Cowan called for a regionally nuanced approach to the Scottish Reformation, pointing out that early support for Protestantism was distinctly localised.[43] And we do have one outstanding urban history – Michael Lynch's *Edinburgh and the Reformation* (1981), which gives a very downbeat account of the slow Protestantisation of the capital – alongside several rather less sophisticated regional studies, the best of which looks at the Protestant stronghold of Ayrshire.[44] A valuable 1985 doctoral thesis on Perth has, unfortunately, never been published.[45] Worst of all, there is no modern study of Scotland's most Protestant town, Dundee.[46] One of the most prolific recent Scottish Reformation historians, Jane Dawson, has ventured into the Highlands, giving a (so far) unparalleled account of the impact of the Reformation in a region of Scotland often seen as a Catholic stronghold. Her work suggests that a real rural Calvinism was built in the Gaelic-speaking regions, slowly but eventually very effectively. It is a remarkable example of the successful Protestantisation of a society whose literary culture was still more oral than written.[47] However, Dawson's pathbreaking research has so far had few imitators. If scholars are drawn to the British Reformations by ease of access to the languages, most of us are likewise daunted by Gaelic. The result is that neither Highland studies, nor other local studies, have yet refreshed Scottish Reformation history as much as we might have hoped.

Irish Reformation historiography has been gripped even more firmly by confessional tensions, tensions which have persisted far more strongly in Ireland than in England or even Scotland. Irish history was long a form of sectarian trench warfare, in which historians from all sides of the confessional divides nursed memories of the injuries they had received, while justifying or downplaying those they had inflicted. Paradoxically, this has meant that modern academic historians of Ireland have been tentative and self-consciously balanced; they cannot afford their English counterparts' cheerful mudslinging. As a result, studies of the Irish Reformation were for decades relatively scarce. This logjam was broken in the 1970s by Brendan Bradshaw, but the subsequent debates on the 'failure' of the Irish Reformation (on which more below) have focused more on politics and institutions than on the religious experience of Irish people in the sixteenth century. In itself, this is a healthy response to the old, common assumption that Irish Catholicism was a 'congenital consequence of the character, culture or "conservatism" of the inhabitants'.[48] Bradshaw did question the widespread assumption that the pre-Reformation Church in Ireland was in a state of collapse, and the recent work of Henry A. Jefferies suggests an even more positive view of late medieval religion in Ireland.[49] The sluggishness of Irish Protestantism's missionary effort has also been linked by Bradshaw, and by his student Alan Ford, to Protestant doctrine: the pessimistic Protestant view of human nature, and the fatalism of predestinarian theology, contributed to the sense that the Irish were beyond redemption.[50] Serious studies of religious culture in early modern Ireland, however, are a very new phenomenon. Two important 1997 books, by Raymond Gillespie and Samantha A. Meigs, have given us something of the lived flavour of early modern Irish religion, with Meigs especially emphasising the power of the bardic tradition of Gaelic Ireland to preserve and transmit a vividly orthodox Catholicism.[51] This is an exciting departure in Irish Reformation studies, but it is only a beginning. The contrast with Dawson's work on Scotland is particularly striking. Why should it be that one Gaelic culture made Calvinism its own, while another put up a stubborn resistance to Protestantism unparalleled across Europe? We do not, yet, know.

nations and nationalism

Each of the British nations has strongly nationalist historical traditions, and all of these have coloured Reformation scholarship. Geoffrey Elton's heady vision of an exceptional England formed under Tudor rule perhaps never convinced a majority of English historians, but it did help to bolster a distinctly insular historiographical tradition. Ironically, the resurgent 'Catholic' interpretation reinforced this, as it tried to distinguish the lacklustre English Reformation from its more fanatical Continental (and Scottish) counterparts. The argument that the Reformation created a new sense of Englishness refuses to die: William Haller's *Foxe's Book of Martyrs and the Elect Nation* (1963), which saw Foxe as a nationalistic writer, has been widely discredited, but it does seem that some English readers used Foxe's text to nationalistic ends.[52] Some recent scholars have, however, been trying to reconnect the English Reformation to its European context. Richard Rex's marvellous study of John Fisher, Henry VIII's bishop of Rochester, has emphasised that he was a theologian of European stature (rare for any Englishman).[53] Continental Protestant individuals and communities fled to England, and English Protestants and Catholics alike sought foreign refuges.[54] However, the study of the international connections of the English Reformation is only just beginning to become respectable again. It is a part of the field where much remains to be done.

The Reformation histories of the other British nations are impoverished in other ways. The most drastically curtailed is that of Wales. Welsh church historians have devoted most of their attention to the eighteenth century; the principality's sixteenth-century experience was chiefly one of conquest, and the key date is not the break with Rome in 1534 but the enforced legal union with England in 1536. Yet there were Welsh Protestants, and also English clerics in Wales who tried, with varying success, to reform the resource-starved church in which they ministered. Moreover, there were some evangelical texts printed in Welsh from as early as 1546, which may have helped both the Welsh Reformation, and the Welsh language, to survive. Study of these questions has been overwhelmingly dominated by one man, Glanmor Williams. In a series of essays and books spanning the period from 1953 to 1997, Williams laid out a view of a Welsh Reformation which was imposed from England, which took root in the localities slowly, but whose progress was steady and inexorable; his view was largely unaffected by the Catholic turn of English scholarship.[55] For all the value of his work, the field remains seriously under-explored. Several historians working on the English Reformation have dipped their toes into Welsh waters, myself included, but there is scope for a fresh generation of Welsh Reformation research – especially from historians literate in the Welsh language.

Scotland's Reformation historiography is much livelier, but again strongly channelled by nationalistic as well as confessional assumptions. In a heady and influential mixture of the two, John Knox claimed of Scotland that 'thair is no realme this day upoun the face of the earth that hath [religion] in grettar puritie'.[56] Scottish nationalism's distinctive mixture of romance and bitterness has also fostered a distinctive historiographical tradition. So too has the subject's institutional shape: Scottish history is predominantly practised in Scottish universities (with some support from Canada and less from the United States), and often quarantined in separate departments of Scottish history. This is beginning to change;

even a few English universities have now noticed that Scotland has a history. International (not merely British) comparisons are becoming more routine. In a 1996 study, Michael Graham compared the theology and practice of ecclesiastical discipline in Scotland with that in several Continental Reformed societies.[57] The greatest achievement of this new trend is Margo Todd's *The Culture of Protestantism in Early Modern Scotland* (2002). Todd, having previously worked on English history, used a detailed examination of the Scottish church's disciplinary records to argue that the new church was remarkably successful in transforming Scottish culture, and in creating a genuinely popular Calvinism. However, she ascribes this achievement in part to the Scottish reformers' prudence in permitting some continuities with the old religion – much as Marsh has argued for England. Although its geographical scope is limited, and it leaves many questions unanswered, this is one of the first pieces of Scottish history which deserves to be read by historians of the European Reformation in general.

Irish Reformation history has been even more determined by nationalistic battles. Both Catholic-nationalist and Protestant-unionist historiographies long agreed that it was inevitable that most of the Irish population would remain Catholic. But we now recognise how odd it is that the Irish should have rejected their rulers' religion, yet remained (uneasily) subject to those rulers. (The phenomenon is not quite so unique as is commonly asserted: Lutheran Brandenburg rejected its princes' Calvinism.) In 1974, Brendan Bradshaw argued that the legislative Reformation in the 1530s and thereafter was actually remarkably successful. Only from the late 1550s onwards, when English policy towards Ireland became more aggressive and colonialist, did the Irish elites firmly adopt Catholicism. Through the Elizabethan period, Bradshaw traces a division in English religious policy in Ireland, between advocates of 'coercion' and 'persuasion' as methods of winning converts. The Irish Reformation's outcome was, in other words, in the balance until late in the day, although few scholars are willing to accept Nicholas Canny's claim that the matter was not finally decided until the nineteenth century. The current consensus locates the hinge of the Irish Reformation in the 1590s.[58]

This debate's focus has been squarely on the political and institutional histories of the Tudor regime and the church it sponsored. The Reformation in Ireland failed, it seems, principally because the regime failed to make it happen. The Elizabethan 'conquest' of Ireland so alienated the Anglo-Irish elites that they adopted a stubborn Catholicism as a form of resistance; and the regime was content for them to do so. A 'persuasive' policy of conversion might well have succeeded, if it had firmly embraced the Irish language and distanced itself from political subjugation; this, as Steven Ellis has pointed out, was more or less what happened in Wales.[59] Or equally, a 'coercive' policy of conversion might have succeeded, had it been pursued with brutal vigour.[60] Yet successive English regimes scrimped and dithered, giving the Catholic Reformation in Ireland time and space to establish and entrench itself. In particular, as Ellis' work has shown, the Protestant church in Ireland was systematically starved of funds.[61] For the English, it seems, the Irish Reformation was simply not a policy priority. In Ronald Hutton's words, 'all that any English government ever intended for Ireland was to keep it from being significant'.[62] This is plausible enough, but is perhaps too neat to be wholly convincing. Another offshoot of the Church of England – the 'King's Church' in colonial North America – was likewise under-supported and widely opposed, but survived rather better.[63] To give the Irish a merely reactive role in their own religious history is not so much to explain their abortive Reformation as to explain it away.

The undoubted intermeshing of the English and Irish Reformations, however, has helped to spur one of the most intriguing recent historiographical projects: the self-styled 'new British history', which attempts to examine collectively the relationships between the peoples of Britain and Ireland without reducing them to a story of English expansionism.[64] This makes most obvious sense for the period after 1603, when one ruler claimed sovereignty over the whole of both islands, but it has also been applied to the sixteenth century, with mixed success. Jane Dawson's biography of the fifth earl of Argyll, a Scottish lord with Irish interests and England ambitions, is a model of how such 'British' history can be written.[65] Some surveys of the period have also deliberately cast themselves in a 'British' idiom. Felicity Heal has produced a splendid study of the Reformations in all three kingdoms, and Susan Brigden's textbook on the Tudors usefully integrates the Irish and English stories.[66] This approach may not revolutionise the subject, however. There are few figures who, like Argyll, link all three kingdoms, and while it is refreshing to compare the three Reformations, the national level remains the most obvious scale on which to examine them. Moreover, while Scottish historians have on the whole welcomed the 'British' project, and have used the two national Reformations to illuminate each other very effectively,[67] their Irish counterparts are more divided.[68] This is partly because incorporating Ireland into something 'British' is not uncontroversial. More seriously, however, given the breadth and complexity of international connections in the early modern world, there seems no reason to privilege the relationships between these three neighbouring kingdoms for collective study. All three belonged to a Europe-wide religious culture, and that seems the natural context in which to study them, especially before 1603.

politics, personality and myth

The historiographies of all of Britain and Ireland's Reformations have been sustained and corrupted by powerful myths: confessional and nationalistic, but also political. In popular histories, the Reformations feature – if at all – as backdrops to the personal dramas of the Tudor and Stuart monarchs. In particular, Henry VIII, Elizabeth I and Mary, Queen of Scots have a firm place in popular culture, both in Britain and beyond it. This is a point of some importance, partly because popular history is unavoidably right in emphasising the importance of politics. The Scottish Reformation, for example, was clearly a political act, in which the high nobility deposed a regent and seized power for themselves. That the earl of Moray was genuinely committed to Protestantism, and that Mary of Guise appears not to have understood the potential power of religious dissidence, are both important facts.[69] The activist political culture of the Scottish nobility is clearly vital to explaining the Reformation process. I have argued that their conversion to the Protestant cause in significant numbers, and their subsequent willingness to fight for it, can only be understood as the result of a sequence of contingent political events.[70] And the role of Queen Mary herself is even more important. Jenny Wormald has argued that Mary was a disastrous political incompetent who was unable to undo her subjects' religious revolution and who, in the end, dramatically destroyed herself.[71] A more recent biography by John Guy is more sympathetic.[72] Given publishers' eagerness for books with the words 'Mary, Queen of Scots' in the title, these debates will go on, but their popularity does not make them unimportant.

Henry VIII's even more political Reformation has been the focus of the most intense scholarly debate. The early Protestant view was of a king who was a plaything of powerful factions: this allowed Henry to be lauded for his Reformation while excused from blame for its shortcomings. As the memories of his tyranny faded, this view was gradually superseded by a loyal monarchism, in which Henry became a larger-than-life, masterful king. This was questioned in the 1970s by Eric Ives, who resurrected something akin to the old Protestant view: the English court was again seen as riven by factionalism. This view won considerable support, largely because it explained the apparent inconsistency of so much of Henry's behaviour. Henry's religious policies, according to Diarmaid MacCulloch, amounted merely to 'a ragbag of emotional preferences'. But others – notably Glyn Redworth and George Bernard – disagree vehemently, citing Henry's (intermittent) close interest in religious politics and claiming to discern some underlying unities to his views.[73] I have also argued that some rather rough-and-ready theological consistency can be found in Henry VIII.[74] The most important intervention in this debate, however, has come from Richard Rex. Rex detects a solid centre to Henry's religion in the unbending ethic of obedience he sought to impose. He has also tried to moderate a debate on faction in which both sides' positions were coming to seem unnecessarily absolute. In Rex's formulation, the king 'could not be pushed around. But he could be worked on.'[75] And there, for the moment, the debate rests; but like his Scottish great-niece, Henry VIII is too tempting a publishing prospect to be left alone for long. The last really substantial biography of the old tyrant was published in 1968, and has aged extremely well, but we are probably overdue for another.[76]

The popular fascination with kings and queens, however, does more than remind us of those rulers' real importance. It reflects a sixteenth-century preoccupation with personalities: with monarchs and ministers, with preachers and persecutors. The great martyrological historians of the Reformations understood this, for they told history of Thomas Carlyle's kind, assembled from the biographies of heroes and of villains. It is not a kind of history which is respectable nowadays, despite the proliferation of heavyweight biographies, but it matters to this extent: it was how most early modern people understood the events that they were living through.

It has become clear since the 1980s that if we are to understand the Reformations we need to understand the controversialists who attacked and defended them at the time; and they did so largely through martyr-stories. This is why one of the most significant recent developments in English Reformation historiography is the resurgence of interest in John Foxe and his colleagues. Since the mid-1990s, the British Academy John Foxe Project has been creating what is, remarkably, the first critical edition of Foxe's vast, complex and shifting text.[77] It has also become the centre of a small industry of Foxe studies.[78] This has had two great achievements. One is to allow us to delve into the worlds which Foxe and the other martyrologists studied. We can see more of subjects such as Marian Protestantism,[79] the Scottish martyrs,[80] or the emerging Tudor hero of women's history, the early Protestant martyr Anne Askew.[81] Secondly, the Foxe Project has helped us to see how martyrology defined and sharpened post-Reformation world views. Although the oft-repeated canard that

Foxe's *Book of Martyrs* was placed in every English church has now been debunked,[82] the book's influence is unmistakable. The very different ways in which Catholic martyrologists told their stories – usually to an international, rather than an English audience – have helped us to understand their different perspective on the events of the sixteenth century.[83] Foxe's friend Knox has not yet been so well integrated into this picture (although Foxe's own work on the Scottish Reformation has been[84]), perhaps because Knox was so much less competent as a historian.

The stories which early modern people told themselves about the religious upheavals they were living through have continued to dominate our histories down to the present. Some of the battles they were fighting have never quite gone away. Distancing ourselves from these struggles has certainly helped to produce a subtler and more humane understanding. Yet some of the finest recent histories of the British and Irish Reformations have embraced unabashedly partisan positions. The resurgence of interest in martyrology is a reminder that the stories which were told at the time are themselves woven into the events which they describe. No doubt we will never know what 'really' happened in the Reformation. Perhaps, if we learn something about what the people who were caught up in the events believed was happening, it will be enough.

further reading

The only convincing attempt to survey the Reformations in Britain and Ireland as a whole is Felicity Heal, *Reformation in Britain and Ireland* (Oxford, 2003). On the 'British' project, see J. G. A. Pocock, 'British History: a Plea for a New Subject', *Journal of Modern History*, 47 (1975), 601–28; Brendan Bradshaw and John Morrill (eds), *The British Problem c. 1534–1707: State Formation in the Atlantic Archipelago* (Basingstoke, 1996).

The best recent historiographical survey of the English Reformation is Peter Marshall, *Reformation England 1480–1642* (London, 2003). The best general surveys lean towards the Catholic view: Christopher Haigh, *English Reformations* (Oxford, 1993) and Eamon Duffy's monumental *The Stripping of the Altars* (New Haven and London, 1992). See also Haigh's collection *The English Reformation Revised* (Cambridge, 1987); and Richard Rex, *Henry VIII and the English Reformation* (Basingstoke, 1993). The classic, and now dated, 'Protestant' treatment is A. G. Dickens, *The English Reformation* (London, 1964, 2nd edn 1989). For recent 'Protestant' treatments of the early Reformation, see especially Diarmaid MacCulloch, *Tudor Church Militant* (London, 1999); Ethan Shagan, *Popular Politics and the English Reformation* (Cambridge, 2002); Alec Ryrie, *The Gospel and*

Henry VIII (Cambridge, 2003); and Peter Marshall and Alec Ryrie (eds), *The Beginnings of English Protestantism* (Cambridge, 2002).

On the later English Reformation, Diarmaid MacCulloch, *The Later Reformation in England, 1547–1603*, 2nd edn (Basingstoke, 2001) is a good survey. The period is still dominated by Patrick Collinson: see especially his *The Religion of Protestants* (Oxford, 1982) and *The Birthpangs of Protestant England* (Basingstoke, 1988); cf. Peter Lake, *Moderate Puritans and the Elizabethan Church* (Cambridge, 1982). The 'post-revisionist' view of the later period has been pioneered by Tessa Watt, *Cheap Print and Popular Piety* (Cambridge, 1991) and especially Alexandra Walsham, *Providence in Early Modern England* (Oxford, 1999). On popular religion more generally, see Christopher Marsh, *Popular Religion in Sixteenth-Century England* (Basingstoke, 1998).

The best English local studies are Christopher Haigh, *Reformation and Resistance in Tudor Lancashire* (Cambridge, 1975) and Susan Brigden, *London and the Reformation* (Oxford, 1989). For Wales, see Glanmor Williams, *Wales and the Reformation* (Cardiff, 1997).

There are only two modern surveys of the Scottish Reformation: Gordon Donaldson, *The Scottish Reformation* (Cambridge, 1960), and Ian Cowan, *The Scottish Reformation* (London, 1982). See also David McRoberts (ed.), *Essays on the Scottish Reformation* (Glasgow, 1962). For the pre-1560 period, see Alec Ryrie, *The Origins of the Scottish Reformation* (Manchester, 2006); on the later Reformation, see James Kirk, *Patterns of Reform: Continuity and Change in the Reformation Kirk* (Edinburgh, 1989); Alan R. MacDonald, *The Jacobean Kirk, 1567–1625: Sovereignty, Policy and Liturgy* (Aldershot, 1998). The best local study is Michael Lynch, *Edinburgh and the Reformation* (Edinburgh, 1981). There is a substantial literature on John Knox, best approached through the essays in Roger Mason (ed.), *John Knox and the British Reformations* (Aldershot, 1998).

On Ireland, see Brendan Bradshaw, *The Dissolution of the Religious Orders in Ireland under Henry VIII* (Cambridge, 1974); and his 'Sword, Word and Strategy in the Reformation in Ireland', *Historical Journal*, 21 (1978), 475–502. See also Nicholas Canny, 'Why the Reformation Failed in Ireland: *Une Question Mal Posée*', *Journal of Ecclesiastical History*, 30 (1979), 423–50; and Karl S. Bottigheimer, 'The Failure of the Reformation in Ireland: *Une Question Bien Posée*', *Journal of Ecclesiastical History*, 36 (1985), 196–207. The debate was moved on by Stephen Ellis, 'Economic Problems of the Church: Why the Reformation Failed in Ireland', *Journal of Ecclesiastical History*, 41 (1990), 239–65. On popular religion in Ireland, see Raymond Gillespie, *Devoted People: Belief and Religion in Early Modern Ireland* (Manchester, 1997); Samantha A. Meigs, *The Reformations in Ireland: Tradition and Confessionalism, 1400–1690* (Basingstoke, 1997).

notes

1. Diarmaid MacCulloch, 'Early Tudor England: a Peripheral Realm in Christendom', lecture delivered at St Mary's College, Strawberry Hill, 17 April 2001.
2. In 1993, remarkably, Christopher Haigh felt it necessary to declare that he was *not* a Catholic: *English Reformations* (Oxford, 1993), pp. vii–viii.
3. See especially J. J. Scarisbrick, *The Reformation and the English People* (Oxford, 1984); Eamon Duffy, *The Stripping of the Altars* (New Haven and London, 1992).
4. Christopher Haigh, 'The Recent Historiography of the English Reformation', in his *The English Reformation Revised* (Cambridge, 1987).
5. Beat Kümin, *The Shaping of a Community: the Rise and Reformation of the English Parish, c. 1400–1560* (Aldershot, 1996); Clive Burgess, 'London Parishioners in Times of Change: St. Andrew Hubbard, Eastcheap, c.1450–1570', *Journal of Ecclesiastical History*, 53 (2002), 38–63.
6. Christopher Haigh, 'Anticlericalism and the English Reformation', in Haigh, *English Reformation Revised.*
7. Christopher Haigh, *Reformation and Resistance in Tudor Lancashire* (Cambridge, 1975); John Bossy, *The English Catholic Community* (London, 1975).
8. Michael Questier, *Conversion, Politics and Religion in England, 1580–1625* (Cambridge, 1996).
9. Alexandra Walsham, *Church Papists: Catholicism, Conformity and Confessional Polemic in Early Modern England* (Woodbridge, 1993).
10. Eamon Duffy and David Loades (eds), *The Church of Mary Tudor* (Aldershot, 2005); William Wizeman, *The Theology and Spirituality of Mary Tudor's Church* (Aldershot, 2005).
11. Duffy, *Stripping of the Altars*, p. 524.
12. Thomas F. Mayer, *Reginald Pole: Prince and Prophet* (Cambridge, 2000); Lucy Wooding, *Rethinking Catholicism in Reformation England* (Oxford, 2000).
13. Jennifer Loach, 'The Marian Establishment and the Printing Press', *English Historical Review*, 101 (1986), 135–48.
14. I am grateful to Dr Freeman for discussing this forthcoming piece with me.
15. John F. Davis, *Heresy and Reformation in the South-East of England 1520–59* (London, 1983); Richard Rex, *The Lollards* (Basingstoke, 2002).
16. Margaret Aston, *Lollards and Reformers* (London, 1984); Anne Hudson, *The Premature Reformation* (Oxford, 1988).
17. Margaret Spufford (ed.), *The World of Rural Dissenters* (Cambridge, 1995), especially the articles by Derek Plumb; Alec Ryrie, *The Gospel and Henry VIII* (Cambridge, 2003), pp. 79–81, 232–7.
18. Nicholas Tyacke (ed.), *England's Long Reformation 1500–1800* (London, 1998).
19. Patrick Collinson, *The Elizabethan Puritan Movement* (London, 1967). See also his seminal lectures on the later Reformation: *The Religion of Protestants* (Oxford, 1982) and *The Birthpangs of Protestant England* (Basingstoke, 1988).
20. Kenneth L. Parker and Eric J. Carlson, *'Practical Divinity': the Works and Life of Revd Richard Greenham* (Aldershot, 1998); Christopher Haigh, 'Success and Failure in the English Reformation', *Past and Present*, 173 (2001), 28–49; Eric

Josef Carlson, 'Good Pastors or Careless Shepherds? Parish Ministers and the English Reformation', *History*, 88 (2003), 423–36.

21. Haigh, *English Reformations*, p. 280.
22. Peter Lake, *Moderate Puritans and the Elizabethan Church* (Cambridge, 1982).
23. See especially Ian Green, *The Christian's ABC: Catechisms and Catechizing in England c.1530–1740* (Oxford, 1996).
24. Tessa Watt, *Cheap Print and Popular Piety* (Cambridge, 1991), p. 327.
25. Peter Marshall, *Beliefs and the Dead in Reformation England* (Oxford, 2002); Peter Lake with Michael Questier, *The Antichrist's Lewd Hat: Protestants, Papists and Players in Post-Reformation England* (New Haven and London, 2002).
26. Norman Jones, *The English Reformation: Religion and Cultural Adaptation* (Oxford, 2002).
27. Haigh, *Reformation and Resistance.*
28. Muriel McClendon, *The Quiet Reformation: Magistrates and the Emergence of Protestantism in Tudor Norwich* (Stanford, CA, 1999).
29. This use of wills has been very effectively attacked by J. D. Alsop, 'Religious Preambles in Early Modern English Wills as Formulae', *Journal of Ecclesiastical History*, 40 (1989), 19–27; and Christopher Marsh, 'In the Name of God? Will-making and Faith in Early Modern England', in G. H. Martin and Peter Spufford (eds), *The Records of the Nation* (Woodbridge, 1990). For one of the most sophisticated recent attempts to extract religious data from wills, see Caroline Litzenberger, *The English Reformation and the Laity* (Cambridge, 1997), pp. 179–87.
30. This scholarship is summarised in Anthony Fletcher and Diarmaid MacCulloch, *Tudor Rebellions*, 5th edn (Harlow, 2004).
31. *Ibid.*, pp. 65–89; Ethan Shagan, 'Protector Somerset and the 1549 Rebellions', *English Historical Review*, 114 (1999), 34–63.
32. Christopher Marsh, *Popular Religion in Sixteenth-Century England* (Basingstoke, 1998).
33. Haigh, 'Introduction', in his *English Reformation Revised*, p. 15; Ethan Shagan, *Popular Politics in the English Reformation* (Cambridge, 2002); Alec Ryrie, 'Counting Sheep, Counting Shepherds: the Problem of Allegiance in the English Reformation', in Peter Marshall and Alec Ryrie (eds), *The Beginnings of English Protestantism* (Cambridge, 2002), pp. 98–105.
34. Diarmaid MacCulloch, *Thomas Cranmer: a Life* (New Haven and London, 1996); idem, *Tudor Church Militant: Edward VI and the Protestant Reformation* (London, 1999); Catharine Davies, *A Religion of the Word: the Defence of the Reformation in the Reign of Edward VI* (Manchester, 2002).
35. M. L. Bush, *The Government Policy of Protector Somerset* (London, 1975).
36. Ryrie, *Gospel and Henry VIII*; Peter Marshall, 'Evangelical Conversion in the Reign of Henry VIII', in Marshall and Ryrie (eds), *Beginnings of English Protestantism*.
37. Diarmaid MacCulloch, 'The Myth of the English Reformation', *Journal of British Studies*, 30 (1991), 1–19; idem, *Tudor Church Militant*, pp. 157–222; idem, 'Richard Hooker's Reputation', *English Historical Review*, 117 (2002), 773–812.
38. Norman Jones, *Faith by Statute: Parliament and the Settlement of Religion, 1559* (London, 1982); W. H. Hudson, *The Cambridge Connection and the Elizabethan Settlement of 1559* (Durham, NC, 1980).

39. Stephen Alford, *The Early Elizabethan Polity: William Cecil and the British Succession Crisis, 1558–1569* (Cambridge, 1998).
40. Nicholas Tyacke, *Anti-Calvinists: the Rise of English Arminianism* (Oxford, 1987).
41. Susan Doran, 'Elizabeth I's Religion: the Evidence of her Letters', *Journal of Ecclesiastical History*, 51 (2000), 699–720.
42. James Kirk, *Patterns of Reform: Continuity and Change in the Reformation Kirk* (Edinburgh, 1989).
43. Ian Cowan, *Regional Aspects of the Scottish Reformation* (London, 1978); idem, *The Scottish Reformation* (London, 1982).
44. Margaret H. B. Sanderson, *Ayrshire and the Reformation* (East Linton, 1997).
45. Mary Black Verschuur, 'Perth and the Reformation: Society and Reform 1540–1560' (Ph.D thesis, University of Glasgow, 1985).
46. The best makeweight is Iain E. F. Flett, 'The Conflict of the Reformation and Democracy in the Geneva of Scotland, 1443–1610' (M.Phil dissertation, University of St Andrews, 1981).
47. Jane Dawson, 'Calvinism and the Gaidhealtachd in Scotland', in Andrew Pettegree, Alistair Duke and Gillian Lewis (eds), *Calvinism in Europe 1540–1620* (Cambridge, 1994).
48. Karl S. Bottigheimer, 'The Reformation in Ireland Revisited', *Journal of British Studies*, 15 (1976), 141.
49. Brendan Bradshaw, *The Dissolution of the Religious Orders in Ireland under Henry VIII* (Cambridge, 1974); Henry A. Jefferies, *Priests and Prelates of Armagh in the Age of Reformations, 1518–1558* (Dublin, 1997).
50. Brendan Bradshaw, 'Sword, Word and Strategy in the Reformation in Ireland', *Historical Journal*, 21 (1978), 497–9; Alan Ford, *The Protestant Reformation in Ireland, 1590–1641* (Frankfurt, 1985).
51. Raymond Gillespie, *Devoted People: Belief and Religion in Early Modern Ireland* (Manchester, 1997); Samantha A. Meigs, *The Reformations in Ireland: Tradition and Confessionalism, 1400–1690* (Basingstoke, 1997).
52. Patrick Collinson, 'John Foxe and National Consciousness', in Christopher Highley and John N. King (eds), *John Foxe and His World* (Aldershot, 2002).
53. Richard Rex, *The Theology of John Fisher* (Cambridge, 1991).
54. Andrew Pettegree, *Foreign Protestant Communities in Sixteenth-Century London* (Oxford, 1986); idem, *Marian Protestantism*; Peter Marshall, 'Catholic Exiles', in his *Religious Identities in Henry VIII's England* (Aldershot, 2005).
55. See *Welsh Reformation Essays* (Cardiff, 1967), containing several earlier pieces; *The Welsh Church from Conquest to Reformation* (Cardiff, 1962); *Wales and the Reformation* (Cardiff, 1997).
56. John Knox, *The Works of John Knox*, ed. David Laing, vol. 2 (Edinburgh, 1848), pp. 263–4.
57. Michael Graham, *The Uses of Reform: 'Godly Discipline' and Popular Behaviour in Scotland and Beyond, 1560–1610* (Leiden, 1996).
58. Bradshaw, *Dissolution of the Religious Orders*; idem, 'Sword, Word and Strategy'; Nicholas Canny, 'Why the Reformation Failed in Ireland: *Une Question Mal Posée*', *Journal of Ecclesiastical History*, 30 (1979), 423–50; Ford, *Protestant Reformation*.
59. Steven G. Ellis, 'Economic Problems of the Church: Why the Reformation Failed in Ireland', *Journal of Ecclesiastical History*, 41 (1990), 259–63.

60. Canny, 'Why the Reformation Failed', 446; Karl S. Bottigheimer, 'The Failure of the Reformation in Ireland: *Une Question Bien Posée*', *Journal of Ecclesiastical History*, 36 (1985), 199–200.
61. Ellis, 'Economic Problems'.
62. Ronald Hutton, *The British Republic 1649–1660* (Basingstoke, 1990), p. 49.
63. James B. Bell, *The Imperial Origins of the King's Church in Early America, 1607–1783* (Basingstoke, 2004).
64. J. G. A. Pocock, 'British History: a Plea for a New Subject', *Journal of Modern History*, 47 (1975), 601–28; Brendan Bradshaw and John Morrill (eds), *The British Problem c. 1534–1707: State Formation in the Atlantic Archipelago* (Basingstoke 1996).
65. Jane Dawson, *The Politics of Religion in the Age of Mary, Queen of Scots: the Earl of Argyll and the Struggle for Britain and Ireland* (Cambridge, 2002).
66. Felicity Heal, *Reformation in Britain and Ireland* (Oxford, 2003); Susan Brigden, *New Worlds, Lost Worlds: the Rule of the Tudors* (London, 2000).
67. See especially Jane Dawson, 'Anglo-Scottish Protestant Culture and Integration in Sixteenth-Century Britain', in Steven G. Ellis and Sarah Barber (eds), *Conquest and Union: Fashioning a British state, 1485–1725* (New York 1995); Clare Kellar, *Scotland, England and the Reformation 1534–1561* (Oxford, 2003).
68. See the sceptical view taken by Nicholas Canny, 'Writing Early Modern History: Ireland, Britain and the Wider World', *Historical Journal*, 46, 3 (2003), 723–47.
69. Maurice Lee, *James Stewart, Earl of Moray* (New York, 1953); Pamela Ritchie, *Mary of Guise in Scotland, 1548–1560* (East Linton, 2002).
70. Alec Ryrie, *The Origins of the Scottish Reformation* (Manchester, 2006).
71. Jenny Wormald, *Mary Queen of Scots: a Study in Failure* (London, 1988), reprinted as *Mary Queen of Scots: Politics, Passion and a Kingdom Lost* (London, 2001).
72. John Guy, *My Heart is My Own: the Life of Mary, Queen of Scots* (London, 2004).
73. Diarmaid MacCulloch, 'Henry VIII and the Reform of the Church', in his *The Reign of Henry VIII* (Basingstoke, 1995), p. 178; George W. Bernard, 'The Making of Religious Policy, 1533–1546: Henry VIII and the Search for the Middle Way', *Historical Journal*, 41 (1998), 321–49; Glyn Redworth, 'Whatever Happened to the English Reformation?', *History Today*, 37 (October 1987), 30–33. Ives usefully summarises the debates over faction in 'Stress, Faction and Ideology in Early-Tudor England', *Historical Journal*, 34 (1991), 193–202.
74. Alec Ryrie, 'Divine Kingship and Royal Theology in Henry VIII's Reformation', *Reformation*, 7 (2002), 49–77.
75. Richard Rex, 'The Crisis of Obedience: God's Word and Henry's Reformation', *Historical Journal*, 39 (1996), 894.
76. J. J. Scarisbrick, *Henry VIII* (London, 1968 and subsequent revisions).
77. <www.hrionline.ac.uk/foxe/>.
78. David Loades has edited three volumes of essays: most valuably *John Foxe and the English Reformation* (Aldershot, 1997), and also *John Foxe: an Historical Perspective* (Aldershot, 1999) and *John Foxe at Home and Abroad* (Aldershot, 2004). See also Highley and King (eds), *John Foxe and his World*; Susan Wabuda, 'Henry Bull, Miles Coverdale, and the Making of Foxe's *Book of Martyrs*', in Diana Wood (ed.), *Martyrs and Martyrologies* (Studies in Church History 30: Oxford, 1993).

79. Thomas S. Freeman, 'Publish and Perish: the Scribal Culture of the Marian Martyrs', in Julia C. Crick and Alexandra Walsham (eds), *The Uses of Script and Print, 1300–1700* (Cambridge, 2004); Brett Usher, '"In a Time of Persecution"', in Loades (ed.), *John Foxe and the English Reformation*.
80. Jane Dawson, 'The Scottish Reformation and the Theatre of Martyrdom', in Wood (ed.), *Martyrs and Martyrologies*.
81. Sarah Elizabeth Wall, 'Editing Anne Askew's *Examinations*: John Bale, John Foxe, and Early Modern Textual Practices', in Highley and King (eds), *John Foxe and his World*; Elaine V. Beilin (ed.), *The Examinations of Anne Askew* (Oxford, 1996).
82. Elizabeth Evenden and Thomas S. Freeman, 'Print, Profit and Propaganda: the Elizabethan Privy Council and the 1570 Edition of Foxe's "Book of Martyrs"', *English Historical Review*, 119 (2004), 1288–307.
83. Anne Dillon, *The Construction of Martyrdom in the English Catholic Community, 1535–1603* (Aldershot, 2001).
84. Thomas S. Freeman, '"The reik of Maister Patrik Hammyltoun": John Foxe, John Winram and the Martyrs of the Scottish Reformation', *Sixteenth Century Journal*, 27 (1996), 43–60.

part two
themes in reformation history

7
renaissance humanism and the reformation

craig d'alton

humanisms and reformations

Would Lorenzo Valla, Leonardo Bruni or Francesco Petrarch have recognised their concerns for good letters and the glories of Ancient Greece and Rome in the intellectual projects of Martin Luther, Huldrych Zwingli, Jean Calvin, or even Desiderius Erasmus?

To link the humanists of late fourteenth- and fifteenth-century Italy with the sixteenth-century reformers of religion is a difficult task; there is so much that is different. Never let it be said, however, that such difficulties have been seen by scholars as anything other than a challenge to which they might rise. The point of connection between what were undoubtedly figures pursuing very different discursive, intellectual and social enterprises is generally made with reference to 'Christian Humanism', and in particular to the work of the doyen of Northern letters, Erasmus of Rotterdam.[1]

Many humanist scholars of the sixteenth century were particularly fond of irony. It is thus somewhat appropriate that one of the central barriers to our understanding their work is our misuse of language. Language was at the heart of humanist concerns, and it is the impact made upon language and its uses that drove humanist scholars and scholarship both into alliance with, and opposition to, the religious reform movements of the sixteenth century. It is equally true, however, that the reformers themselves had substantial, indeed, seminal impact upon the language not merely of religion, but of scholarship and of society at large. If the re-emergence of Greek and Hebrew scholarship made the Reformations possible, then the reformers' championing of the vernacular changed forever the central place of the favourite humanist tongue – Latin – particularly within those societies that embraced reformist ideals. Berndt

Moeller once famously wrote 'No humanism, no Reformation'.[2] Might we refashion his words by adding 'No Reformation, no triumph of the vernacular'?

Before walking down such fraught scholarly paths, it is necessary at the outset to define our terms. So first, what is, or was, Renaissance Humanism?

Words ending in '-ism' almost invite deconstruction. Scholars and popular writers alike tend to see movements and ideologies where only methods or conditions exist. For example, in recent years many have spoken with ease about 'postmodernism' and 'postmodernists' as though they were discussing an ideology and its promoters. This is despite the fact that postmodernity is a condition rather than an ideology: to claim that the processes at work within the postmodern condition constitute an '-ism' is akin to coining 'medievalism' as an ideology which somehow explains with clarity what happened to society and culture in Europe between the fall of Rome and the beginnings of modernity. Similarly, 'humanism' is one of the most problematic terms employed by scholars of sixteenth-century Europe. Humanism was not an ideology either, but nor was it a condition; rather, it was a hugely diverse literary, historical and general academic method. Since the invention of the term in the nineteenth century, 'humanism' has been deployed to describe the activities of widely differing groups of scholars, clerics, bureaucrats, poets and artists who, over several centuries, developed methodologies of language, writing and research which, if they stood as a whole in dichotomous relationship to anything, rejected the logic and dialectic of medieval scholasticism. When such a catch-all term is further linked with the equally broad and problematic concept of 'The Reformation' we enter a quagmire of language and confused concepts.

For the second term in the title of this chapter is as problematic as the first. 'The Reformation' as a coherent, definable and positive term to describe what happened to religion in sixteenth-century Europe has become less than fashionable in recent decades. Carter Lindberg provides a good summary of the advantages of moving to the plural in his *The European Reformations*, and the question is addressed elsewhere in the present volume.[3] Most scholars are now more comfortable with the plural 'Reformations', often as a broadly inclusive term compassing events in regions that remained Catholic as well as in regions that became Protestant. Thus the plural term will generally be employed in what follows here. Where Reformation is used in the singular it will be in reference to a particular geographical manifestation of widely divergent change. Generally we will speak not of Humanism and Reformation, but of humanists and reformers, acknowledging that these were very often the same people.

For the purposes of this chapter, then, we may define humanism in its early sixteenth-century form as the pursuit of learning and the search for truth facilitated by an appeal to and the study of ancient authorities and sources, which occurred across confessional boundaries. Humanism was a method rather than a philosophy, and was employed by religious reformers and traditionalists alike to justify and advance their beliefs and religio-political positions.

The key to understanding humanism's impact upon the reformation of religion is that it was a *method*, a set of tools. It was used sometimes as a weapon and sometimes as armour, in the theological literary warfare which enabled the parties to fight if not always on identical discursive battlegrounds, then at least with similar discursive munitions. Once we make the transition from regarding humanism as an ideology to a scholarly method it becomes easier to discern both who employed it, and what impact it had.

Humanism thus defined was not the invention of the sixteenth century. Paul Oskar Kristeller's definition of a Renaissance humanist acts as a helpful counterpoint to the somewhat technical definition I have proposed above. He speaks of

> a broad class of Renaissance intellectuals . . . who were active as teachers and secretaries, writers, scholars and thinkers; who exercised a wide and deep influence on all aspects of Renaissance civilization; and who left to posterity, along with the record of their lives and activities, vast writings that might be roughly classified as literature, historical and philological scholarship, and moral thought, but which often deal with such diverse subjects as philosophy and sciences, literary and art criticism, education, government and religion.[4]

This 'broad class' began their work in late fourteenth- and fifteenth-century Italy, and their influence and methods expanded throughout Europe. Their style of writing – both their literature and their script – was disseminated by scholars who had studied in Italy, and by Italians going abroad to teach. Peter Burke offers a helpful reflection on the spread of Italian humanism in the fifteenth and early sixteenth centuries. He notes the rapid but uneven spread of this style of scholarship throughout Europe, that some areas were influenced more rapidly than others and that, importantly for our purposes, 'the [humanist] package [was] broken up in transit and . . . some of its contents were put to uses which might well have surprised Leonardo Bruni or even Lorenzo Valla'.[5] As we examine the spread of humanist method, we discover that the teaching of Greek, for example, met with a mixed reception. In newer university foundations in particular, such as the trilingual college at Louvain, at the new university of Wittenberg, or at Alcalá in Spain it rapidly took hold, leading to its application in theological fields. The Complutensian Polyglot Bible overseen by Cardinal Ximenes at Alcalá was one of the crowning glories of non-Italian humanist enterprise. By contrast, older, more established institutions greeted Greek and the new humanities with a scepticism that occasionally overflowed into outright hostility. Often the leaders of such opposition were scholastic theologians who (rightly) saw a threat to the coherence of their curriculum. At Paris and Cologne opposition was considerable, and in Oxford, reaction to the establishment of lectureships in Greek and 'new' theology led to a minor riot.[6]

The humanist versus scholastic arguments were perhaps most entertainingly written up in 1515 by Ulrich von Hutten (1488–1523) in his bitingly satirical *Letters from Obscure Men*. The argument between the two groups was not so much over their philosophies of life or religion as over how one approached fundamental questions, and what tools one employed to answer them. For the schoolmen, the traditional method of dialectic, disputation and exposition of sentences was enough. For the humanists this needed to be tempered by reference to and production of the best texts, by conversation and critical reading, by historical precedent and looking for 'right' rather than 'correct' answers.[7]

Prior to the substantial dismantling of confessional historiography in the 1960s and beyond, the link between humanists and reformers was traditionally presented in somewhat black and white terms. Humanists were seen either as reformers *manqué*, or they were presented as those who remained within the Catholic fold whilst the reformers corrupted their ideas and turned against them. Jacob Burckhardt first argued with force for the idea that the Renaissance and the Reformation, humanism and Protestantism, were opposing discourses, with the latter being an outcome of the medieval mindset, whilst the former represented a return to classical ideals. This position, reinforced by Wilhelm Dilthey and Ernst Troeltsch, was followed until quite recently by scholars as highly regarded as Louis Spitz.[8] Now, the majority position in contemporary scholarship on this issue is that the Reformations were, possibly in large part, an outcome of humanist ideas and methods.

Yet problems arise immediately the Reformations are defined as an outcome even of 'Christian' humanism. Erasmus lived until 1536 yet, famously, remained a Catholic in Roman obedience. So did many who shared at least some of his scholarly vision. Thomas More, John Fisher and many others died in defence of the old faith. Jacques Lefèvre d'Etaples, Jacob Wimpheling, Johannes Reuchlin and many of the *spirituali* of Italy and France resisted the temptation to break with Rome despite their concerns at a perceived moral decay within the Church. Other humanists, however, took the plunge into schism and, to some minds at least, heresy. Although some regard Martin Luther's own humanist pedigree as debatable,[9] the humanist credentials of Huldrych Zwingli, Martin Bucer, Peter Martyr, Jean Calvin and many others are not. Having chosen their position, humanists on each side of the confessional divide wrote against each other, sometimes vehemently, drawing from the same storehouse of rhetorical and theological techniques to press home their arguments. Even the most cursory glance at the corpus of writers with views as divergent as Thomas More and Huldrych Zwingli reveals that the humanist and the theologian were often the same person, and that they used every rhetorical method at their disposal no matter what their confessional stance. There was no 'evangelical style' of writing pitted against a 'humanist style'. Peter Matheson highlights the dangers of claiming, for example, dialogue as the irenic humanist alternative to the polemic of the theologian.[10] If humanism was a precursor of the Reformation, then it was also a precursor of Counter-Reformation, as the Fathers of Trent, including humanist scholars of the calibre of Reginald Pole, reinvented the Church by employing humanist as well as scholastic methods, scholarship and rhetoric.

What, then, is the nature of the relationship between even the 'Christian' (or perhaps better, 'Northern') form of Renaissance humanism and the sixteenth-century Reformations? To paraphrase Matheson again, was one in any way inclusive of the other, did the one precede the other, were they totally separate, or were they so intimately entangled that it is impossible to separate them?[11]

The extent to which the work of the reformers relied upon humanist foundations has been well rehearsed. The classic story is perhaps most easily demonstrable in the case of Scripture, and what Peter Matheson has described as the 'most astonishing phenomenon' of the rise of 'Biblical humanism'.[12] The publication of Erasmus' Greek New Testament (1516) and its impact upon the Biblical scholarship of Martin Luther is the stuff of legend and cliché, but it did not come out of the blue. We have already noted the work at Alcalá to produce the first polyglot Bible. Erasmus was also able to draw heavily upon the work of other scholars. John Colet, Jacob Wimpheling, Jacques Lefèvre d'Etaples, Johannes Reuchlin, and a good many others were studying Scripture in the primary tongues of

Greek and Hebrew long before Erasmus himself acquired the necessary facility to work in either. We should not underestimate the impact of humanist approaches to the text of Scripture upon developing religious understandings. Just as Luther's re-reading of Romans relied heavily upon Erasmus' work and Huldrych Zwingli was positively a disciple of the famed scholar-monk, so the version of Scripture which captured the popular imagination – Luther's German Bible – was as much a product of the humanist scholarly factory as Erasmus' *Novum Instrumentum*. The tone and cadence of the Reformation in England relies directly upon Tyndale's translation of the Scriptures, for which he used not the Vulgate, but Erasmian and Lutheran editions. The same is true in different ways throughout Europe.

Even within those jurisdictions which remained under Roman obedience there were challenges to established Scriptural hermeneutics which were a direct outcome of humanist concerns. Erasmus, of course, famously remained in Roman obedience whilst pushing the margins of orthodox hermeneutics in his *Paraphrases* and commentaries. Jacques Lefèvre d'Etaples and the circle of Meaux turned French Biblical scholarship on its head.[13] Gasparo Contarini and Reginald Pole could as easily have defected to Luther as become the voice of the papacy at Regensburg or one of the Tridentine Fathers.[14] The *spirituali* as a group have been labelled everything from proto-Protestants to steadfast Catholics.[15] Although such labels may serve as much to muddy the waters as to make them clear, what is beyond dispute is that lines of religious demarcation were fluid in their construction and that being a practitioner of humanist literary or scholarly methods was neither a guarantee of sympathy towards evangelical views nor a necessary precondition for holding them. We cannot speak with authority of the impact of Humanism on The Reformation. What, then, might we more helpfully do?

Recent scholarship has emphasised that there were as many reformations as there were reformers. So too we might argue that there were many humanists, but no one Humanism. Most of the remainder of this chapter therefore examines in brief the scholarship and religious aims of ten key figures who proposed varieties of reform – theological, religious and social – in early sixteenth-century Europe. The list is very far from exhaustive, but may serve to illustrate both the extent of humanist impact, the variety of its practitioners, and the dangers both of equating humanists with reformers or of placing them in opposition to each other. For convenience sake the ten may be divided into: Erasmus, plus three groups of three. The first group comprises what might be labelled the 'Wittenberg reformers': Luther himself, Philip Melanchthon, and Andreas

Bodenstein von Karlstadt. Then, we examine three key figures in the 'Reformed' tradition: Huldrych Zwingli, Martin Bucer, and Jean Calvin. The final group takes us through three generations of English humanist scholar-politicians, only one of whom became an evangelical – Thomas More, Thomas Cranmer, and Reginald Pole. A simple question is put regarding each figure: what was the relationship between their education and literary method and their religious beliefs and practices? The variety of answers demonstrates both the variety and scale of humanist scholarship in the early sixteenth-century, as well as the enormously varied impact it had upon religion and church. The reason for taking this approach is that, unlike so much of broader Reformation discourse, humanism was the project of cultural elites, and thus of particular privileged and well-educated individuals. By eschewing generalisations and concentrating on specific examples we are able to illustrate the variety and complexity of humanist–reformist interconnection. As the chapter concludes we move from discussing how different scholars' approaches to language and text influenced the religious landscape of Western Europe, to outlining how the broader religious change they inspired itself impacted upon the world of letters.

ten humanists

Humanists intimately interested in religious matters did not always end up developing opinions which placed them in irrevocable conflict with the Church. Before surveying some of the key reformers who left Roman obedience and asking what their relationship was to Renaissance humanism, let us highlight the most prominent figure in the humanist tradition who, despite his interests in ecclesiastical reform, remained in obedience to the Roman church: Desiderius Erasmus of Rotterdam (1469–1536).[16]

In spite of, and perhaps even because of, his extraordinary capacity for self-promotion, virtually all scholars would agree that Erasmus was the key figure in Northern European humanism. Erasmus' interests and contacts are too various to rehearse at length. Instead what will be highlighted here is his interest in theology and his contribution as a theologian to ecclesial and theological reform.

Erasmus obtained his doctorate in theology at the University of Turin in 1506, keen to establish his qualifications in an area in which some had perceived him to be little more than a talented amateur. He described his most basic scholarly aim as 'the revival of the genuine science of theology',[17] and John Olin argues that for Erasmus scholarship was not an end in itself, but was the means by which Christianity might be revived and revitalised.[18] His first explicitly religious publication was the 1503 *Enchiridion militis Christiani* (Handbook of the Christian Soldier), which set the tone for much of what was to follow. One of the central arguments of the book – that knowledge of scripture is, alongside prayer, the great weapon of the Christian – was programmatic for his subsequent work as a Biblical translator and commentator, and inspired other more radical reformers, such as Luther and Zwingli. So too did the other theme of the *Enchiridion* – that the key to religious experience was interior rather than external; a doctrine which lent at least some weight to Luther's repudiation of works as a means to salvation.

Erasmus' first work of direct Biblical scholarship was the 1505 *Annotationes*, but the work which sealed his place in the reformers' canon was the 1516 annotated Greek and revised Latin New Testament. Interestingly, this *Novum Instrumentum* was well received not only in those circles which would use it to re-read fundamental Christian doctrines, but by the pope himself (Leo X), who provided a commendatory epistle to preface the second edition. Its impact upon subsequent reformers has already been discussed. Erasmus continued to publish on Scriptural and related subjects throughout his career, including translations of the Fathers and original works. His monumental edition of Jerome was published in the same year as the New Testament. Translations of commentaries by Cyprian and Arnobius followed, as did Augustine's *City of God*, which presaged a full edition of that bishop's works in 1529. Other editions of St John Chrysostom, Irenaeus, Ambrose, Basil and (published posthumously) Origen further expanded a growing library of the Greek and Latin fathers. Original works on religious subjects included commentaries, especially on the psalms, and his last major work, the *Ecclesiastes*, which dealt with the art of preaching.

Through all this work Erasmus had to tread a fine line of refuting accusations of sympathy with the reformers whilst maintaining the integrity of his own theological scholarly project. What is important to remember here, perhaps, is that his circle of influence included both evangelical and Counter-Reformation humanists, and provides one of the most tangible links between them.

Perhaps the single most important factor contributing to the perception that humanists and evangelical reformers were different animals is the dispute between Erasmus and Luther over free will. After considerable pressure had been put to bear, Erasmus finally entered the fray against Luther in 1524, with his publication of *De libero arbitrio* (A Diatribe on Free Will), arguing that Luther was wrong in asserting that humans are fully corrupted and unable to contribute anything to their salvation. Erasmus instead argued that the Fall was not absolute, that humanity retained free will, and that human beings had a role to play in their personal salvation by responding appropriately to the grace of God. Luther responded with *De servo arbitrio* (Of the Slavery of the Will) which, in Diarmaid MacCulloch's terms, argued from 'the very heart of the Reformation's reassertion of the darkest side of Augustine: a proclamation that the humanist project of reasonable reform was redundant'.[19]

But was Luther declaring humanism redundant, or merely condemning the application of civic humanism to religious matters? Just as Erasmus (quite sensibly) couched many of his arguments in the Augustinian terms which would have spoken most readily to Luther, and presented a position which John Olin describes as essentially that of Thomas Aquinas,[20] so Luther's reply could hardly be described as either scholastic or an unreconstructed medieval approach, despite his attacks on Erasmus' 'Lucianism'. This was not a fight between a humanist and a reformer, but between a humanist with a doctorate in theology and a theologian with humanist training.

This begs the obvious question: is it reasonable to label Luther a humanist? At least in the years prior to 1521, most other humanists who knew him seemed to think so.[21] From his early schooling in Eisenach he was exposed to and enjoyed Latin classics and history. Importantly, he also learned Greek. At Wittenberg, following the publication of Erasmus' edition of the New Testament in Greek, Luther's desire to work *ad fontes* in his teaching of scripture led him to develop a revisionist theology curriculum wholly at odds with the scholastic theologians of his day. There can be no doubt that Luther took full advantage of humanist tools to analyse scripture, but it is also true that his debt to his Augustinian heritage ensured a reluctance to engage in what one might label 'secular humanist' activities. He was a prolific translator – one of the best tests for humanism – but his translations were always religious in nature, often into the vernacular rather than classical Latin, and included no classical works. On balance, perhaps it is best to argue that Luther was a direct product of fifteenth- and early sixteenth-century humanism, but that humanism was not his only, nor even his most important, academic grounding. That accolade, surely, must go not to the *studia humanitatis* but to Augustine.

Luther's Wittenberg colleagues are somewhat easier to define in terms of their style of learning. Few scholars of the sixteenth century could claim a better humanist pedigree than Philip Melanchthon (1497–1560). The grandnephew of Johannes Reuchlin, trained in the classics, who had graecised his name from Schwarzerd, was the author of one of the most popular Greek primers of the period. His contribution to the Wittenberg reformation was undoubtedly primarily a scholarly one, and unlike Luther, there is no ambiguity in his influences. Given his primary role in authoring the Augsburg Confession of 1530, it can easily be argued that, even if Luther was not a humanist of the pure school, the Lutheran movement, under Melanchthon's influence, retained a humanist flavour in its most foundational document.

Andreas Bodenstein von Karlstadt (c.1480–1541), on the other hand, was probably the least humanist-influenced of the core Wittenberg group. Prior to his conversion to Luther's reading of Augustine in 1516, he was an unreconstructed Thomist with doctorates in theology and both laws. Although his conversion by Luther came about as a result of studying the writings of Augustine in the best new edition, there is no indication that the conversion amounted to anything more than a shift from Thomist to Augustinian theology. German mysticism rather than Renaissance humanism was his most profound non-Augustinian influence. The Puritan and pietist tendencies identified by his modern biographers are the marks of a scholastic mystic rather than a humanist scholar.[22]

Thus in Wittenberg even the three key figures of reformation demonstrate the diversity of the impact of humanist scholarship upon the theological reform movement. Luther's fundamental discoveries were predicated on his ability to make use of the tools produced by Erasmus and his confrères. Melanchthon was able fully to integrate his well-honed skills in the humanist disciplines of philology, grammar and rhetoric into the new theology he enthusiastically supported as being in tune with his own readings of the Greek scriptures. Karlstadt, on the other hand, was a supporter of reform despite its humanist origins rather than because of them.

The key figures in what has been broadly called the 'Reformed' tradition were similarly diverse in their application of humanist methods to the task of reform.

Huldrych Zwingli (1484–1531) is probably the closest we come among the anti-Roman reformers to a straightforward disciple of Erasmus. Exposed to the Latin classics from early youth in Basle, Berne and Vienna, it was on his return to Basle that his humanist education reached a heightened stage of development. If Luther's theological grounding was firmly Augustinian, Zwingli's was Thomistic, but he appears to have had little difficulty harmonising his two fields of interest and method. Moving from the university into parish ministry, Zwingli maintained and developed his interest in the humanities, learning Greek, and studying the classics, the Fathers and Scripture. The turning point of his influence came about as a direct result of his reputation as an expositor of Scripture through humanistic preaching. In 1518 he was appointed to the position of people's priest at the Great Minster in Zurich. Zwingli's programme of didactic preaching on entire Scriptural books began on 1 January 1519, and through careful, extended and systematic exposition of key Scriptural texts he gradually succeeded in turning Zurich into a centre of reformist views. Like Luther, Zwingli's *sola scriptura* approach was predicated on the humanist notion of going back to the text, *ad fontes*, in order to determine the best way to proceed both in the reform of the Church and in the Christian life. Zwingli's argument with Luther over Eucharistic theology was in fact a classic humanist dispute: what is the correct way to translate and explain a key Greek phrase, 'This is my body'? The debate traversed arguments on the uses and dangers of Platonic dualism in Christian theology as well as the key issue of figurative versus literal interpretation of Scripture. The key point for our purposes is that this was an argument between reformers over translation, and over how the rhetoric of Scriptural authors is to be interpreted. It was a humanist argument over fundamental theological method.

Throughout his career Zwingli freely acknowledged his debt to Erasmus, and Erasmus was equally effusive in his compliments on Zwingli's scholarship. Some scholars have been happy to name Erasmus as the real founding father of Reformed theology.[23] This is debatable, but there is little doubting that the theology of the Zurich programme, even after Zwingli's untimely death in 1531, owed a great debt to humanist methods.

Martin Bucer (1491–1551) was the key reformer in Strasburg, one of the defining influences on the Reformation in the Franco-German region, and had a profound impact upon the reforming methods of Jean Calvin. Bucer was a Dominican whose humanist interests saw him move into the circles of Jacob Wimpheling at Sélestat and the Heidelberg group which included Oecolampadius, Brenz and Melanchthon. Eventually his preparedness to examine and incorporate new ideas into his thinking led him to accept many of Luther's key doctrines.

If one of the signs of humanist influence on a debate is moderation, then Bucer was an exemplary humanist. He was perhaps the first great ecumenist, making efforts to bring together Luther and Zwingli at Marburg, and Evangelicals and Catholics at Regensburg. Bucer first encountered Luther's teachings at Heidelberg in 1518, and by mid-1523 arrived at Strasburg as an independent disciple of Luther's theology.

In Strasburg, Bucer was a comparatively open-minded and tolerant reformer, who none the less regarded good discipline as one of the signs of the true Church. His preparedness to make room for radicals, so long as they did not disturb good social order, made some uncomfortable, and led to what MacCulloch has called 'highly individual versions of Reformation' developing in Strasburg.[24]

Perhaps the most interesting, and final, period of Bucer's career followed an invitation by Archbishop Thomas Cranmer to come to England in 1549. There, Bucer assisted Cranmer in the process of liturgical revision, and resumed an academic career, lecturing in Cambridge and composing *De Regno Christi*, which put forth his ideas on how the reign of Christ might be mirrored on earth. He died in early 1551, and one cannot help but wonder how much greater his impact may have been on the course of English reform had he lived longer.

Jean Calvin (1509–1564) was at the vanguard of the second generation of reformers, but like many in the first generation, was a humanist before his religious conversion. He studied Latin, Greek and Hebrew, and his first published work (1532) was a translation of Seneca *On clemency*. As a secular academic, neither a cleric nor a monk, Calvin stands out from other evangelical reformers. By adopting a didactic catechetical style in his magnum opus, the *Institutes*, Calvin provided a systematic theology firmly grounded in Biblical humanist method that was accessible even to those without humanist training.

During his second, long period as reformer in Geneva (1541–64), Calvin drew upon his experiences of gradual reform in Strasburg under Bucer. The great disaster of his first time in Geneva (1536–38), when he and Guillaume Farel were expelled from the city, was the result of a failure to negotiate the political climate of the Town Council. This second time around he made no such mistake. The Ecclesiastical Ordinances of 1541 were a triumph of diplomacy and of what one might term 'Christian civic humanism' as well as of ecclesiology. The extent to which the Genevan reformation was to be led by sound Scriptural learning is constantly underlined, with the Council – thanks to its judicious editing of Calvin's draft and Calvin's acceptance of their changes – as the ultimate authority in matters of discipline and appointments.

Calvin's writings, and the model of the godly city which he established in Geneva and which became so programmatic for the Reformations in other parts of Europe, were predicated both on his knowledge of the classics and on his belief that the Scriptures surpassed all else.[25] The pious opinions of early twentieth-century scholars may require some mitigation, yet Neuenhaus' observation that Calvin 'preserved to the end the reputation of an excellent humanist' is fair comment.[26]

The final trio to be examined here, More, Cranmer and Pole, would probably have objected violently to being labelled a 'group' (although Reginald Pole may have been happy enough to be linked with Thomas More). These three English proponents of very different sorts of reform were all humanist-trained, all wrote theological tracts, and were all employed in the highest echelons of England's Tudor government. Two of the three would die for their faith.

Thomas More, Lord Chancellor from 1529 to 1533, was executed in 1535 because of his commitment to the 'common faith and belief of the whole church'. More was the most prominent English humanist of his generation and, along with Bishop John Fisher, was the most vociferous opponent of the theological reforms suggested by Luther, Zwingli and the rest. More's writings are a classic case in point of a humanist turned theologian who used every tool in his rhetorical armoury to argue against change. Some scholars have seen startling discontinuities between More's early humanist writings (which included such secular works as translations of Lucian and, of course, *Utopia*) and his voluminous polemical output from the 1520s and early 1530s. Alistair Fox, for one, has seen this second period as evidence that the clash between Reformation discourses and More's humanist training and sympathies led him to the verge of breakdown, and that 'calm was regained' only after More was arrested and retired from polemics, instead composing equally voluminous devotional treatises which contained much that was highly traditional in flavour, albeit expressed in beautiful humanist Latin and elegant English prose.[27]

Why did More not become an evangelical?

One myth about More is that, despite his early explicitly humanist publications, once he entered royal service he ceased to be a proponent of reform. This was not so. In the legal arena, which was his primary area of professional expertise, John Guy has demonstrated that More was a reformer of almost radical proportions.[28] Within the religious sphere, however, More had always composed quite conventional devotional works, even contemporaneously with his most adventurous humanist writings. Part of his early poetry falls into this category, as does the unfinished treatise on the Four Last Things, written in English c.1522, which Alistair Fox has described as a repudiation of the view of providence attained in *Utopia*.[29] By far his greatest theological output, however, was anti-Lutheran English polemic. Bearing in mind Peter Matheson's warning that all literary genres were employed by all sides in debate, this is not surprising. Perhaps the most interesting of these works, the *Dialogue Concerning Heresies* (1529), is in fact an extended example of the breakdown of humanist methods of reform in the face of the insidious nature of heresy.[30] For More, as for many other humanists, reform ceased to be reform when it became heresy. When theological propositions were 'novelties' rather than a return to the common faith of the Church, or where divine revelation was seen to be ignored by scholars filled with hubris, some humanists, More amongst them, were prepared to draw a line in the sand and say 'enough'. Such action is reminiscent of Luther's almost reactionary stance upon returning to Wittenberg after his stay in the Wartburg, when he reinstated many practices his more radical followers had discarded. Both for More and for Luther change, if divinely inspired, was never radical or violent. More remained a Catholic, because in his view the Continental reformers and the English evangelicals had moved from reform into heresy. Even amongst the Utopians, for whom religious tolerance was allowable, those who attacked the established religion of others, who used violence, or who were too vehement in expressing their unorthodox views could not be tolerated. Moreover, those who denied fundamental principles of providence and the salvation of souls forfeited their humanity.[31]

Thomas Cranmer, archbishop of Canterbury from 1533 until the accession of Queen Mary in 1553, was the key establishment figure of religious reform in Tudor England. Cranmer first came to royal attention because of his learning, and Diarmaid MacCulloch has demonstrated the extraordinary extent of his networking with reformers and non-reforming humanists alike throughout Europe, even as far distant as Cracow in Poland.[32] In the early 1520s he preferred Erasmus to Luther, and counted Robert Ridley, a humanist protégé of Cuthbert Tunstall and a vociferous opponent of Tyndale, as a mentor. MacCulloch concludes that Cranmer was 'certainly a humanist, but of a very conservative and eclectic sort'.[33]

Cranmer, of course, did what More did not, and moved into the reformers' camp in the late 1520s and 1530s. It is clear that he did not do so purely for political reasons, as his views were frequently, indeed usually, more advanced than those of the king. Did this conversion, then, turn him from his humanist leanings, and if not, then what impact did they have on his style of reform?

Certainly in his disputes with Stephen Gardiner over eucharistic theology, Cranmer demonstrated an impressive depth of learning of the Fathers, and did not eschew references to the pre-Christian classical authors, particularly when searching for biting metaphors.[34] With

Cranmer as with Luther, however, we see the most enduring legacy of his humanist literary training not in his polemical works, but in his new vernacular liturgies. Like Luther, Cranmer was meticulous in trying to conform his changes in the Prayer Books of 1549 and 1552, as well as in other less dramatic forms, to early liturgical and theological texts – peeling back, as he thought, the layers of medieval accretion in order to regain the pure order of worship. In this way Cranmer followed reformers throughout Europe in applying the humanist principle of *ad fontes* not merely to obtaining the best Scriptural and theological texts, but to ensuring the best possible public worship of and private devotion to God. His actions in this regard, it must be noted, were those of the conservative humanist rather than a radical; leading to sometimes vigorous critique from more 'advanced' reformers on the Continent.

After the death of King Edward VI in 1553 and the accession of his Catholic half-sister Mary, Cranmer was labelled a heretic and condemned. Ironically, as he languished in prison in Oxford, it was a work by Thomas More, the *Dialogue of Comfort*, written in the Tower in 1535, which appears, albeit temporarily, to have reconverted Cranmer back to the Catholic faith.[35] His successor at Canterbury was yet another humanist-trained politician – Cardinal Reginald Pole. John Guy has commented that, as papal legate under Queen Mary, Pole concentrated on behaviour rather than belief, discipline before preaching, order before regeneration.[36] Others have been more kind, and have seen Pole as one whose work was undercut by his allies rather than his enemies.[37] Perhaps by the time of his return to England, Pole's opportunity to have a lasting impact upon Catholic reform had already passed. Pole, a relative of King Henry VIII and a potential heir to the throne, had left England in 1532 at the height of debates over the succession, in order to pursue humanist studies in Italy. His breeding and his learning combined to ensure that he moved in the highest circles from the outset. From the safety of exile, Pole was able vigorously to oppose Henry's break with Rome, to defend papal primacy, and to become one of the primary architects of reform from within the Church.[38] As one of the Fathers of Trent, Pole was at the vanguard of those who identified the need for the Protestant threat to be countered not with force but with learning. The training of clergy, in particular, was to be the highest priority, and it was Pole himself who came up with the idea of seminaries. He was also, however, the legate with probably the greatest sympathy for Lutheran views on justification. Building on Contarini's views, there is some evidence to suggest that whilst Pole always outwardly conformed to official teaching, his departure from Trent after the tenth session was linked to the Council's complete repudiation of Luther's position in favour of doctrine which Pole regarded as somewhat Pelagian.[39]

Those who have argued that Pole somehow left behind his humanism when he became archbishop need to read his actions as legate and archbishop in the light of the constant financial and administrative crisis of Mary's reign. Pole needed at least to impose the outward signs of a restored Catholicism, in order that in the coming years (which he never lived to see) the process of re-education of the populace and clergy could properly be mounted. Pole's death, within hours of that of Queen Mary herself, was probably a blessing for him. It certainly was for the new queen Elizabeth, who was able to install her own candidate as archbishop – another humanist politician, Matthew Parker – in his stead, without the need for messy arrests or treason trials.

This brief survey of some key Reformation-period humanists and reformers has been far from comprehensive. One could speak of countless other humanists who remained in Roman obedience – Jacques Lefèvre d'Etaples, Gasparo Contarini, Vittoria Colonna, Marguerite de Navarre or John Fisher, just to name a few. One could add a legion of voices of evangelical reform from throughout Europe – Peter Martyr Vermigli, Johannes Bugenhagen, Andreas Osiander, Johannes Brenz, Heinrich Bullinger, Guillaume Farel or Argula von Grumbach. One could go on to speak of the radical reformers, particularly Thomas Müntzer.[40] One could explore the towns and cities of the German states, the Netherlands, France, Poland, Sweden and beyond for humanists who were or were not reformers, and reformers who were or were not humanists. What this small sample has shown, however, is that when discussing humanism one is forced to examine individuals rather than classes of person or genres of belief and action. The emergence of the individual as a category of person is surely one of the lasting legacies of both humanism and evangelical reform.

humanism's legacy

We have explored in this chapter the influence of humanism on several key religious reformers of the sixteenth century. A question almost never asked until very recently, however, has been to what extent the reformations of religion, which developed into a broad social movement affecting every stratum of society, impacted upon humanist scholarship, which remained the province of a relatively small, educated elite.

The most useful and original work in recent years on humanism and the Reformations is that of Erika Rummel, and the impact of reform upon humanism is her major concern. Rummel's *The Confessionalization of Humanism in Reformation Germany* highlights the two-way traffic of intellectual development between humanist and reformist scholars of the early sixteenth century. Rummel's work deals substantially only with the German states, but many of her conclusions might usefully be tested and perhaps applied to Reformations elsewhere.[41]

When examining the idea that humanists and reformers were allies in the same intellectual project, Rummel notes 'a certain overlap' in the causes each promoted. That overlap is consonant with our analysis of key reformers, and Rummel argues that it is sufficient to suggest that the Reformation debate may be understood as 'a continuation of an ongoing debate over curriculum between scholastic theologians and teachers of language arts'.[42] Humanist scholars crossed into theology, and theologians, whether evangelical or traditional, made use of and acknowledged the importance of texts in primary languages and of humanist philology. In Rummel's terms, the conceptual link was made between humanism and reform.[43]

Rummel is correct in arguing against the idea that humanism was somehow superseded by the 'new' discourse of reform. Traditions of reform, theological as well as moral, can be traced throughout the history of the Church, and although it is unfashionable to refer to Wyclif and Hus as 'Forerunners of the Reformation', it remains true that they and other streams of what the *magisterium* of the Church characterised as 'heresy' did set a precedent both for how those questioning official theology might proceed, and for how the Church might deal with them.[44] Moral reformers, sometimes labelled 'Catholic reformers', such as John Colet or even Savonarola, were equally important as forerunners of the concept of reform of religious practice that was the logical consequence, and often the impetus for, theological change. Equally, sixteenth-century humanists had their ancestors, from Petrarch to Poggio to Pico; from Nicholas of Cusa to John Tiptoft to Johannes Reuchlin. The two movements – humanist and reformist – were contemporary, and the one did not overrun the other. Instead, as Rummel has demonstrated, the two progressed for some time 'in lockstep', before the reformers' became the more dominant discourse, selectively suppressing or enhancing the parallel development of the very humanist interests in which most of them themselves had been trained:

> The subordinate role that humanism played because its appeal was limited to the educated classes meant that its champions could not impose their views on the religious protagonists at will; rather the

extent and area of their contribution to the debate was determined by the religious movement.[45]

Charles Nauert notes the apparent disappearance of humanism as a distinctive intellectual strain in the later Renaissance: 'It seemed to melt away.'[46] That may well have been the case, but given that we have identified that humanism was a method rather than an ideology, and that its practice varied as widely as its practitioners, ultimately perhaps Nauert, and even Rummel, are not asking quite the right question. Rather than asking 'What happened to Humanism after the Reformation?', perhaps we should ask whether the sixteenth-century interaction between religious thought and classical rhetoric, philology, ethics and politics in fact made possible the emergence of new, secular modes of thought – modes in which the notion of 'returning to the sources' was superseded by the modernist concept that the new could be better than the old?

further reading

John P. Dolan (ed.), *The Essential Erasmus* (New York and London, 1964)

John C. Olin (ed.), *Christian Humanism and the Reformation: Selected Writings of Erasmus, with the life of Erasmus by Beatus Rhenanus* (New York, 1975). See especially Olin's introduction

Peter G. Bietenholtz (ed.), *Contemporaries of Erasmus: a Biographical Register of the Renaissance and Reformation*, 3 vols (Toronto, 1985–87)

Maria Dowling, *Humanism in the age of Henry VIII* (London, 1986)

Anthony Goodman and Angus MacKay (eds), *The Impact of Humanism on Western Europe* (London, 1990)

James Kirk (ed.), *Humanism and Reform: the Church in Europe, England and Scotland, 1400–1643* (Oxford, 1992)

James K. McConica, *English Humanists and Reformation Politics under Henry VIII and Edward VI* (Oxford, 1965)

Arthur F. Kinney, *Humanist Poetics: Thought, Rhetoric and Fiction in Sixteenth Century England* (Amhurst, MA, 1986)

Peter Matheson, *The Rhetoric of the Reformation* (Edinburgh, 1998)

Charles Nauert, *Humanism and the Culture of Renaissance Europe* (Cambridge, 1995)

James Overfield, *Humanism and Scholasticism in late Medieval Germany* (Princeton, 1984)

Erika Rummel, *The Confessionalization of Humanism in Reformation Germany* (Oxford, 2000)

notes

1. The two classic biographies of Erasmus in this mode are Jan Huizinga, *Erasmus of Rotterdam* (London, 1952) and Ronald H. Bainton, *Erasmus of Christendom* (New York, 1969).
2. Bernd Moeller, *Imperial Cities and the Reformation: Three Essays*, ed. and trans. H. C. Erik Midelfort and Mark U. Edwards, Jr. (Durham, NC, 1982), p. 36.
3. Carter Lindberg, *The European Reformations* (Oxford, 1996), pp. 8–23. See p. 3 above.
4. Paul Oskar Kristeller, *Renaissance Thought II: Papers on Humanism and the Arts* (London, 1965), p. 23.
5. Peter Burke, 'The Spread of Italian Humanism', in Anthony Goodman and Angus MacKay (eds), *The Impact of Humanism on Western Europe* (London, 1990), p. 20. Burke appends two very useful tables to this essay, of Italian humanists abroad and foreign scholars in Italy.
6. Charles Nauert, 'The Struggle to Reform the University of Cologne, 1525–1535', in James V. Mehl (ed.), *Humanismus in Köln / Humanism in Cologne* (Cologne, 1991); Craig D'Alton, 'The Trojan War of 1518: Politics, Patronage and the Rise of Humanism', *Sixteenth Century Journal*, 28 (1997), 727–38. On the university of Paris, see Augustin Renaudet, *Préréforme et humanisme à Paris pendant les premières guerres d'Italie 1494–1517* (Paris, 1953). James K. Farge, *Orthodoxy and Reform in Early Reformation France: the Faculty of Theology of Paris, 1500–1543* (Leiden, 1985) provides a useful analysis, but concentrates on religious rather than pedagogical reform.
7. James Overfield, *Humanism and Scholasticism in Late Medieval Germany* (Princeton, 1984).
8. Jacob Burckhardt, *The Civilization of the Renaissance in Italy*, trans. S. G. C. Middlemore (London, 1990, cf. first edition 1860); Louis Spitz (ed.), *The Reformation: Basic Interpretations* (Lexington, MA, 1972), pp. 11–43.
9. Peter Matheson, 'Humanism and Reform Movements', in Anthony Goodman and Angus MacKay (eds), *The Impact of Humanism on Western Europe* (London, 1990), pp. 36–9.
10. Peter Matheson, *The Rhetoric of the Reformation* (Edinburgh, 1998), p. 3.
11. Matheson, 'Humanism and Reform Movements', p. 23.
12. *Ibid.*, p. 28.
13. R-J. Lovy, *Les Origines de la Reforme Francaise: Meaux, 1518–1546* (Paris, 1959).
14. Peter Matheson, *Cardinal Contarini at Regensberg* (Oxford: Clarendon Press, 1972); David C. Steinmetz, *Reformers in the Wings: from Geiler von Kaysersberg to Theodore Beza*, 2nd edn (Oxford, 2001), pp. 23–31.
15. Barry Collet, *Italian Benedictine Scholars and the Reformation: the Congregation of Santa Guistina of Padua* (Oxford, 1985).
16. Scholarship on Erasmus is enormous. See especially the still incomplete *Collected Works of Erasmus* (Toronto, 1974–) (hereafter *CWE*); Cornelius Augustijn, *Erasmus, his Life, Works and Influence* (Toronto, 1991); James McConica, *Erasmus* (Oxford, 1991); Richard Schoeck, *Erasmus Grandescens: the Growth of a Humanist's Mind and Spirituality* (Nieuwkoop, 1988); Erika Rummel, *Erasmus and his Catholic Critics*, 2 vols (Nieuwkoop, 1989).
17. *CWE*, vol. I, letter 139.

18. John C. Olin (ed.), *Christian Humanism and the Reformation: Selected Writings of Erasmus, with the life of Erasmus by Beatus Rhenanus* (New York, 1975), p. 8.
19. Diarmaid MacCulloch, *Reformation: Europe's House Divided 1490–1700* (London, 2003), p. 151; cf. Marjorie O'Rourke Boyle, *Rhetoric and Reform: Erasmus' Civil Dispute with Luther* (Cambridge, MA, 1983).
20. Olin, *Christian Humanism*, p. 28.
21. Leif Grane, *Martinus Noster: Luther in the German Reform Movement, 1518–1521* (Mainz, 1994).
22. Ronald Sider, *Andres Bodenstein von Karlstadt: the Development of his Thought 1517–1525* (Leiden, 1974).
23. See Lindberg, *European Reformations*, p. 177.
24. MacCulloch, *Reformation*, p. 182.
25. Jean Calvin, Institutes, I.8.1, cited in François Wendel, *Calvin: the Origins and Development of his Religious Thought*, trans. P. Mairet (London, 1963), p. 34.
26. Johannes Neuenhaus, 'Calvin als Humanist', *Calvinstudien* (1909), 2.
27. Alistair Fox, *Thomas More: History and Providence* (London, 1982).
28. John Guy, *The Public Career of Sir Thomas More* (New Haven and London, 1980).
29. Fox, *History and Providence*, p. 107.
30. Craig D'Alton, 'Charity or Fire? The Argument of Thomas More's 1529 *Dyaloge*', *Sixteenth Century Journal*, 33 (2002), 51–70.
31. Edward Surtz and J. H. Hexter (ed.), *The Complete Works of St Thomas More, Volume 4: Utopia* (New Haven and London, 1965), pp. 220–1.
32. Diarmaid MacCulloch, *Thomas Cranmer: a Life* (New Haven and London, 1996), pp. 33–7.
33. *Ibid.*, p. 31.
34. *Ibid.*, p. 488.
35. *Ibid.*, pp. 587–605.
36. John Guy, *Tudor England* (Oxford, 1988), pp. 236–7.
37. Thomas F. Mayer, *Reginald Pole: Prince and Prophet* (Cambridge, 2000); Dermot Fenlon, *Heresy and Obedience in Tridentine Italy: Cardinal Pole and the Counter-Reformation* (Cambridge, 1972).
38. See his *De Unitate*, published as *A Defense of the Unity of the Church* (Westminster, 1965).
39. Reginald Pole, *A Treatie of Justification* (Scolar Press, 1976).
40. Matheson, 'Humanism and Reform Movements', pp. 39–41.
41. Erika Rummel, *The Confessionalization of Humanism in Reformation Germany* (Oxford, 2000).
42. *Ibid.*, p. 10.
43. *Ibid.*, pp. 11–23.
44. Heiko Oberman, *Forerunners of the Reformation: the Shape of Late Medieval Thought Illustrated by Key Documents* (New York, 1966); idem, *The Dawn of the Reformation: Essays in Late Medieval and Early Reformation Thought* (New Haven and London, 1989).
45. Rummel, *Confessionalization*, p. 150.
46. Charles Nauert, *Humanism and the Culture of Renaissance Europe* (Cambridge, 1995), p. 195.

8
print and print culture

andrew pettegree

print and the reformation

The printed book has always been – and no doubt always will be – a cornerstone of research on the Reformation. This is true in more than the obvious sense that sixteenth-century books provide one of the most important windows on the religious changes of the period. Books also represent a cornerstone of our understanding of the early modern period. For those brought up with an essentially progress-orientated view of the Renaissance the printing press is firmly established as one of the critical technologies that justify our sense of fundamental change. Print was an essential component of the surge towards modernity. This view, most eloquently articulated for the English-speaking world in the work of Elizabeth Eisenstein, in fact underpins most of the general studies of the origins and growth of printing.[1] According to Eisenstein, print played a crucial role in the three main revolutionary developments of early modern Europe: the Renaissance, the Reformation and the Scientific Revolution. Print was thus essential to the democratisation of information: a profusion of information allowed for a new climate of intellectual cross-fertilisation. In this presentation print was (and still to a large extent is) seen as a crucial stage in the process of intellectual enlightenment and political empowerment.

It is also widely seen as having a particular affinity with the new movements of evangelical change. Certainly Martin Luther himself promoted the power of the book as energetically as he seized on the new medium to promote his own theological agenda. For Luther print was a gift of a beneficent God, a view shared and repeatedly articulated by other Protestant preachers and spokesmen. In Luther's words, printing was 'God's highest and extremist act of Grace, whereby the business of the

Gospel is driven forward'. For John Foxe, author of the famous English martyrology, it was printing that made the Reformation possible. 'The Lord began to work for his Church not with sword and target to subdue his exalted adversary, but with printing, writing and reading ... so that either the Pope must abolish knowledge and printing or printing must at length root him out.'[2] This sense of the power of print had also the beneficent result that from the first years of the Reformation Protestant books were systematically preserved and collected: even the sort of short ephemeral pamphlets, or single-sheet broadsheets, that might otherwise have been expected to have been destroyed, were carefully gathered up. Thus books are not only an indispensable window on the culture of the Reformation – they are also a sturdy and dependable source. Rates of survival are generally good, and especially so for Protestant books.

That said, it is as well to understand the limitations of our knowledge of this crucial communication medium. Much less work has been done on what we might call the contextual history of Reformation print – the place of the book in the broader information culture of the period. Nor has there been the same interest shown in the way in which the book industry functioned commercially, either in specialised local markets for vernacular print, or in a persuasive explanation of the pan-European market in Latin works. Study of the place of print and print culture in the Reformation has, as with much else, started with Martin Luther and his transformation of theological writing in the 1520s. How far this can stand as a metaphor for the wider impact of print has not yet been fully explored.

In the first decade of the Reformation the impact of the book was both profound and to many contemporaries deeply shocking. Early critics within the church were first scandalised and then alarmed by Luther's blatant courting of a broad public through printed tracts and sermons. Contemporaries were well aware that in harnessing the previously rather formal, stolid world of the book to serve these ends, evangelical critics of the Church had achieved something fundamentally new. A new form of book, the *Flugschriften*, short pamphlets in quarto format, came rapidly to dominate the output of German print-shops. Demand for books expanded very rapidly after 1517, as religious debate engaged the interest of a new, largely non-clerical audience. An exceptionally high proportion of these books addressed the new controversies: and in Luther, a writer of genius and extraordinary facility, Germany's publishers had found their ideal partner. Luther could write with phenomenal speed and quickly developed an extraordinary range, from the homiletic sermon, through excoriating satire, to careful, systematic exposition of complex theological issues.

The impact of the book in these years has been extensively explored in a range of bibliographical and analytical studies reaching back to Josef Benzing's milestone bibliography of Luther's works, published in 1966.[3] The wider context of Luther's writing has also been illuminated with the fundamental work of the Tübingen *Flugschriften* project, which over a space of years collected and quantified a vast mass of the writings published by Luther and others sympathetic to the Reformation.[4] Mark Edwards draws heavily on this material for his *Printing, Propaganda and Martin Luther*, a fine and perceptive survey of the impact of Reformation literature.[5] Further statistical analysis is provided in articles by Richard Crofts, and by the work of Miriam Usher Chrisman on Strasburg.[6] Chrisman's work has been particularly influential, because she offered a vision of how statistical analysis – represented by her pioneering bibliography of books printed in Strasburg – could be combined with analytical work based on her detailed knowledge of the Strasburg archives.[7]

Meanwhile, the late Robert Scribner offered in his highly influential monograph, *For the Sake of Simple Folk*, a vision of how the Reformation message might resonate with a larger, largely non-literate public.[8] For the repertoire of illustrated books and broadsheets that figure largely in this project he was able to draw both on his own researches, and on a four-volume collection dedicated to *The German Single Leaf Woodcut*; a publication that both brought a new appreciation for the exquisite quality of much German woodcut art, and helped fix the association of the reform movement with exploitation of the visual media.[9] From this range of studies, and our sense of the vibrancy of the book in the service of the German Reformation, has developed a dominant view of the role that the printed media played in the process of Reformation. This model, what we might call the German paradigm, consists of several elements. First, and most importantly, it has engendered a clear sense of the dominance of print by evangelicals. We also understand that the Reformation achieved its impact partly through the rapid spread of print to multiple printing centres, as popular texts were spread by local reprints. This was accentuated by the difficulty authorities faced in controlling the spread of dissident ideas, especially in Germany. Lastly we associate the Reformation as a critical phase in the victory of the vernacular over Latin in European intellectual culture; and we affirm the importance of illustration in spreading the Reformation message to those who could not read for themselves.

It is important to recognise for what follows the extent to which this paradigm has been developed essentially from the German model of the first evangelical generation; and how little it has been tested against the experience of other European cultures, where the Reformation enjoyed more halting progress, where the medium of print was often used effectively by the defenders of the old order, or where the Reformation sometimes succeeded before the development of an indigenous print culture.[10] The fact is that the German paradigm has been allowed to stand as normative partly because we have lacked the basic data for other national cultures to subject it to systematic, sceptical analysis. There was, until recently, no equivalent to the Tübingen *Flugschriften* project for other parts of Europe. This deficiency is now in the process of being addressed, not least through the general ongoing transformation of the resource base through major national bibliographical projects and the availability of major online resources (including library catalogues). These make possible not only a new generation of analytical studies, sketched below, but a more general refinement of our understanding of the relationship between the book and the Reformation. This can be characterised in three main developments: a more complete sense of how printing served (or failed) Protestantism outside Germany; a developing understanding of how the book was also successfully exploited by the defenders of traditional religion; leading, finally, to an understanding that the overwhelming domination of print by evangelicals suggested by our knowledge of Germany in the 1520s is by no means characteristic of the situation over the whole Reformation century, and in all countries.

the bibliographical task

Given that so much attention has been given to the power of the printed word, it is perhaps surprising that our knowledge of the global world of sixteenth-century print has rested on such insecure foundations. Many studies have focused on the origins of print, and the great innovative publishers of the first generations, but attempts to map the expanding world of print (such as that attempted by Febvre and Martin for their seminal text, *The Coming of the Book*) have rested on remarkably shaky statistical foundations.[11] In the 1960s an attempt was made to address this deficiency with the *Index Aureliensis*, a project that was to publish a list of all books published in the century, organised by author.[12] Progress with this massive enterprise was inevitably slow, and even now has progressed only as far as the letter D. The repertoire of information was based, particularly in the earliest volumes, on the largest collections (such as London and Paris) for which printed catalogues already existed; the project has now in large measure been overwhelmed, and superseded, by the vast amount of data available through online catalogues, and it cannot be expected ever to be concluded.

The insufficiency of the global bibliographical data has to some extent been disguised for students of the English Reformation by the exceptionally high quality of bibliographical resources for English-language printing. For almost one hundred years now students of the book in England have disposed of a complete listing of books published in England and in English abroad in Pollard and Redgrave's *Short Title Catalogue*, a model for all subsequent ventures of this type.[13] We owe this good fortune to particular circumstances, not least the fact that the English book world was in absolute terms relatively small, and heavily concentrated in one well-regulated centre of production (London) – this was not the normative case for other European language zones. The accident of collecting, with especially rich concentrations of books in only three centres, Oxford, Cambridge and London, also assisted Pollard and Redgrave in their labours. Books in collections further afield, particularly in the United States, were then incorporated into a second edition, completed under the direction of Kathleen Pantzer at Harvard and published in three volumes between 1976 and 1991, a work which also incorporated the fruits of much specialist bibliographical work accomplished in the intervening half century. This resource has been further developed with the production of the whole STC on microfilm, a boon for those working at a distance from any but the largest research libraries. The onward march of technology has produced, in recent years, an online searchable version of the STC, and finally EEBO (Early English Books online), a near-complete selection of texts available in digital facsimile.[14]

New books unknown to the compilers of the STC continue to surface, such as a broadsheet summary of the essentials of the Protestant faith discovered in the binding of a book purchased by the Scheide Library in Princeton, and two French-language London imprints recently discovered in Halle.[15] But, in the main, students of the English Reformation are quite exceptionally well served by this fundamental resource base.

The situation is very different for France, where a truly comprehensive listing of sixteenth-century publications is still lacking. This situation reflects both the greater complexity of the task, and the very different tradition of French bibliographical research. France was from the beginning one of the largest centres of European book production, based largely in the precocious book metropolis of Paris, with Lyons as a second major centre of production. During the course of the sixteenth century, book publication also became established in over one hundred provincial towns. Patterns of collecting have also been very different. While most students of sixteenth-century French culture rely mostly on the great collections in Paris, the confiscations of the property of the religious houses during the French Revolution also brought enormous numbers of books into the possession of the French Bibliothèques Municipales – where they remain to this day. The enormous task of charting the history of Paris printing was undertaken by two great bibliographers, Brigitte Moreau and Philippe Renouard. Neither completed their work. Moreau, working on a chronological survey of Paris printing in the post-incunabula age, had reached only 1535 at the time of her death, when the history of the French book was only beginning to be transformed by contemporary events.[16] Renouard adopted a different approach, embarking on a printer-by-printer survey of each publishing house. Published volumes cover only printers whose names begin with the first two letters of the alphabet, although Renouard's card fiche catalogues, deposited in the Réserve of the Bibliothèque Nationale, are made available to researchers on a generous basis.[17] Lyons print was surveyed at the beginning of the century in twelve volumes, again arranged by printer, by H.-L. and J. Baudrier.[18] The Baudriers based their studies on their own substantial private collection of local imprints, and on the holdings of the libraries of Lyons, Aix and Grenoble; an attempt to rework and complete this work, this time on a chronological basis, is ongoing.[19] The work of the Baudriers inaugurated a tradition of local bibliographical scholarship continued in the *Répertoire bibliographique des livres imprimés en France au seizième siècle*, a town-by-town listing of all the books published in each locality, consigned to the responsibility of a local specialist.[20] The volumes are of varying quality, many based only on inspections of local collections, collated with the major collections in Paris; all inevitably miss dispersed items, located in other parts of France or in libraries abroad. The lack of a global survey of French vernacular print, so essential to a real understanding of the impact of the Reformation, is presently being addressed by the St Andrews French Vernacular Book project, due for publication in 2007.[21]

For Germany the listing and analysis of the pamphlet literature of the Reformation proceeded largely in advance of systematic bibliographical research. Like France, German-language publication suffered from the problem of fragmentation, with no single dominant centre of production. In the case of Germany this problem was further exacerbated by the dispersal or destruction of major collections of printed books in the wars of the twentieth century. The mid-century division of Germany, with many of the largest libraries located in the communist German Democrat Republic, also posed a major obstacle to a systematic study of all books published in the German-speaking lands. In the 1980s an attempt was made to address this deficiency with the *VD16*, a global survey of all sixteenth-century German print.[22] This project based its published volumes on the two largest collections then available to book specialists, at Munich and Wolfenbüttel, and since its completion in 2000 has been an indispensable research tool. But inevitably the use of only two main libraries, however vast, has resulted in the omission of large numbers of editions, a problem exacerbated by the decision, now somewhat perplexing, to omit all single-sheet publications (broadsheets, and such like). We still await a second edition of the *VD16*, incorporating many thousands of new items unearthed in other libraries.

The religious controversies raised by Luther made an immediate impact in the Netherlands, stimulating a wave of local Dutch publications of his works, and an immediate hostile response from the church authorities. The bibliographical footprint of these conflicts, and the great outpouring of print that accompanied the later events of the Dutch Revolt, may now be followed in three major works. The three volumes of the *Nederlandsche Bibliographie* by Nijhoff and Kronenberg document the publication of books in the Low Countries until 1540, encompassing the great outpouring of religious print of the first generation after Luther.[23] This work can be supplemented from the catalogue of the great Knuttel collection of pamphlets in the Royal Library, The Hague, now also available in its entirety on microfilm.[24] But it is only recently that the work of *NK* has been continued through to the end of the century, in two separate projects: one, the *Belgica Typographica*, dealing with books published in present day Belgium, the other, the *Typographia Batava*, covering books published in the present day Netherlands.[25] Leaving aside the complications that arise from a division based on modern national boundaries that have no basis in sixteenth-century reality, both projects are in their own way problematic. The *Belgica Typographica* is a three-volume work, the first registering the holdings of the Royal Library in Brussels, with the holdings of other Belgian libraries covered in two further volumes. Books published in the Southern Netherlands that survive in libraries elsewhere are not surveyed. The *Typographia Batava* has cast its net wider to identify books published in the northern Netherlands (a much smaller number), but it is essentially the work of a single bibliographer, Paul Valkema Blouw, and reflects his own specific interests. Thus, while a huge amount of scholarly work is embedded in the project, particularly in the identification of printers responsible for works published anonymously, the *Typographia Batava* lacks information normally regarded as standard in a work of this sort (such as formats). Students of the religious book culture of the Netherlands are badly in need of an initiative that would consolidate and harmonise this diverse bibliographical material. For the moment we may follow the progress of two major ongoing projects: the Short Title Catalogue Netherlands (STCN) and the STC Vlaanderen. The STCN is a retrospective bibliography of all books published in the Netherlands, and those in Dutch printed abroad for the period 1540–1800. So far the data files include the entire collections of the Royal Library in The Hague, and the university libraries of Amsterdam, Leiden and Utrecht up to the end of the seventeenth century.[26] The STC Vlaanderen is a complementary project for Dutch-language material printed in Flanders, focused thus far entirely on the seventeenth century.[27]

For the historians of religion in Italy and Spain the state of fundamental bibliographical data is even more provisional. For Italy there is a large ongoing project, *Edit 16*, which will in due course provide a complete survey of books published on the peninsula in the sixteenth century.[28] The small number of volumes published thus far is in fact misleading, since the project has continued as a solely electronic resource since this date. Such an undertaking appears not yet even to be contemplated for Spain.

These six major markets – France, Germany, Italy, Spain, the Netherlands and England – were responsible, together with the international market in Latin publishing, for over 95 per cent of all books published in the sixteenth century. Outlying markets, such as Scandinavia, Poland and Scotland, were small, and have been the subject of mostly local investigations. The current and recently published bibliographical projects surveyed above present the possibility of a much more complete and integrated understanding of the European book world, along the lines of what is already available for the fifteenth century (albeit with much smaller global numbers) with the I-STC.[29] In this the transformation of technology in the digital age will undoubtedly play its part, with the provision of new online resources and composite library catalogues. Historians of the Reformation also have many reasons to be grateful for commercial publishers of major microfilm and fiche series, such as the Dutch-based company IDC, that have made available a large number of sixteenth-century books in thematically organised collections.[30] The tendency of libraries to catalogue these microfiche as individual items without distinguishing sufficiently clearly from sixteenth-century originals presents a hazard for bibliography; for scholars, however, the multiplication of available texts is an enormous boon. All of this has helped to stimulate a new wave of scholarship on the Reformation and the book, though, as we shall see, this scholarship has advanced more rapidly in some areas than in others.

recent studies on reformation-era print

For Germany, scholars of the Reformation have long been in possession of an unrivalled range of high quality resources, from complete editions of the works of Luther, in both German and English, to a huge range of German Reformation *Flugschriften* in the published microfiche version of the Tübingen project.[31] The groundbreaking bibliographical work of Benzing has been enhanced by the patient surveys of dispersed collections of German Reformation pamphlets by the British bibliographer Michael Pegg.[32] Given the quality of these resources, and the very high rate of survival of Luther's works, it is all the more astonishing that there has been so little work on the print industry of Wittenberg, a subject that still awaits systematic study. Specialist monographs (based on exhibition catalogues) have now been devoted to illustrated title-page design in Wittenberg, and to the Luther Bible.[33] There is also a fine reproduction edition of the early catalogue of Wittenberg University Library.[34] Strasburg, the second city of Protestant publishing in the Empire, has fared better, thanks largely to Miriam Usher Chrisman's pioneering study of Strasburg imprints, and subsequent monograph treatment of the same subject.[35] But there is no comparable study of Nuremberg, Augsburg, or of Protestant print in the Low German language zone of northern Germany. The role of music in the dissemination of the Reformation is explored in two important monographs, by Rebecca Wagner Oettinger and Alexander Fisher.[36] This apart, much of the most original and groundbreaking work has been devoted to the previously ignored subject of Luther's Catholic opponents, who while undoubtedly cautious of the implications of public controversy, did engage the polemical debate with a vigour not always recognised. To the work of David Bagchi on Luther's Catholic opponents must now be added important monographs by Frank Aurich and Christoph Volkmar.[37] The importance of print to traditional religion, before but also during the Reformation, is the subject of work on published sermon epitomes by Anne Thayer.[38] Attention has also turned to the provision of published literature for the Lutheran peoples of Europe in the second half of the sixteenth century. Pending the publication of an important monograph study of the pamphlet literature of the Madgeburg resistance movement to

Charles V, much of this work is embedded in major ongoing collaborative research projects, such as the Database Sources Confessionalisation, 1548–1577/80 of the University of Mainz.[39] The radical Reformation is also well represented in bibliographical projects, not least in the series Bibliotheca Dissidentium.[40]

In the Swiss Confederation Zurich never, strangely, developed a print culture commensurate with the towering status of its two major reformers, Huldrych Zwingli and Heinrich Bullinger. Bullinger's influence was felt largely through his correspondence, but also in his printed works, the subject of a fine and painstaking bibliography by Joachim Staedtke.[41] These works formed the mainstay of the published editions of Zurich's pre-eminent (and for much of the time, only) publishing house, Froschover's, analysed in a recent article by Urs Leu.[42] There is also a serviceable bibliography of Zurich publications.[43] Urs Leu has also recently published a painstaking reconstruction of Bullinger's private library.[44] That Zurich publishing did not develop beyond these limits may partly be attributed to the established position of Basle in the international publishing world, a subject that badly requires a modern study.[45]

In the francophone world one should first of all treat the particular case of Geneva. Geneva has always been especially well served by its bibliographers, who have charted the blossoming from exceptionally modest beginnings of one of Europe's major publishing centres, fuelled by the international popularity of the works of Jean Calvin, and of the French Protestant Bible. The bibliography of Genevan print published by Chaix, Dufour and Moekli is now being replaced by Jean-François Gilmont, already responsible for the fine bibliography of the works of Jean Calvin published in three volumes between 1991 and 2000.[46] Gilmont has currently accumulated information on over four thousand Genevan imprints, in French and Latin. This work will also substantially replace Gardy's venerable bibliography of Théodore de Bèze, at least for works published in Geneva: for de Bèze the St Andrews French Vernacular Book project has also contributed a large number of previously unknown editions published in France at the beginning of the French Wars of Religion.[47] This fundamental bibliographical work is the basis of a body of interpretative scholarship, including Gilmont's own *Jean Calvin et le livre imprimé* and an important collection of essays.[48] The economics of the Genevan book industry are considered in two older monographs, and in the more recent work of Ingeborg Jostock.[49]

Looking beyond Geneva our knowledge of French Protestant printing has been transformed, particularly for the early decades of the sixteenth century, by the work of Francis Higman. His discoveries of previously unknown French translations of Luther's works are set out in a series of essays now available in a collected volume, *Lire et découvrir*.[50] For French Bibles the standard work remains Bettye Chambers's *Bibliography of French Bibles*, reinforced by a specialist study of the maps and diagrams developed for these iconic publications.[51] There is also an ongoing project to establish a parallel bibliography of the French Protestant Psalter, perhaps the archetypal book of militant French Protestantism.[52] For the period of the religious wars, earlier bibliographical studies of printing at La Rochelle and Orleans can be supplemented by new work on the Orleans press of Eloy Gibier by Jean-François Gilmont, and on Protestant printing in Caen and Lyons by Andrew Pettegree.[53] Important French Protestant authors such as Simon Goulart, Philippe du Plessis Mornay, Jean de l'Espine and Antoine de La Roche Chandieu are also the subject of recent and ongoing study.

In France, Protestant domination of the printing press was never uncontested, yet the skill with which Catholic authors exploited the press has been slow to be recognised. An older tradition of scholarship, in which the robust polemical style of these Catholic authors was widely deprecated, is represented by Frank Giese's study of Artus Désiré, who was in fact a subtle and effective defender of traditional values.[54] A more sensitive evaluation of the skill shown by Catholic authors (often doctors of the Sorbonne, writing away from their natural medium of Latin) is provided in recent monograph studies by Larissa Taylor, Christopher Elwood and Luc Racaut.[55]

For the Netherlands the standard work on Luther remains Visser's *Luthers geschriften in de Nederlanden*, sadly never translated into English. More recent discoveries of German Protestant influences are explored by Andrew Johnston.[56] Early Dutch Bibles have also been the subject of systematic study.[57] The emergence of Calvinism in the Netherlands is charted in the bibliographic investigations of Willem Heijting (on Catechisms and Confessions of Faith), and Andrew Pettegree (on the exile printing centre Emden, the northern Geneva).[58] But the most fundamental work is that of Paul Valkema Blouw, who succeeded, in a series of intricate bibliographical case studies pursued in parallel with his work on the *Typographia Batava*, in establishing the corpus of works published by a series of largely obscure or even wholly unknown printers. These men often published in conditions of great secrecy in relatively obscure corners of the Netherlands away from the main centres of population – their rediscovery represents bibliographical detective work at its most masterly. These articles, translated by Alastair Hamilton, are mostly to be found in the Dutch bibliographical journal *Quaerendo*.[59]

Alastair Hamilton has made his own significant contribution to the field through editions of the major works of Dutch dissident thinkers such as Hendrik Niklaes.[60] In this way the enduring Anabaptist strand of Dutch dissent has not been neglected. For the period of the Dutch Revolt considerable attention has also been devoted to satirical prints and woodcuts that spread like a virus through the cities of Flanders and Holland, pouring scorn and opprobrium on the Spanish ruling power. Here we have both valuable collections of the prints (often in exhibition catalogues) and the beginnings of analytical work, represented most accessibly in recent articles published by Alastair Duke.[61]

In Scandinavia, the introduction of a state-sponsored Lutheran Reformation did not stimulate a robust vernacular print culture. The limited output of the Danish and Scandinavian presses is reviewed by Anne Rijsing and Remi Kick in the articles in the accessible and valuable collection *The Reformation and the Book*.[62] Scotland was another place where a successful Protestant Reformation largely preceded the establishment of an indigenous publishing industry. Even in the later stages of the century a close, largely dependent relationship with printers in London and continental Europe could not entirely be avoided.[63] In the case of Italy, the cause of evangelical reform inevitably became tangled with the larger conflicts over authority in the Roman church, and the debate, conducted at the highest levels of the hierarchy, over how the church should respond theologically to the challenge of the Reformation. Interest in the evangelical doctrines found its echo in a tentative attempt to make the works of the northern reformers known through vernacular Italian editions. These works were published under the auspices, sometimes even disguised as their work, of the group of reform minded writers and churchmen known as the *spirituali*. These rare and interesting editions are surveyed in the article contributed by Ugo Rozzo and Silvana Seidel Menchi to Jean-François Gilmont's collection on *The Reformation and the Book*.[64]

Charting the difficult and dangerous work of bringing the evangelical gospel to Spain was the life work of Gordon Kinder, pursued through a series of bibliographical and biographical studies of the major figures of Spanish evangelism.[65] In truth, the perils of the enterprise were such that evangelical publishing made only a limited impact on the Iberian peninsula. Most Spanish dissidents made their most substantial contribution to the evangelical movement while abroad, in the cities of the Swiss Confederation, in the Netherlands, or in England.[66]

The role of print in the English Reformation has not been a leading feature of the debates that have consumed scholars since the concept of the rapid and largely painless triumph of Protestantism was first subjected to sustained criticism some thirty years ago. The attempt to chart the slow progress of Protestant penetration in English county communities led to a different type of monograph study. It is possible too that the early success of the *Short Title Catalogue* in establishing the corpus of English print has meant that the absence of a sense of fundamental discovery had led to a comparative lack of interest in the role of English print. This relative neglect may now be in the process of being corrected. The dependence of English Protestantism on Continental imports in the first generation of reform had been remarked, as had the extraordinary transformation of the London printing industry under Edward VI, though this still awaits systematic study. The recognition of the importance of the publications of the Marian exiles in keeping the Protestant faith alive has not been matched by a similarly systematic exploration of Catholic printing in England during the same years.[67] The raw materials for such a study certainly exist, not just in the E-STC and EEBO, but in the numerous individual printer biographies embedded in the *Oxford Dictionary of National Biography*, and in the systematic mapping of book collecting in the successive volumes of the English Renaissance Libraries series.[68]

In bibliographical terms the second half of the sixteenth century has fared rather better. The debate over the English Reformation concentrated attention on the reign of Elizabeth, when historians now believe Protestantism established a real resonance with a substantial mass of the population. The role of the printed book in this process has received a substantial amount of attention, not least in Ian Green's monumental (and ongoing) three-volume study of English Protestant book culture.[69] The role of cheap print in disseminating the principles of Protestantism among the broader population is one subject of a monograph by Tessa Watt; Adam Fox has offered a sensitive exploration of the interface between print, manuscript and oral culture.[70] At the other end of the theological spectrum the influence of Calvin's writings in England (and in English) is analysed in articles by Francis Higman and Andrew Pettegree; full details of the editions can be found in Jean-François Gilmont's exhaustive bibliography.[71] The impact of different Continental authors on English readership can also be gauged in the inventories of books compiled from Cambridge inventories by Elisabeth Leedham-Green.[72]

The organisation of the London print industry can be followed in the eighty individual contributions contained in the two relevant volumes of the Cambridge History of the Book in Britain.[73] One of the editors of this project, David McKitterick, has also charted the uncertain fortunes of academic publishing in England in his history of the Cambridge

University Press.[74] Other specialist aspects of the book trade in England are studied in Krummel's monograph on music printing, and various articles on the publication history of Foxe's book of martyrs scattered through the serial volumes of essays sponsored by the British Academy John Foxe project.[75] Finally the early halting beginnings of the Book in Colonial America are the subject of a fine volume edited by Hugh Amory and David D. Hall, the first volume of the Cambridge History of the Book in America.[76]

For the Reformation, the book was always a potent instrument; for that reason the dissemination of the word in print was never without strict controls. For the most part – and this is a point that needs emphasis – these controls were developed and maintained by the industry itself. The publication of almost any book in the sixteenth century involved an element of financial risk (to this the early publications of Luther in the 1520s were almost certainly a shining exception, which is why they were so popular with printers). So the publisher of religious works, whether these were a lucrative catechism or a stately Latin Biblical commentary, required a guarantee that their market would not be spoiled by an opportunist competitor. For this reason all of Europe's book markets developed systems of privilege to protect the interests of publishers – in the established book markets these systems were generally in place before the Reformation.[77]

The competition of ideas with the dawn of Protestantism brought a new edge to this desire to control markets, where not only economic rivals, but dissident theologies were seen to threaten social harmony. In Germany the very dispersed nature of political power made the circulation of books especially hard to control; elsewhere the governing powers could build naturally on existing mechanisms of control to impose strict censorship. The pioneer in this regard was the regime of Charles V in the Netherlands, but France soon followed suit, and soon an Index of Prohibited Books was a standard feature of all European Catholic cultures.[78] Protestant regimes also developed systems of regulation to ensure that only works in harmony with the prevailing local confession could be published. This prior inspection of texts was especially firmly enforced in Geneva.[79] In England, the establishment of the Stationers' Company brought together the twin goals of industry regulation and intellectual control. The records generated by these two forms of regulation – prior control and prohibition of books already on the market – in fact form a valuable source of information for books that may have existed but now appear to be lost.[80] Other sources of information on such lost books are the early encyclopaedic bibliographical projects of Antoine du Verdier or Du Croix du Maine, and the early booksellers' catalogues, such as those produced for the Frankfurt Fair.[81]

It is for historians one of the great scholarly conundrums to come to a realistic estimate of how large a proportion of the books published in the Reformation era may now be totally lost. The best approach may be to postulate highly differential rates of survival. It is likely that almost all editions of works by a man as famous as Luther survive in at least one copy, since they would have been preserved as highly collectable, such was Luther's renown, even in his own lifetime. Large-scale books, such as folio Bibles and Latin commentaries are also unlikely to have disappeared entirely. But small books by anonymous authors have probably survived far less well. At the end of the nineteenth century a small cache of Dutch evangelical books was unearthed in a castle in the Netherlands. Of the seven books in the cache only two were previously known through a surviving library copy. Only a few years ago a second discovery, this time in the rafter of a house being renovated, produced a further three previously unknown editions (of five books in the bundle). It is an enticing thought what new discoveries may await historians of the Reformation as the systematic logging of books in large and small libraries around the world increases the corpus of information at our disposal.

further reading

The classic exposition of the impact of print is Elizabeth Eisenstein, *The Printing Press as an Agent of Change* (Cambridge, 1982). See also Lucien Febvre and Henri-Jean Martin, *The Coming of the Book: the Impact of Printing, 1450–1800* (London, 1976; cf. French edition, 1958).

For the 'German paradigm' of print's role in the Reformation, see Mark U. Edwards, *Printing, Propaganda and Martin Luther* (Berkeley, 1994); Richard A. Crofts, 'Books, Reform and Reformation', *Archiv für Reformationsgeschichte*, 71 (1980), 21–36; idem, 'Printing, Reform and the Catholic Reformation in Germany (1521–1545)', *Sixteenth Century Journal*, 16 (1985); Miriam Usher Chrisman, *Conflicting Visions of Reform: German Lay Propaganda Pamphlets, 1519–1530* (Boston, 1996); and especially Robert W. Scribner, *For the Sake of Simple Folk: Popular Propaganda for the German Reformation* (Cambridge, 1981). As counterweights to this focus on Protestant print, see David Bagchi, *Luther's Earliest Opponents: Catholic Controversialists, 1518–1525* (Minneapolis, 1991); Anne T. Thayer, *Penitence, Preaching and the Coming of the Reformation* (Aldershot, 2002).

The wider applicability of the 'German paradigm' is challenged in Andrew Pettegree and Matthew Hall, 'The Reformation and the Book: a

Reconsideration', *Historical Journal*, 47 (2004), 785–808. Several strong European case studies can be found in Jean-François Gilmont (ed.), *The Reformation and the Book*, trans. Karin Maag (Aldershot, 1998).

On France, see Andrew Pettegree, Paul Nelles and Philip Connor (eds), *The Sixteenth-Century French Religious Book* (Aldershot, 2001); Francis Higman, 'Theology for the Layman in the French Reformation', *The Library*, 6th ser. 9 (1987), 105–27; Christopher Elwood, *The Body Broken: the Calvinist Doctrine of the Eucharist and the Symbolisation of Power in Sixteenth-Century France* (Oxford, 1999); Luc Racaut, *Hatred in Print: Catholic Propaganda and Protestant Identity During the French Wars of Religion* (Aldershot, 2002); Larissa Juliet Taylor, *Heresy and Orthodoxy in Sixteenth-Century Paris* (Leiden, 1999).

On the Netherlands, see Andrew G. Johnston, 'Lutheranism in Disguise: the Corte Instruccye of Cornelis vander Heyden', *Nederlands archief voor Kerkgeschiedenis*, 68 (1988), 23–9, and his essay in Gilmont (ed.), *The Reformation and the Book*; Andrew Pettegree, *Emden and the Dutch Revolt: Exile and the Development of Reformed Protestantism* (Oxford, 1992).

On England, see Ian Green, *The Christian's ABC: Catechisms and Catechizing in England, c. 1530–1740* (Oxford, 1996); idem, *Print and Protestantism in Early Modern England* (Oxford, 2000); Tessa Watt, *Cheap Print and Popular Piety, 1550–1640* (Cambridge, 1991); Adam Fox, *Oral and Literate Culture in England, 1500–1700* (Oxford, 2000); Andrew Pettegree, 'The Latin Polemic of the Marian Exiles', in his *Marian Protestantism: Six Studies* (Aldershot, 1996); and volumes III and IV of *The Cambridge History of the Book in Britain* (1999, 2002). On Scotland, see Alastair J. Mann, *The Scottish Book Trade, 1500–1720: Print Commerce and Print Control in Early Modern Scotland* (East Linton, 2000).

On Catholic attempts to control print, see George Putnam, *The Censorship of the Church of Rome*, 2 vols (London, 1906); Paul F. Grendler, *The Roman Inquisition and the Venetian Press, 1540–1605* (Princeton, 1977).

notes

1. Elizabeth Eisenstein, *The Printing Press as an Agent of Change* (Cambridge, 1982).
2. Quoted *ibid.*, pp. 150, 151.
3. Josef Benzing, *Lutherbibliographie* (Baden-Baden, 1966).
4. On the Tübingen *Flugschriften* project see Hans-Joachim Köhler (ed.), *Fluschriften als Massenmedium des Reformationszeit* (Stuttgart, 1981).
5. Mark U. Edwards, *Printing, Propaganda and Martin Luther* (Berkeley, 1994); idem, 'Statistics on Sixteenth-Century Printing', in Philip N. Bebb and Sherrin

Marshall (eds), *The Process of Change in Early Modern Europe* (Athens, OH, 1988).

6. Richard A. Crofts, 'Books, Reform and Reformation', *Archiv für Reformationsgeschichte*, 71 (1980), 21–36; idem, 'Printing, Reform and the Catholic Reformation in Germany (1521–1545)', *Sixteenth Century Journal*, 16 (1985), 369–81.
7. Miriam Usher Chrisman, *Bibliography of Strasbourg Imprints, 1480–1599* (New Haven and London, 1982); idem, *Conflicting Visions of Reform: German Lay Propaganda Pamphlets, 1519–1530* (Boston, 1996).
8. Robert W. Scribner, *For the Sake of Simple Folk: Popular Propaganda for the German Reformation* (Cambridge, 1981).
9. Max Geisberg, *The German Single-Leaf Woodcut, 1500–1550*, rev. and ed. Walter L. Strauss, 4 vols (New York, 1974); Walter L. Strauss, *The German Single-Leaf Woodcut, 1550–1600: a Pictorial Catalogue*, 3 vols (New York, 1975); Dorothy Anderson with Walter L. Strauss, *The German Single-Leaf Woodcut, 1600–1700: a Pictorial Catalogue*, 2 vols (New York, 1977).
10. Andrew Pettegree and Matthew Hall, 'The Reformation and the Book: a Reconsideration', *Historical Journal*, 47 (2004), 785–808.
11. Lucien Febvre and Henri-Jean Martin, *The Coming of the Book: the Impact of Printing, 1450–1800* (London, 1976; cf. French edition, 1958).
12. *Index Aureliensis: Catalogus Librorum Sedecimo Saeculo Impressorum*, 13 vols (Baden-Baden, 1965–2003).
13. A. W. Pollard and G. R. Redgrave, *A Short Title Catalogue of Books Printed in England, Scotland and Ireland, and of English Books Printed Abroad, 1475–1640* (London, 1926).
14. E-STC available from the Research Libraries Group at <www.rlg.org/estc.html>. Early English Books Online: <http://eebo.chadwyck.com/home>.
15. *L'Epistre du roy d'angleterre aux princes & peuple chrestien touchant le concile à venire* (London [Thomas Berthelet], 1539); *La protestation et advis du roy d'Angleterre touchant le Concile qui se debuoit tenir à Mantue* (London [Thomas Berthelet], 1539). Halle ULB: AB 154113 (15 & 16).
16. Brigitte Moreau, *Inventaire Chronologique des éditions parisiennes du XVIe siècle*, 4 vols. (Paris, 1972–1992). A further volume, covering 1536–1540, has been recently published under the editorial supervision of Geneviève Guilleminot-Chrestien (Paris, 2004).
17. Philippe Renouard, *Imprimeurs & libraires Parisiens du Xve siècle*, 5 vols, A- BON (1964–91).
18. H.-L. Baudrier and J. Baudrier, *Bibliographie lyonnaise. Recherches sur les imprimeurs, libraires, reliers et fondeurs de lettres de Lyon au XVIe siècle*, 12 vols (Lyon, 1895–1921).
19. Sybille von Gültingen, *Bibliographie des livres imprimés à Lyon*, 4 vols (Baden-Baden, 1992–96). Thus far this covers Lyons print between 1500 and around 1540.
20. *Répertoire bibliographique des livres imprimés en France au seizième siècle*, 30 vols (Baden Baden, 1968–80).
21. Information on the current state of the project and libraries surveyed available on the project website at <www.st-andrews.ac.uk/~www_rsi/book/book.htm>.

22. *VD16. Verzeichnis der im deutschen Sprachbereich erscheinenen Drucke des XVI. Jahrhunderts*, 25 vols (Stuttgart, 1983–2000).
23. Wouter Nijhoff and M. E. Kronenberg, *Nederlandsche Bibliographie van 1500 tot 1540*, 3 vols (The Hague, 1923–42).
24. *Dutch Pamphlets, 1486–1853: the Knuttel Collection* (IDC, 1982–1995).
25. Elly Cockx-Indestege, Geneviève Glorieux and Bart Op de Beeck (eds), *Belgica typographica 1541–1600: catalogus librorum impressorum ab anno MDXLI ad annum MDC in regionibus quae nunc Regni Belgarum partes sunt* (Nieuwkoop, 1968–94); Paul Valkema Blouw, *Typographia Batava, 1541–1600: a Repertorium of Books Printed in the Northern Netherlands between 1541 and 1600* (Nieuwkoop, 1998). Together the three bibliographies log some 21,000 editions of books published in the Low Countries during the sixteenth century.
26. <www.kb.nl/stcn/index-en.html>.
27. <www.stcv.be/eng/home.html>.
28. <http://edit16.iccu.sbn.it/>.
29. <http://picarta.pica.nl/login/DB=3.29/LNG=EN/>. The I-STC records more than 27,000 editions published before 1501, including around 1,500 items of single-sheet printing. The expectation is that the total will eventually settle at around 28,000 editions. Full information about the project can be found at <www.bl.uk/collections/hoinc.html>. Lotte Hellinga and John Goldfinch, 'Ten Years of the Incunabula Short-Title Catalogue (ISTC)', *Bulletin du Bibliophile* (1990), 125–31; idem, *Bibliography and the Study of Fifteenth Century Civilization* (London, 1987); Paul Needham, 'Counting Incunables: the IISTC CD-ROM', *Huntington Library Quarterly*, 61 (1998–2000), 457–529.
30. <www.idc.nl/>.
31. *Sixteenth Century Pamphlets: Flugschriften des 16. Jahrhunderts* (IDC). Part I (1500–1530) contains 5,000 German and Latin pamphlets. Part II (1531–1600), still in progress, contains between 3,500 and 5,000 pamphlets.
32. Michael A. Pegg, *Bibliotheca Lindesiana, and Other Collections of German Sixteenth-Century Pamphlets in Libraries of Britain and France* (Baden-Baden, 1977); idem, *Catalogue of German Reformation Pamphlets (1516–1550) in Swiss Libraries* (Baden-Baden, 1983); idem, *A Catalogue of German Reformation Pamphlets (1516–1550) in Swedish Libraries* (Baden-Baden, 1995).
33. *Cranach im Detail. Buchschmuck Lucas Cranachs des Älteren und seiner Werkstatt* (Austsellung Lutherhalle Wittenberg, 1994); Heimo Reinitzer (ed.), *Biblia deutsch. Luthers Bibelübersetzung und ihre Tradition* (Herzog August Bibliothek Wolfenbüttel, 1983).
34. Sachiko Kusokawa, *A Wittenberg University Library Catalogue of 1536* (Cambridge, 1995).
35. Chrisman, *Bibliography of Strasbourg Imprints*; idem, *Conflicting Visions*.
36. Rebecca Wagner Oettinger, *Music as Propaganda in the German Reformation* (Aldershot, 2001); Alexander J. Fisher, *Music and Religious Identity in Counter-Reformation Augsburg, 1580–1630* (Aldershot, 2004).
37. David Bagchi, *Luther's Earliest Opponents: Catholic Controversialists, 1518–1525* (Minneapolis, 1991); Frank Aurich, *Die Anfänge des Buchdrucks in Dresden. Die Emserpresse, 1524–1526* (Dresden, 2000); Christoph Volkmar, *Die Heiligenerhebung Bennos von Meissen (1523–1524)* (Aschendorff Münster, 2002).

38. Anne T. Thayer, *Penitence, Preaching and the Coming of the Reformation* (Aldershot, 2002).
39. <www.litdb.evtheol.uni-mainz.de/databank/index.php>.
40. *Bibliotheca Dissidentium. Répertoire des non-conformistes religieux des seizième et dix-septième siècles*, 23 vols (Baden-Baden, 1980–2003).
41. Joachim Staedtke, *Heinrich Bullinger Bibliographie. Band 1: Beschreibendes Verzeichnis der gedruckten Werke von Heinrich Bullinger* (Zurich, 1972).
42. Urs Leu, 'Die Zürcher Buch- und Lesekultur, 1520–1575', in Emidio Campi (ed.), *Heinrich Bullinger und seine Zeit* (*Zwingliana*, 31 (2004)), 61–90.
43. Manfred Vischer, *Bibliographie des Zürcher Druckschriften des 15. und 16. Jahrhunderts* (Baden-Baden, 1991); idem, *Zürcher Einblattdrucke des 16. Jahrhunderts* (Baden-Baden, 2001).
44. Urs B. Leu and Sandra Weidmann (eds), *Heinrich Bullingers Privatbibliothek* (Heinrich Bullinger Werke. Abt. 1: Bibliographie Heinrich Bullinger. Band 3, Zurich, 2004).
45. Peter G. Bietenholz, *Basle and France in the Sixteenth Century: the Basle Humanists and Printers in their Contacts with Francophone Culture* (Geneva, 1971).
46. P. Chaix, A. Dufour, and G. Moeckli, *Les Livres Imprimés à Genève de 1550 à 1600* (Geneva, 1966); Jean-François Gilmont and Rodolphe Peter, *Bibliotheca Calviniana. Les oeuvres de Jean Calvin publiés au XVIe siècle*, 3 vols (Geneva, 1991–94).
47. F. Gardy, *Bibliographie des oeuvres théologiques, littéraires, historiques et judidiques de Théodore de Bèze* (Geneva, 1960).
48. Jean-François Gilmont, *Jean Calvin et le livre imprimé* (Geneva, 1997); idem, *Le livre et ses secrets* (Geneva, 2003).
49. P. Chaix, *Recherches sur l'imprimerie à Genève de 1550 à 1564. Etude bibliographique, économique et littéraire* (Geneva, 1954); Hans Joachim Bremme, *Buchdrucker und Buchhändler sure it der Galubenskämpfe. Studien zur Genfer Druckgeschichte, 1565–1580* (Geneva, 1969); Ingeborg Jostock, 'La censure au quotidian: le contrôle de l'imprimerie à Genève, 1560–1600', in Andrew Pettegree, Paul Nelles and Philip Connor (eds), *The Sixteenth-Century French Religious Book* (Aldershot, 2001).
50. Francis Higman, 'Luther et la piété de l'église gallicane: le Livre de vraye et parfaicte oraison', *Revue d'Histoire et de philosophie Religieuses*, 63 (1983), 91–111; idem, 'Theology for the Layman in the French Reformation', *The Library*, 6th ser. 9 (1987), 105–27; idem, *Lire et découvrir. La circulation des idées au temps de la Réforme* (Geneva, 1998).
51. B. T. Chambers, *Bibliography of French Bibles: Fifteenth- and Sixteenth-Century French-Language Editions of the Scriptures* (Geneva, 1983); Catherine Delanoo-Smith and Elizabeth Morley Ingram, *Maps in Bibles, 1500–1600: an Illustrated Catalogue* (Geneva, 1991).
52. An ongoing project under the direction of Jean-Daniel Candaux and Jean-Michel Noailly, with the cooperation of Bettye Chambers. *Bibliographie des Psaumes imprimés en vers français* (to be published by Droz, Geneva).
53. Eugénie Droz, *L'imprimerie à La Rochelle. 1: Barthélémy Berton, 1563–1573* (Geneva, 1960); idem, *L'imprimerie à La Rochelle. 3. La veuve Berton et Jean Portau, 1573–1589* (Geneva, 1960); Louis Desgraves, *L'imprimerie à La Rochelle. 2: Les Haultin, 1571–1623* (Geneva, 1960); idem, *Elie Gibier imprimeur à Orléans (1536–1588)* (Geneva, 1966); Andrew Pettegree, 'Protestantism, Publication

and the French Wars of Religion: the Case of Caen', in Robert J. Bast and Andrew C. Gow (eds), *Continuity and Change: the Harvest of Late-Mediaeval and Reformation History* (Leiden, 2000); idem, 'Protestant Printing during the French Wars of Religion: the Lyon Press of Jean Saugrain', in Tom Brady and James Tracy (eds), *Essays Presented to Heiko Oberman on his Seventieth Birthday* (Brill, 2003); Jean-François Gilmont, 'La première diffusion des Mémoires de Condé par Éloi Gibier en 1562–1563', in P. Aquilon and H.-J. Martin (eds), *Le livre dans l'Europe de la Renaissance. Actes du XXVIIIe Colloque international d'études humanistes de Tours* (Paris, 1988), pp. 58–70. Cf. revised and updated version of this article in Gilmont, *Le livre & ses secrets* (Louvain-la-Neuve, 2003).

54. F. S. Giese, *Artus Désiré: Priest and Pamphleteer of the Sixteenth Century* (Chapel Hill, 1973).
55. Christopher Elwood, *The Body Broken: the Calvinist Doctrine of the Eucharist and the Symbolisation of Power in Sixteenth-Century France* (Oxford, 1999); Luc Racaut, *Hatred in Print: Catholic Propaganda and Protestant Identity During the French Wars of Religion* (Aldershot, 2002); Larissa Juliet Taylor, *Heresy and Orthodoxy in Sixteenth-Century Paris: François Le Picart and the Beginnings of the Catholic Reformation* (Leiden, 1999).
56. C. Ch. G. Visser, *Luthers geschriften in de Nederlanden tot 1546* (Assen, 1969); Andrew G. Johnston, 'Lutheranism in Disguise: the Corte Instruccye of Cornelis vander Heyden', *Nederlands archief voor Kerkgeschiedenis*, 68 (1988), 23–9; idem, 'Printing and the Reformation in the Low Countries, 1520– c. 1555', in Jean-François Gilmont (ed.), *The Reformation and the Book*, trans. Karin Maag (Aldershot, 1998).
57. Bart A. Rosier, *The Bible in Print: Netherlandish Bible Illustrations in the Sixteenth Century*, 2 vols (Leiden, 1997); A. A. Den Hollander, *De Nederlandse Bijbelvertalingen 1522–1545* (Nieuwkoop, 1997).
58. Willem Heijting, *De catechisme en confessies in de Nederlandse Reformatie tot 1585*, 2 vols (Nieuwkoop, 1989); Andrew Pettegree, *Emden and the Dutch Revolt: Exile and the Development of Reformed Protestantism* (Oxford, 1992).
59. See, among other pieces, Paul Valkema Blouw, 'Printers to Henrik Niclaes: Plantin and Augustijn van Hasselt', *Quaerendo*, 14 (1984), 247–72; idem, 'The Secret Background of Lenaert der Kinderen's Activities, 1562–67', *Quaerendo*, 17 (1987), 83–127.
60. Alastair Hamilton (ed.), *Documenta anabaptistica Neerlandica. Deel 6, Cronica. Ordo sacerdotis. Acta HN* (Leiden, 1988).
61. Daniel R. Horst, *De opstand in zwart-wit. Propagandaprenten uit de Nederlandse Opstand, 1566–1584* (Walburg Pers, 2003); Alastair Duke, 'Posters, Pamphlets and Printers: the Ways and Means of Disseminating Dissident Opinion on the Eve of the Dutch Revolt', *Dutch Crossing*, 27 (2003), 23–44.
62. Anne Rijsing, 'The Book and the Reformation in Denmark and Norway, 1523–1540' and Remi Kick, 'The Book and the Reformation in the Kingdom of Sweden', in Gilmont, *Reformation and the Book*.
63. Alastair J. Mann, *The Scottish Book Trade, 1500–1720: Print Commerce and Print Control in Early Modern Scotland* (East Linton, 2000).
64. Ugo Rozzo and Silvana Seidel Menchi, 'The Book and the Reformation in Italy', in Gilmont, *Reformation and the Book*, pp. 319–67.

65. A. Gordon Kinder, *Spanish Protestants and Reformers in the Sixteenth Century: a Bibliography* (London, 1983).
66. A. Gordon Kinder, *Casiodoro de Reina: Spanish Reformer of the Sixteenth Century* (London, 1975); idem, *Confessión de fe Christiana: the Spanish Protestant Confession of Faith (London, 1560/61) / Casiodoro de Reina; Edited from the Sole Surviving Copy of the Bilingual Edition* (Exeter, 1988).
67. E. Baskerville, *A Chronological Bibliography of Propaganda and Polemic Publications Published in English between 1553 and 1558* (Philadelphia, 1979); Andrew Pettegree, 'The Latin Polemic of the Marian Exiles', in his *Marian Protestantism: Six Studies* (Aldershot, 1996).
68. R. J. Fehrenbach and E. S. Leedham-Green (eds), *Private Libraries in Renaissance England: a Collection and Catalogue of Tudor and Early Stuart Book-Lists* (New York, 1992–).
69. Ian Green, *The Christian's ABC: Catechisms and Catechizing in England, c. 1530–1740* (Oxford, 1996); idem, *Print and Protestantism in Early Modern England* (Oxford, 2000).
70. Tessa Watt, *Cheap Print and Popular Piety, 1550–1640* (Cambridge, 1991); Adam Fox, *Oral and Literate Culture in England, 1500–1700* (Oxford, 2000).
71. Francis Higman, 'Calvin's Works in Translation', in Andrew Pettegree, Alastair Duke and Gillian Lewis (eds), *Calvinism in Europe, 1540–1620* (Cambridge, 1994); Andrew Pettegree, 'The Reception of Calvinism in Britain', in Wilhelm H. Neuser and Brian G. Armstrong (eds), *Calvinus Sincerioris Religionis Vindex: Calvin as Protector of the Purer Religion* (Kirksville, MO, 1997).
72. E. S. Leedham-Green, *Books in Cambridge Inventories*, 2 vols (Cambridge, 1986).
73. Lotte Hellinga and J. B. Trapp (eds), *The Cambridge History of the Book in Britain. Volume III: 1400–1557* (Cambridge, 1999); John Barnard and D. F. Mckenzie (eds), *The Cambridge History of the Book in Britain. Volume IV: 1557–1695* (Cambridge, 2002).
74. David McKitterick, *A History of Cambridge University Press. Volume I: Printing and the Book Trade in Cambridge, 1534–1698* (Cambridge, 1992).
75. Donald W. Krummel, *English Music Printing, 1553–1700* (London, 1975); Jeremy L. Smith, *Thomas East and Music Publishing in Renaissance England* (Oxford, 2003). On Foxe studies, see above, p. 139.
76. Hugh Amory and David D. Hall, *A History of the Book in America. Volume I: The Colonial Book in the Atlantic World* (Cambridge, 2000).
77. Elizabeth Armstrong, *Before Copyright: the French Book-Privilege System 1498–1526* (Cambridge, 1990).
78. Franz Heinrich Reusch, *Der Index der verboten Bücher; ein Beitrag zur Kirchen- und Literaturgeschichte*, 2 vols (Bonn, 1883–85); George Putnam, *The Censorship of the Church of Rome*, 2 vols (London, 1906); Paul F. Grendler, *The Roman Inquisition and the Venetian Press, 1540–1605* (Princeton, 1977).
79. Chaix, *Recherches sur l'imprimerie à Genève*.
80. J. M. de Bujanda (ed.), *Index des livres interdits*, 10 vols (Sherbrooke, 1985–96).
81. *La bibliotheque d'Antoine dv Verdier* (Lyon, 1585); *Premier volume de la bibliothèque de la Croix du Maine* (Paris, 1584); *Die Messkatalogue Georg Willers. Herbstmess 1564 – Herbstmesse 1600*. Faksimiledrucke ed. Bernhard Fabian, 5 vols (New York, 1972–2001).

9
the catholic reformation

trevor johnson

As the presence of a chapter on the Catholic Reformation in this book demonstrates, historians tend not to interpret the rise of Protestantism and the changes within Catholicism in the sixteenth century as isolated events but as interdependent developments, together defining an epoch in European history. Like that on its Protestant counterpart, the literature on the Catholic Reformation is vast. As befits a movement which encompassed much of Europe and touched many parts of the wider world, this literature exists in many languages and reflects separate national historiographical traditions as well as varied ecclesiastical positions. What follows, therefore, cannot be an exhaustive summary, but is rather intended to offer a glimpse of some of the more important tendencies in older and more recent historical research on the theme.

historiographical origins and the problem of definition

Books on aspects of the Catholic Reformation have been written since the movement's origins. In 1619, for example, the Italian Servite friar Paolo Sarpi (1552–1623) published a famous *History of the Council of Trent*. However, Sarpi's *History* was less a work of scholarly analysis (as understood today) than a brilliant piece of propaganda by a noted defender of the ecclesiastical autonomy of the Republic of Venice against papal authority. Other 'histories' composed in the period (including a riposte to Sarpi by the Jesuit, Pallavicini) were similarly reflections of contemporary concerns. Meaningfully, therefore, the historiography of the Catholic or Counter Reformation can in fact properly be said to have begun only in the late eighteenth and early nineteenth centuries, when these very terms were coined and systematic large-scale characterisations of a defined chapter in the history of the Church were attempted.

Historians may have agreed as to many of the specific events, personalities, institutions and processes to be investigated (the Council of Trent, Philip II, the Roman Inquisition, the Society of Jesus, and so on), but they have often differed severely over naming the movement which supposedly united them. The labels in use today have a complicated and contested history, which is worth reviewing since the choice of names reveals fundamental assumptions and approaches.[1] 'Counter-Reformation' seems to have been the first such designation to appear on the historiographical map, invented as it was in the 1770s by a German Lutheran jurist named Johann Stephan Putter (1725–1807) in order to describe instances of the forced reconversion of Protestant regions and communities to Catholicism in the sixteenth and seventeenth centuries. Of course, there is something of an irony here, in that the terminology, and therefore in a sense the field itself, was created by a representative of Catholicism's religious opponents. Indeed, it was other German Protestant scholars who, after Putter, first enlarged the meaning of his term 'Counter-Reformation' to encompass an entire period and movement. Most influential among these was Leopold von Ranke, author of a *History of the Popes* (1834–) and a six-volume history of Germany 'in the Age of the Reformation' (1842–47). For Ranke the Counter-Reformation was primarily a negative and defensive reaction against the rise of Protestantism and was epitomised by repressive measures and institutions like the Inquisition and the Index of Prohibited Books. Ranke's approach stressed the importance of statecraft, politics and warfare, in particular the wars of religion which led to the return of some Protestant areas to Catholicism. With some terminological looseness he still wrote, as had Putter, of 'Counter-Reformations' (*Gegenreformationen*, or even *Antireformationen*) in the plural to denote specific instances of such Catholic re-conquest, whilst simultaneously using the singular compound noun to define the movement in general. 'The Counter-Reformation gained a fresh driving force and a new field', wrote Ranke of the Bavarian annexation of the Upper Palatinate in the early phase of the Thirty Years War, for example, in his *History of the Popes*. A few pages later he identified other Catholic conversions in Germany in the 1620s:

> In this period the Anti-Reformations were established in Upper Germany with a new zeal ... so many decrees, resolutions, decisions, recommendations, and all of them in favour of Catholicism! The young Count of Nassau-Siegen, the younger Counts Palatine of Neuburg, the Grandmaster of the Teutonic Order, all these undertook

> new Reformations [sic]; in the Upper Palatinate the nobility was now compelled to adopt Catholicism.[2]

A generation after Ranke, Moritz Ritter penned a multi-volume history of Germany in the period of 'the Counter Reformation and the Thirty Years War', covering the period 1555–1648, whilst Eberhard Gothein, in a work of 1895, located the Counter-Reformation not so much in the confessional battlegrounds of Germany but rather in Spain, highlighting the importance of the contribution made by Ignatius Loyola and the early Jesuits.

This shift of geographical focus can be found among other late-nineteenth-century German Protestant historians. With it came also a new, subtly different, designation when Wilhelm Maurenbecher (1838–1892) brought out a volume on what he termed the 'Catholic Reformation' in 1880. Because it indicated internal reform which was already underway before 1517 and therefore a movement which was more than simply a negative reaction to Protestantism, this new title found approval from Catholic scholars, most notably from Ludwig von Pastor, who used the phrase 'Catholic Reformation and Restoration' in his multi-volume *History of the Popes from the Close of the Middle Ages* (1886–1933), a work conceived as a riposte to Ranke. This more positive framing of the field influenced writers beyond Germany. With an alternative nomenclature now available, the strategic choice of designation became an indicator of an author's entire approach and likely evaluation. In late nineteenth-century Italy, for example, clerical authors tended to favour the term 'Catholic Reformation'; lay liberal and nationalist scholars, who were critical of the repressive aspects of the Church of their own day as well as those exhibited by it in the past, preferred to stick with 'Counter-Reformation'.[3]

Following on from the great synthetic surveys and sweeping histories of the late nineteenth century, but without putting aside arguments over terminology, early twentieth-century scholarship delved more deeply into the various individual themes, personalities or events which collectively comprised the Catholic Reform movement. For example, the French historian Marcel Bataillon's classic studies investigated the shifting fortunes in Spain of the ideas of the Netherlandish humanist, Desiderius Erasmus, from their enthusiastic reception in the early 1520s to harsh reaction against them in the 1530s, while the German scholar Karl Brandi produced an exhaustive biography of Emperor Charles V.[4] Post-1945 scholarship on the Catholic Reformation was long dominated by another German, Hubert Jedin (1900–1980), author of the definitive history of the Council of Trent and senior editor of a handbook of Church history which appeared between 1962 and 1979, the fifth volume of which was entitled 'Reformation and Counter Reformation'. Its tone was neutral, indicating, as the preface to the English edition declared, 'a step in the right direction' of freeing scholarship from what Jedin saw as the accumulated bias and misconceptions that had continued to cloud both Catholic and Protestant interpretations of the movement.[5]

Jedin had already signalled his approach in an influential essay first published in 1946 in which he debated the relative merits of the terms 'Catholic Reformation' and 'Counter-Reformation'. Seeking to transcend this polarisation, Jedin argued that wide-ranging institutional reform predated both the Protestant Reformation and the Council of Trent, but that such various localised reformist currents (what he called 'reform from below') could only converge into a powerful torrent when the papacy itself gave them active and centralised direction, which did not occur until after the rise of Protestantism imposed on the Church an 'external impetus' to reform. Giving legislative form to these new priorities in the decrees of the Council of Trent (1545–63), represented a third stage in the Catholic Reformation. A fourth stage saw the implementation of the decrees under the tutelage of successive post-Tridentine (i.e. post-Council of Trent) popes through a range of mechanisms: the activities of papal nuncios, episcopal visitations and synods, the foundation of seminaries for the formation of clergy, the revision of liturgical texts and a reorganisation of the papal administration. As for terminology, Jedin proposed 'Catholic Reform' as a more neutral and acceptable alternative to 'Catholic Reformation'. However, Catholic Reform was the precondition for the 'Counter-Reformation', a term which Jedin retained but applied in a restricted fashion solely to the measures taken to defend the Catholic church against the Protestant Reformation, such as the publication of works of theological controversy, the revival of the Inquisition, the creation of an Index of Prohibited Books or the employment of the secular arm against heresy. 'It is impossible to speak of Catholic reform *or* Counter-Reformation; rather one must speak of Catholic reform *and* Counter-Reformation', he concluded.[6]

Jedin's compelling framework held much subsequent scholarship in its sway and until recently it seemed to have settled the problem of definition, whilst substantially freeing historical debate on the Catholic Reformation from the legacy of confessional polemic. However, his approach to Church history remained rigidly institutional. A different approach developed out of the French intellectual tradition, in particular from the *Annales* school of historians and from the religious sociology of Gabriel le Bras. This sought to recapture the experiences of ordinary lay people, rather than limiting itself to the careers of bishops or the formulation and defence of doctrines. Another complementary approach was that advanced in England by H. Outram Evennett, whose 1951 lectures (published posthumously in 1968) tried to distil what he termed the 'spirit' of the Counter-Reformation, stressing its devotional culture, spirituality and religious experience, including mysticism, and suggesting that this spirit was actually more important in defining the movement than were its institutional reforms. Unsurprisingly, the character and piety of St Ignatius Loyola and their expression in his influential spiritual guide, the *Spiritual Exercises*, here assumed prominence.[7] Evennett's emphasis also echoed the large study undertaken by Henri Bremond, and published as a *Literary History of Religious Sentiment in France*, which had much to say on mystical experience in the seventeenth century.[8] For these approaches the politically defined cut-off dates favoured by historians from Ranke to Jedin, which limited the Catholic Reformation to the sixteenth century, or which allowed it to last only until 1648 (the end of the Thirty Years War) were no longer meaningful. However, the main contribution of such non-institutional ways of regarding Church history has been to open up the field to a vastly increased range of tendencies and approaches, which in the last two or three decades have produced radically different views of the Catholic Reformation.

the catholic reformation and the institutional church

Nevertheless, if the Catholic Reformation is seen to have been about many things, none would deny the defining importance to the movement of institutional reform: the attempt to curb perceived abuses in the lifestyle and professional practice of the clergy – from humble parish priests, through the bishops and the religious orders, up to and including the papacy itself. As Eric Cochrane's description of this period as one of 'Tridentine Reformation' (after the Latin word for the Italian Alpine city of Trent) implies, it was the Council of Trent which was and is seen as setting the benchmark of institutional reform in its fullest contemporary expression.[9] The modern historiography of the Council is dominated by Jedin's monumental history (1949–75). Original in its conception (it was the first comprehensive history of the Council since the works by Sarpi and Pallavicini in the seventeenth century), the scale and depth of Jedin's analysis may be grasped immediately from the first of his six volumes, which is entirely taken up with background to the convening of the Council in 1545 and which in the English translation runs to over 580 pages.[10] As a reconstruction of the origins and course of the Council, and in particular of the entanglement of doctrinal debate with ecclesiastical politics and European diplomacy, Jedin's masterpiece of erudition and sensitive analysis of detail remains definitive. However, since Jedin, scholars have concentrated on fleshing out and interpreting the careers of Trent's principal participants, uncovering the natures of the factions and groupings and (by far the most favoured approach) exploring and assessing the Council's impact on Catholic institutions and society. In short, Trent's history has become much more socially and culturally contextualised.

A substantial body of work has focused on Rome and that most obvious of Catholic institutions, the papacy. Such work includes biographies of individual popes, but also structural histories which have explored and questioned the apparent resurgence of papal authority over the Church after Trent, amid the renewal of Roman administrative structures and practices. Scholars such as Paolo Prodi have emphasised that every Catholic Reformation pope continued to play a twofold role, as both governor of the universal Church and also territorial monarch of the Papal States. Prodi has analysed the papacy as more or less successful state-builder, and reminded us of the seriousness with which any post-Tridentine pontiff saw himself as a 'papal prince'.[11] Popes did not rule the Papal States alone, but at the apex of an apparatus, and their style of governance has been examined by a number of historians, including Nicole Reinhardt, whose analysis, employing the approaches of social anthropology and microhistory, focuses on patronage and clientelism to highlight the necessity of pragmatic compromises between old and new elites in local administration.[12] The strengths and weaknesses of the popes as territorial rulers are further evoked in Giampiero Brunelli's study of the papal army in the late sixteenth and early seventeenth centuries.[13] Meanwhile the larger political role of the papacy within the post-Reformation world is underscored by scholarship on papal diplomacy, students of which can profit from such outstanding editorial projects as the *Nuntiaturberichte aus Deutschland*, the published correspondence between Rome and successive papal nuncios at the Imperial court (a series begun in 1892 and still going).

Another way in which the papacy presented itself was through conspicuous cultural consumption and display. The city of Rome itself underwent a resurgence after Trent, when the munificent patronage of successive popes, cardinals and religious orders created the Baroque urban space which could act as a worthy architectural and artistic setting for the enactment of liturgy and as an imposing haven for the thousands of pilgrims who came to venerate its vast repository of relics. In Rome, as elsewhere, art played a key role in the Catholic Reformation.[14]

As for governance over the Church, as historians have stressed, the key was a renewal of administrative practice and a bureaucratic centralisation, facilitated by the creation in the 1580s under Pope Sixtus V of new curial 'congregations', or departments, in Rome which were designed to routinise the Tridentine decrees and to oversee all aspects of the life of the Church, from liturgical reform, saint-making or the coordination of missionary work through to the maintenance of doctrinal orthodoxy. Heads of these congregations were the senior ecclesiastics, the Cardinals. A series of revisionist studies of the sixteenth- and seventeenth-century cardinalate, mostly by Italian historians, has modified an older view of their decline under the heel of papal reforms, stressing their continued independence and indeed the new opportunities for assertiveness provided by their role within the new organs of curial bureaucracy.[15] Alongside such structural histories, studies of individual 'Princes of the Church' abound, particularly for the generation of reformers raised to the purple by Pope Paul III.[16] The story has become one not of sharply defined groups of reformers and traditionalists, or centralisers and resisters, still less of clear-cut victory or defeat, but rather of complex competition for patronage and webs of power-brokerage – which gives the papacy the typical appearance of any sixteenth-century court.

Nonetheless, the papacy's spiritual responsibility to the Church beyond Rome made it special. One of the important legacies of the Council of Trent was the greater uniformity in the practice of liturgy, or public worship, which the papacy succeeded in imposing on Catholic Europe. In an approach combining institutional with intellectual and cultural history, and using the example of one early modern Italian diocese, Simon Ditchfield has shown how local churches were gradually compelled to respond to the liturgical revisions and greater standardisation decreed by the newly created curial body, the Congregation of Rites – not least by revising the saints' lives included in their prayer books in line with new Roman measures of authenticity. In the process, a new practice of conceiving and writing history was disseminated. Indeed, such local ecclesiastical histories neatly complemented the grander projects of Church history which were being undertaken by the Oratorian Cardinal, Cesare Baronio, in his *Annales Eccelsiasticae*. There were paralleled in the *Acta Sanctorum*, a collection of meticulously researched saints' lives which began to be published in 1643 by a group of Flemish Jesuit scholars who came to be known as the Bollandists, after their leading figure, Jean Bolland. The new standards of source criticism exhibited here owed much to humanist scholarship, but such ambitious projects were prompted by an urgent need to refute alternative Protestant histories, such as those of John Foxe in England or the 'Magdeburg Centuriators' in Germany.[17]

Although the writing of lives of the saints may have been led by the Bollandists, the institutional process of saint-making and supervision of cults was controlled by Rome as never before. The study of sanctity has lately enjoyed a rich and lively historiography, much of which has examined the complex process whereby Rome officially proclaimed an individual as a saint. Giovanni Papa has studied the organisation, personnel, definitions and case studies of 'canonisation' for the later sixteenth and early seventeenth centuries, when Urban VIII's reforms led to a wholesale tightening up of procedure.[18] As Peter Burke has argued in an influential essay, a study of Catholic-Reformation saint-making casts light not just on an important means by which the papacy strengthened its centralising grip on the Church after Trent, but also, more broadly, on the shifting religious priorities of the age, for which the favoured personifications of sanctity can be seen as barometers. Thus the canonisation of missionaries, bishops and founders of religious orders as opposed to lay men and women can be seen as reflecting the clericalist and masculine values of the Church; whilst the number of new saints chosen from the Mediterranean lands suggests that their societies were 'programmed' to perceive sanctity in certain stereotypical forms.[19] The study of saints therefore touches on local and popular religion. David Gentilcore's examination of the devotional culture of southern Apulia uses the evidence presented at canonisation trials as one important source for exploring the region's spiritual priorities.[20]

Much of the debate at Trent centred on the rights and responsibilities of the episcopate; and it was primarily to the bishops that the Council and subsequently the papacy looked to implement the Tridentine decrees, for example by founding seminaries, conducting visitations, and presenting edifying models of pastoral care, as St Carlo Borromeo was held to have done in his see of Milan. The study of the post-Tridentine episcopate is therefore essentially an historiography of the impact of the Council of Trent at the local level, a history that has usually been written in terms of success or failure.[21]

Among the institutions which the Church embraced alongside its diocesan hierarchy were the religious orders and congregations of priests, monks, friars and nuns. Historians have long credited the orders, especially the many new foundations of the sixteenth and seventeenth centuries, with a key role both in renewing the spirituality, devotional life and religious discipline of Catholic Europe, and also in spearheading missionary work overseas. Admittedly, much of this scholarship has been undertaken by scholars from among their own ranks and has been published in their own historical journals, such as the Jesuits' *Archivum Historicum Societatis Iesu*. Of all the religious orders, indeed, it is the Society of Jesus which has traditionally been most closely identified with the core values of the Catholic Reformation and which has consequently received the greatest attention. In the past these accounts were extremely divergent, ranging from the triumphalism of Jesuit apologists to the negative stereotypes concocted by their enemies.[22] Modern work by Jesuit and non-Jesuit historians has produced a much more dispassionate account of the Society's identity, methods and impact in a variety of fields and contexts. Despite continuing scholarly interest in Ignatius Loyola, the thrust of recent work has been on the structures and context of Jesuit activity.[23] A landmark collection of essays has illuminated the importance of Jesuit artistic, literary, scientific and philosophical endeavours, both in Europe and on the overseas missions. Here the Society is approached as a 'cultural ecosystem' or alternatively as a 'global corporation', its hierarchy, dynamism, multi-functionality and developed communication networks bearing comparison with modern multinational enterprises.[24]

The role of the Jesuits as significant art patrons has long been recognised, but new studies by Gauvin Bailey on the decorative schemes of the Jesuits' Roman churches in the late sixteenth century and by Jeffrey Chipps Smith on their artistic campaigns in Germany reveal the central importance which the Society attached to powerful visual imagery as a favoured vehicle to convey complex spiritual messages.[25] The corporate identity of the order also emerges from research on the traditionally controversial area of Jesuit political theory and practice. Robert Bireley has demonstrated the important, although neither unchallenged nor dominant, role played by Jesuits as confessors to Catholic monarchs; whilst political theory is the theme of Harro Höpfl's new study, suggesting a 'fit' between the order's internal culture of rigid discipline and obedience and the ideas expressed by Jesuit writers on the subject of secular politics. They advocated strong monarchy, opposed any toleration of non-Catholic minorities in Catholic countries and fostered an 'anti-Machiavellian' rhetoric of the compatibility of princely piety with political stability.[26] Given the scope of their activities it is easy to lose sight of the Jesuits' primary identity as promoters of spiritual renewal, despite a large literature on the foundation document of Jesuit spirituality, the *Spiritual Exercises*, reflecting a debate on whether the latter is best characterised as a 'medieval' text rooted in traditional piety, or as an innovative, 'modern' approach to religion.[27] Against a backdrop of older writing which saw the Jesuits as at the forefront of Catholic-Reformation zeal from the start, John O'Malley has stressed the almost accidental way in which the early Jesuits became involved in the systematic educational programme for which they would become justly famed.[28] Indeed, in a provocative essay, aiming in part to redefine the Catholic Reformation more broadly, O'Malley has questioned whether, despite their reputation, the early Jesuits can be considered 'reformers' in any conventional sense at all, stressing their relative independence from the institutional preoccupations of the Council of Trent.[29] The result of such research has not been to diminish the importance of the Jesuits, but rather to discard some of the exceptionalism which coloured older accounts, to contextualise their activities and to cast more light on their internal culture.

An equally decisive transformation can be seen in the historiography of the Roman and more particularly the Spanish and Portuguese Inquisitions, institutions which have also often been closely identified with the spirit of the Catholic Reformation. Inquisition studies have expanded immeasurably since the groundbreaking research of Henry Charles Lea in the early twentieth century. As Henry Kamen makes clear, this has meant dismissing the old 'black legend' of the tribunal as an all-powerful, unscrupulous and murderous machine of ideological and social repression. Certainly the Inquisition was sanguinary in its early period, when its principal victims were from among Spain's population of *conversos*, or Christianised Jews. However, by the mid-sixteenth century it had largely ceased to prosecute either *converso* or Protestant heresy and from the early seventeenth century concentrated on superstition, blasphemy and sexual immorality, which it penalised somewhat lightly compared with other judicial bodies in Spain and elsewhere.[30] Understaffed, the Inquisition made little headway across swathes of rural Spain and was always only one of a number of competing jurisdictions, finding itself, even in cases of witchcraft and superstition, vying with rival diocesan, state and baronial tribunals.[31] The history of institutions designed to defend religious orthodoxy is of course indissolubly connected to a history of religious dissidence or heterodoxy. Scholarship on the *Alumbrados* of sixteenth-century Spain, or the Jansenists of seventeenth- and eighteenth-century France demonstrates that a revisionist approach cannot mask the persecuting face of the Catholic Reformation, even while its aims and instruments are placed more firmly than ever in the general context of an intolerant age.[32]

the catholic reformation, society and belief

Much modern historiography has been concerned less with institutions and more with the relationship between religion and social change, and with everyday religious practice and belief. In part charged with the mission to gauge whether Trent succeeded or failed in making an impact on grassroots Catholicism, and under the influences of Marxism, the *Annales* school, social anthropology, gender studies and, latterly, post-structuralism, a rich new social and cultural history of the Catholic Reformation has developed. The *Annales* school's concept of a 'history of mentalities' has encouraged scholars to view belief not just as a set of doctrines but also as a cosmology and moral economy integral to the cultural framework of European society. This has encouraged the use of statistical analysis, compiled from parish registers, episcopal visitations and trial records, to measure shifting patterns of belief and even religious experience, as in Jean Delumeau's attempt to quantify the fear which Catholic Reformation preachers of hellfire instilled in their flocks.[33] Delumeau exhibits a further *Annaliste* trait, the tendency to view change over a long timescale, the *longue durée*; his best-known work on the Catholic Reformation, significantly bearing the baldly descriptive title *Catholicism between Luther and Voltaire*, spans the best part of three centuries.[34] In anglophone scholarship one of the most influential historians to draw on *Annaliste* currents has been John Bossy, whose work has powerfully demonstrated the profound social significance to early modern Europeans of the sacraments and the liturgy. In a groundbreaking essay of 1970 he argued that (like the Protestant Reformation) the Catholic Reformation constituted a major watershed in the social history of Europe. For Bossy, the clergy-dominated parochial structure mandated by the Council of Trent, with its enhanced regime of sacramental penance (epitomised by the confessional boxes disseminated by Archbishop Carlo Borromeo in late-sixteenth-century Milan), made for an individual and interiorised religion which destroyed traditional ties of kinship and collective sentiment and a communal moral economy, but which largely failed to establish new forms of affective religion to replace them.[35] It is fair to say that much of the work, at least of anglophone scholars, since 1970 has set out to test this nuanced conclusion of Tridentine success and failure.

In part the answers have been determined by the particular phenomena which scholars have chosen to study and the methodologies which they have employed. Historians of 'popular religion', for example, have tended to stress continuities across the period and to argue for the resilience of popular religious culture in the face of the disciplining (or 'acculturating') impulses from the official elite.[36] Perhaps the best example in the modern literature is Carlo Ginzburg's 1976 microhistory, *The Cheese and the Worms*, which uses inquisitorial trial records to reconstruct the idiosyncratic religious world-view assembled from oral tradition and a haphazard collection of written texts by a sixteenth-century North Italian miller, Domenico Scandella, known as Menocchio.[37] Ginzburg's approach to popular religion assumed the validity of a distinction between 'elite' and 'popular' which other historians have questioned, but such detailed case-studies, reconstructing the life of an individual or the social life of a small community either at a snapshot of time or over the *longue durée*, remain a fruitful way of exploring belief, where surviving sources permit. In the regional case study mentioned above, for example, David Gentilcore has used Inquisitorial records to describe the pattern of religious life in southern Italy as a 'system of the sacred'. In this analysis the religious beliefs and practices of the population are seen as encompassing a spectrum from those officially endorsed by the Church hierarchy to the unorthodox, including widespread recourse to magical rituals.[38] Taking the perspective of the locality permits the historian to present local communities not as the passive targets of an external movement, the 'Catholic Reformation', but as actors helping to shape that movement in a dynamic interplay between the (Roman) centre and the periphery. The example of social anthropology has also led to a new appreciation of the spatial dimensions of early modern Catholicism and has led to a renewed interest in the history of sacred sites. The study of wonder-working shrines and their attendant pilgrimages has provided a fertile area for examination of attitudes to the sacred and to beliefs about the body. Shrines, as David Lederer has shown, were centres of 'spiritual physic' and, as William Christian has argued, pre-eminent places for encounters between the everyday world and the sacred.[39] Shrines were important to the individual pilgrim, but since they were also established and maintained in response to collective vows, they functioned as part of the cultural glue providing cohesion in Catholic societies, arguably compensating for the loss of communal institutions and rites discussed by Bossy. Admittedly, after Trent, ecclesiastical authorities sought to control and manage such sites as never before, attempting to 'spiritualise' the experience of pilgrimage. Moreover, as Philip Soergel has shown for Bavaria, medieval pilgrimage shrines which had been waning in popularity at the time of the Reformation were reanimated by Jesuits and other clerics in the later sixteenth century, their alleged miracles providing ammunition in the propaganda campaign fought by the Catholic-Reformation elite against Protestantism.[40]

A similar story of tension between traditional communal religion and a newer, post-Tridentine clericalism emerges from extensive research, particularly in Italy, into religious guilds and confraternities.[41] Such organisations gained their distinctive identities most obviously from their association with such devotions as the Blessed Virgin Mary, the Guardian Angels or the so-called 'Holy Souls' suffering in Purgatory. Louis Châtellier has argued that the numerous confraternities established by the Jesuits, which were known as Marian sodalities or congregations from their dedication to the Virgin, exercised a particular appeal to the upper social strata of the cities of Catholic Europe and played a pivotal role in the inculcation of post-Tridentine religious deportment, extending a bourgeois culture and creating what he called the 'Europe of the devout'.[42] However confraternities also had explicit social, especially charitable, roles and functions and their history therefore slides into the history of Catholic education and welfare.[43] Indeed, the rededication of the Church to practical charity and its reorganisation of the means for delivering it was one hallmark of the Catholic Reformation. Much practice was a continuation of pre-Tridentine trends, but there were new features, including new religious orders, and new ideological imperatives, such as the anti-Semitic prejudices underlying the establishment of *monti di pietà*.[44] In Catholic Europe, charity extended to the dead and Holy Souls confraternities played a role in managing it. Given the salience of indulgences as a trigger for the Lutheran Reformation, it is surprising that the history of the indulgence system and beliefs in Purgatory in the century or so after Trent is still under-researched, although some recent work stresses the continuing importance of charity towards the dead in the moral economy, which found artistic expression in the sponsorship of indulgenced altars dedicated to the Holy Souls.[45]

Perhaps the most important recent departure in the social history of the Catholic Reformation is in questions of gender. With the growth of women's history and gender history since the 1970s, scholars have asked whether and to what extent women and men perceived, experienced and practised their religion differently, and have directed attention to traditionally ignored activities and spaces, including the domestic sphere and sites of female sociability and spirituality. This can mean a fresh approach to the history of persecution and intolerance, since some dissident groups, such as the *Alumbrados*, gave women an authoritative voice.[46] Within the mainstream, the focus has tended to be on the constraints imposed by male elites in the name of religion upon women's lives generally and their spiritual expression in particular. A reassertion of misogynist and patriarchal assumptions has been discerned by most historians, even if simultaneously some aspects of the Catholic Reformation, such as its encouragement of individual piety, private prayer and frequent examination of conscience, arguably resulted in a certain feminisation of its devotional culture.[47]

Practitioners of women's history with an interest in the Catholic Reformation have devoted much attention to the gendered religious experience of nuns, whose individual and institutional records bear testimony to a vigorous and in some respects distinctive spiritual life. This was despite their subjugation to stricter regimentation after the Council of Trent, which decreed that strict enclosure be imposed or reinstated in all convents, a measure which was both enforced and resisted with vigour.[48] Judith Brown's story of the sixteenth-century convent mystic, Sister Benedetta Carlini, shows how one nun's claims to sanctity were not vindicated by broader approval and in so doing highlights the behavioural norms expected of segregated female religious.[49] An important study of the best-known Spanish mystic, St Teresa of Ávila, has focused on the rhetorical strategies she adopted to secure both patronage and acceptance from an often hostile male ecclesiastical establishment.[50] The creativity of less famous nuns is glimpsed in Elissa Weaver's new study of convent theatre in Italy, which shows how drama provided 'spiritual fun' for the sisters and their audiences, who included laywomen and sometimes men, peering through the grilled windows into the nuns' enclosures. Despite the Catholic Reformation's campaign to curtail such fraternisation, it continued into the post-Tridentine period.[51]

Historiography has also investigated the shifting fortunes of gendered religious symbols. Next to Christ the most potent figure in the sixteenth-century Catholic imagination was the Virgin Mary, and the association of resurgent Marian veneration with Catholic-Reformation patriarchy has intrigued scholars. One recent study of sixteenth-century sermons on Mary charts a shift in emphasis in her portrayal from a medieval glorification of Mary's shared flesh with her son towards a greater stress on Mary's spiritual role as Christ's mother, embodying the 'virtues' of silence, passivity and obedience. It is argued that this reconfiguration of Marian symbolism both reflected the changing social position of women and also, given the power of symbols, helped to promote such change.[52] But it is clear that the Virgin was an ambiguous symbol. As the figurehead of the militant Catholic Reformation, her image decorated military standards, and as 'Our Lady of Victory' she was credited with the triumph of Catholic arms at Lepanto (1571) and the White Mountain (1620) against Turks and heretics respectively. Mary remained the most powerful intercessor, the most potent saint, whose shrines proliferated throughout the Catholic world, even as aspects of Marian doctrine, in particular the Immaculate Conception, continued to be debated by Catholic theologians.[53]

the catholic reformation and the state

Histories of the Catholic Reformation are replete with figures of European rulers, like the Habsburg Emperor Charles V or his son, Philip II of Spain, whose policies and actions furthered or impeded the cause of ecclesiastical reform. But the Catholic Reformation's political history is more than simply a chronicle of the deeds of princes. A more structural approach, focusing on the relationship between religious change and state formation, is found in the German historiographical paradigm of 'confessionalisation'. Developed by Ernst Walter Zeeden in the 1950s and subsequently elaborated by Heinz Schilling and Wolfgang Reinhard, this model holds that, even whilst they established competing doctrinal systems and rival visions of Christianity during the Reformation century, the Protestant and Catholic churches and the rulers who supported them adopted similar strategies and methods in order to achieve their goals. Propaganda, indoctrination and censorship not only forged new religious identities (creating, for example, self-conscious Roman Catholics) but also served the interests of the secular state. Confessionalisation fostered the unity of a government's subjects or citizens, instilled in them obedience to authority and 'social discipline' and helped to create the modern state with its expanded fiscal regimes, rationalised bureaucracies and enhanced means of social control. In Germany the Catholic Reformation was indeed often driven forward by the state, epitomised by severe Catholic princes like Duke Maximilian I of Bavaria (ruled 1598–1651) who enforced harsh punishments against adulterers and fornicators; cracked down on 'superstition' (his 'Witchcraft Mandate' of 1611 was the most comprehensive piece of witchcraft legislation anywhere, even if the witch craze in Bavaria was comparatively limited); decreed that every one of his subjects should carry a set of Rosary beads with them at all times; and fined or pilloried them for eating meat on Fridays or failing to genuflect at the tolling of the *Angelus*. Admittedly, as his Jesuit confessor later testified, Maximilian was as happy to flagellate his own hair-shirted back as he was to have his subjects whipped for religious laxity.[54] Naturally, such control was not achieved overnight (if indeed it was achieved at all), so the model posits a long-term analysis, extending into the eighteenth century.[55]

The confessionalisation hypothesis represents an ingenious attempt to interpret religion as a dynamic, modernising force. However it has had its critics, some arguing that it presents a rather one-dimensional image of religion, stripped of spirituality and emotion, others that it marginalises non-mainstream confessions and ignores religion's subversive potential, or the potential for pluralism to provide a basis for a stable polity. Attempts to apply it to areas of Europe beyond the Holy Roman Empire have often proven unsuccessful. Many historians have chosen to stress the limitations, rather than the power of early modern states.[56] So far, confessionalisation has not been attractive to anglophone scholars, even those writing on Germany. Marc Forster's important studies of the Catholic Reformation in the bishopric of Speyer and in southwest Germany stress the weakness of both secular and ecclesiastical authorities and their inability to implement far-reaching change, whether to institutional structures or to patterns of belief and practice. In Speyer, Forster shows resistance coming from local ecclesiastical bodies, such as cathedral chapters and monasteries.[57] In his broader study, power is seen to reside more with the lay communes which controlled the flavour of religious life in their localities, albeit whilst subscribing to an ideology of 'clericalism' which did place the priesthood at the heart of parochial religious life.[58] Generally one might argue that the confessionalisation model works less well for Catholic than for Protestant Europe, given that Catholic ecclesiastical institutions always retained a degree of independence from the state.

the catholic reformation and the world

One recent textbook of the Counter-Reformation is entitled *The World of Catholic Renewal*.[59] As the title implies, R. Po-chia Hsia's book gives almost as much emphasis to the expansion of early modern Catholic Christianity overseas and to its encounter with non-Christian civilisations in America, Africa and Asia as it does to developments within Europe. Whilst scholars (including Ranke) had always included some discussion of the overseas missions in their accounts, this often seemed almost an afterthought. Recent trends, however, stress the formation of early modern Catholicism as the first tradition to achieve the literal status of a *world* religion as well as the role of the missions in shaping Catholic Christianity back 'at home' in Europe. As David Brading puts it, writing of the Franciscan mission to Mexico in the 1520s:

> In Spain the renewal of religious enthusiasm preceded the Protestant Reformation and the conversion of the natives of New Spain must be counted among the chief works of that renewal. In any discussion of

the Catholic Church in the sixteenth century the establishment of the Mexican Church must thus occupy an important place.[60]

One of the first findings to emerge even from the older scholarship was a realisation of the great diversity of methods of attempted Christianisation and their varied success in different missionary contexts. For the Americas, the classic interpretation was the French scholar Robert Ricard's analysis of the evangelising efforts of the mid-sixteenth-century friars, *The Spiritual Conquest of Mexico* (1933).[61] Ricard distinguished between two missionary strategies, the 'tabula rasa', whereby the entire indigenous culture was seen as a corrupting force which had to be eradicated before the new faith could be imposed, and 'opportunistic preparation', whereby certain elements in the indigenous culture were retained by the missionaries as a point of departure for evangelisation. Ricard argued that the Mexican mission used both systems, noting the ethnographies researched by the Franciscans and their aspiration to foster a native clergy, but also their drive to eliminate 'idolatry' by destroying Aztec temples and outlawing traditional ceremonies. Generally, Ricard concluded, the friars held a low opinion of the intellectual and spiritual capacities of the 'Indians', albeit combined with a genuine – if patronising – desire for their well-being, for the preservation of their language, for their separation from Spanish settler society and, often, for their preservation from the harsh excesses of colonial rule.[62]

If Ricard wrote largely from the missionaries' perspective, more recent historiography has tried to recapture the indigenous experience, or to focus on the dynamics of encounter. For Mexico, Inga Clendinnen and Fernando Cervantes, although looking respectively at the sixteenth-century beginnings and the seventeenth-century consolidation of evangelisation, come to similar conclusions about the failure of the friars to inculcate a 'pure' notion of Christianity among the conquered populations. Rather, Christianity was itself partially assimilated into a pre-Columbian belief system, to produce, as Cervantes describes it, a situation where it seemed to be precisely the most ardently Christian 'Indians' who were the most devoted to the Devil.[63] But whilst language and cultural difference created obstacles to evangelisation everywhere, the trajectories of missionary endeavours, and the strategies adopted, varied greatly. Much of the literature on this theme relates to particular localities, but Gauvin Bailey's global study of Jesuit-sponsored art presents the missionary field as one of cultural encounter, producing a 'hybrid' or syncretic art which reflected the differing dynamics of cultural convergence in each mission context, whether in the Amazon, in Japan, in China, in Goa or any other site where the indefatigable Jesuits located their missions.[64]

conclusion

As John O'Malley's historiographical survey recalls, Jedin's neat 'Catholic-*and* Counter-Reformation' label has not settled the problem of definition and nomenclature. Recent English-language textbooks have avoided the older terms in favour of 'Catholic Renewal' (Hsia), the 'Refashioning of Catholicism' (Bireley) and O'Malley's own preferred 'Early Modern Catholicism', whilst Marc Forster invokes 'Catholic revival'.[65] Nonetheless, both Counter-Reformation and Catholic Reformation (and their non-English equivalents), as broad, handy descriptors, are probably likely to remain on the historiographical scene for a while yet. Perhaps the names are less contested these days because of the decline of the confessional point-scoring which distorted older treatments.

The historical movement that all these labels try to define continues to fascinate scholars, who are producing a literature ever broadening in its scope. Some of the new terms themselves indicate an expansion of the geographical range beyond Europe or an extension of the chronology to run throughout the entire early modern period. The *longue durée* appears to have won. Some of them indicate an even more important historiographical development, the broadening of the definition of religion as a subject of historical analysis. The history of the Catholic Reformation may still be fruitfully written as a history of ideas and of institutions, but it is now more commonly approached through a social and cultural history which concentrates on the study of a wide variety of religious experiences, practices and responses and is sensitive to distinctions of place, age, class and gender. As a result, the Council of Trent may appear less of a watershed, the papacy less dominant, local religion more resilient, and early modern Catholicism as a whole less monolithic.

further reading

In addition to John W. O'Malley, *Trent and All That: Renaming Catholicism in the Early Modern Era* (Cambridge, MA, 2000), the following historiographical surveys are valuable points of departure: David Gentilcore, 'Methods and Approaches in the Social History of the Counter-Reformation in Italy', *Social History* 17 (1992), 73–98; Craig Harline, 'Official Religion – Popular Religion in Recent Historiography of the Catholic Reformation', *Archive für Reformationsgeschichte*, 81 (1990), 239–62; and John W. O'Malley (ed.), *Catholicism in Early Modern History: a Guide to Research* (St Louis, 1988). David M. Luebke (ed.), *The Counter-Reformation: the Essential Readings* (Oxford, 1999) reproduces some of the landmark contributions to the historiography. Besides the invaluable textbooks by Bireley and Hsia already cited, which summarise previous research and provide useful bibliographies as well as offering important interpretive frameworks of their own, one may profit from: A. D. Wright, *The Counter Reformation: Catholic Europe and the Non-Christian World* (London, 1982) and Michael A. Mullett, *The Catholic Reformation* (London and New York, 1999). Martin D. W. Jones, *The Counter Reformation: Religion and Society in Early Modern Europe* (Cambridge, 1995) contains a collection of document extracts and commentary which reflect recent historiographical concerns.

notes

1. John W. O'Malley, *Trent and All That: Renaming Catholicism in the Early Modern Era* (Cambridge, MA, 2000).
2. Leopold von Ranke, *Die römischen Päpste in den letzten vier Jahrhunderten.* I have used the edition *Rankes Meisterwerke*, vol. 7 (Munich and Leipzig, 1915), pp. 443, 489.
3. On these developments, O'Malley, *Trent and All That*, esp. chapter 1.
4. Marcel Bataillon, *Erasme en Espagne* (Paris, 1937); Karl Brandi, *The Emperor Charles V* (London, 1939) (cf. German edition, 1937).
5. Erwin Iserloh, Joseph Glazik and Hubert Jedin, *History of the Church. Volume V: Reformation and Counter Reformation*, trans. Anselm Biggs and Peter W. Becker (New York, 1990).
6. H. Jedin, 'Catholic Reformation or Counter-Reformation?', trans. David Luebke, in David Luebke (ed.), *The Counter-Reformation: the Essential Readings* (Oxford, 1999).
7. H. Outram Evennett, *The Spirit of the Counter-Reformation* (Cambridge, 1968).
8. Henri Bremond, *Histoire littéraire du sentiment religieux en France depuis la fin des guerres de religion jusqua'à nos jours*, 11 vols (Paris, 1916–33).
9. Eric W. Cochrane, 'Counter-Reformation or Tridentine Reformation? Italy in the Age of Carlo Borromeo', in John M. Headley and John B. Tomaro (eds), *San Carlo Borromeo: Catholic Reform and Ecclesiastical Politics in the Second Half of the Sixteenth Century* (Washington, 1988).
10. Hubert Jedin, *A History of the Council of Trent*, trans. E. Graf, vol. I (London, 1957).
11. Paolo Prodi, *The Papal Prince*, trans. Susan Haskins (Cambridge, 1987).
12. Nicole Reinhardt, *Macht und Ohnmacht der Verflechtung. Rom und Bologna unter Paul V.* (Tübingen, 2000).
13. Giampiero Brunelli, *Soldati del papa: politica militare e nobiltà nello stato della chiesa (1560–1644)* (Rome, 2003).

14. For Italian art of the Catholic Reformation, including Rome, much of the framework of current discussion was established in Francis Haskell, *Patrons and Painters: a Study in the Relations Between Italian Art and Society in the Age of the Baroque* (London, 1963). For a summary of post-Haskell literature see Simon Ditchfield, '"In Search of Local Knowledge": Rewriting Early Modern Italian Religious History', *Cristianesimo nella storia*, 19 (1998), 255–96, esp. 272–6. More broadly, modern work on the theme ultimately owes its inspiration to Émile Mâle: see his *L'Art religieux de la fin du XVIe siècle du XVIIe siècle et du XVIIIe siècle. Étude sur l'iconographie après le Concile de Trente. Italie, France, Espagne, Flandres* (Paris, 1951). On connections between art and Catholic mysticism: Victor I. Stoichita, *Visionary Experience in the Golden Age of Spanish Art* (London, 1995).
15. Ditchfield, '"In Search of Local Knowledge"', esp. 266–72; cf. Barbara McClung Hallman, *Italian Cardinals, Reform, and the Church as Property, 1492–1563* (Berkeley, 1985).
16. E.g. Thomas F. Mayer, *Reginald Pole: Prince and Prophet* (Cambridge, 2000).
17. Simon Ditchfield, *Liturgy, Sanctity and History in Tridentine Italy: Pietro Maria Campi and the Preservation of the Particular* (Cambridge, 1995).
18. Giovanni Papa, *Le cause di canonizzazione nel primo periodo della congregazione dei riti (1588–1634)* (Rome, 2001). See also Simon Ditchfield, 'Sanctity in Early Modern Italy', *Journal of Ecclesiastical History*, 47 (1996), 98–112.
19. Peter Burke, 'How to be a Counter-Reformation Saint', in Kaspar von Greyerz (ed.), *Religion and Society in Early Modern Europe, 1500–1800* (London, 1984), reprinted in Burke, *The Historical Anthropology of Early Modern Italy* (Cambridge, 1987) and in Luebke, *Counter-Reformation.*
20. David Gentilcore, *From Bishop to Witch: the System of the Sacred in Early Modern Terra d'Otranto* (Manchester, 1992).
21. Headley and Tomaro, *San Carlo Borromeo*; Philip T. Hoffman, *Church and Community in the Diocese of Lyon, 1500–1789* (New Haven, 1984).
22. On the latter, Peter Burke, 'The Black Legend of the Jesuits: an Essay in the History of Social Stereotypes', in Simon Ditchfield (ed.), *Christianity and Community in the West* (Aldershot, 2001).
23. W. W. Meissner, *Ignatius Loyola: the Psychology of a Saint* (New Haven and London, 1992).
24. John W. O'Malley, Gauvin Alexander Bailey, Steven J. Harris, and T. Frank Kennedy (eds), *The Jesuits: Cultures, Sciences, and the Arts, 1540–1773* (Toronto, Buffalo and London, 1999).
25. Gauvin Alexander Bailey, *Between Renaissance and Baroque: Jesuit Art in Rome, 1565–1610* (Toronto, 2003); Jeffrey Chipps Smith, *Sensuous Worship: Jesuits and the Art of the Early Catholic Reformation in Germany* (Princeton, 2002).
26. Robert Bireley, *Religion and Politics in the Age of the Counter-Reformation: Emperor Ferdinand II, William Lamormaini, S.J., and the Formation of Imperial Policy* (Chapel Hill, 1981); Harro Höpfl, *Jesuit Political Thought: the Society of Jesus and the State, c. 1540–1630* (Cambridge, 2004).
27. Javier Melloni, *The Exercises of St Ignatius Loyola in the Western Tradition* (Leominster, 2000).
28. John W. O'Malley, *The First Jesuits* (Cambridge, MA, 1993).
29. John W. O'Malley, 'Was Ignatius Loyola a Church Reformer? How to Look at Early Modern Catholicism', *Catholic Historical Review*, 77 (1991), 177–93; reprinted in Luebke, *Counter-Reformation.*

30. The progress of revisionism can be charted from the successive editions of a study first published in 1965: Henry Kamen, *The Spanish Inquisition* (New Haven, 1998).
31. María Tausiet, *Pozoña en los ojos. Brujería y superstición en Aragón en el siglo XVI* (Zaragoza, 2000).
32. Alastair Hamilton, *Heresy and Mysticism in Sixteenth-Century Spain: the Alumbrados* (Cambridge, 1992); William Doyle, *Jansenism: Catholic Resistance to Authority from the Reformation to the French Revolution* (Basingstoke, 2000).
33. Jean Delumeau, *Sin and Fear: the Emergence of a Western Guilt Culture 13th–18th Centuries*, trans. Eric Nicholson (New York, 1990).
34. Jean Delumeau, *Catholicism between Luther and Voltaire: a New View of the Counter-Reformation*, trans. Jeremy Moiser (London, 1977).
35. John Bossy, 'The Counter-Reformation and the People of Catholic Europe', *Past and Present*, 47 (1970), 51–70; reprinted in Luebke, *Counter-Reformation*.
36. Louis Châtellier, *The Religion of the Poor: Rural Missions in Europe and the Formation of Modern Catholicism, c1500–c1800*, trans. Brian Pearce (Cambridge, 1997).
37. Carlo Ginzburg, *The Cheese and the Worms: the Cosmos of a Sixteenth-Century Miller*, trans. John Tedeschi and Anne Tedeschi (Baltimore, 1980).
38. Gentilcore, *From Bishop to Witch*.
39. David Lederer, 'Reforming the Spirit: Society, Madness, and Suicide in Central Europe 1517–1809' (PhD thesis, New York University, 1995); William Christian, *Local Religion in Sixteenth-Century Spain* (Princeton, 1981).
40. Philip Soergel, *Wondrous in his Saints: Counter-Reformation Propaganda in Bavaria* (Berkeley, 1993).
41. J. P. Donnelly and M. W. Maher (eds), *Confraternities and Catholic Reform in Italy, France and Spain* (Kirksville, MO, 1999).
42. Louis Châtellier, *The Europe of the Devout: the Catholic Reformation and the Formation of a New Society*, trans. Jean Birrell (Cambridge, 1989).
43. Maureen Flynn, *Sacred Charity: Confraternities and Social Welfare in Spain, 1400–1700* (Basingstoke, 1989); Sandra Cavallo, *Charity and Power in Early Modern Italy: Benefactors and their Motives in Turin, 1541–1789* (Cambridge, 1995).
44. C. Breshahan Menning, 'The Monte's "Monte": the Early Supporters of Florence's Monte di Pietà', *Sixteenth Century Journal*, 23 (1992), 661–76.
45. Christine Göttler, *Die Kunst des Fegefeuers nach der Reformation* (Mainz, 1996).
46. Mary E. Giles (ed.), *Women in the Inquisition: Spain and the New World* (Baltimore and London, 1999).
47. Sherrin Marshall (ed.), *Women in Reformation and Counter-Reformation Europe: Private and Public Worlds* (Bloomington and Indianapolis, 1989); Elizabeth Rapley, *The* Dévotes*: Women and Church in Seventeenth-Century France* (Montreal, 1990).
48. On this trend, Gabriella Zarri, 'Gender, Religious Institutions and Social Discipline: the Reform of the Regulars', in J. Brown and R. Davis (eds), *Gender and Society in Renaissance Italy* (London and New York, 1998).
49. Judith Brown, *Immodest Acts: the Life of a Lesbian Nun in Renaissance Italy* (Oxford, 1986).
50. Alison Weber, *Teresa of Avila and the Rhetoric of Femininity* (Princeton, 1990).

51. Elissa Weaver, *Convent Theatre in Early Modern Italy: Spiritual Fun and Learning for Women* (Cambridge, 2001).
52. Donna Spivey Ellington, *From Sacred Body to Angelic Soul: Understanding Mary in Late Medieval and Early Modern Europe* (Washington, 2001).
53. Patrick Preston, 'Cardinal Cajetan and Fra Ambrosius Catharinus on the Controversy over the Immaculate Conception of the Virgin in Italy, 1515–51', in R. N. Swanson (ed.), *The Church and Mary* (Woodbridge, 2004).
54. Dieter Albrecht, *Maximilian I. Von Bayern 1573–1651* (Munich, 1998); Wolfgang Behringer, *Witchcraft Persecutions in Bavaria*, trans. J. C. Grayson and David Lederer (Cambridge, 1997).
55. Wolfgang Reinhard and Heinz Schilling (eds), *Die katholische Konfessionalisierung. Wissenschaftliches Symposion der Gesellschaft zur Herausgabe des Corpus Catholicorum und des Vereins für Reformationsgeschichte 1993* (Münster, 1995); Wolfgang Reinhard, 'Reformation, Counter-Reformation, and the Early Modern State: a Reassessment', *Catholic Historical Review*, 75 (1989), 383–405, reprinted in Luebke, *Counter-Reformation*.
56. Ute Lotz-Heumann, 'The Concept of "Confessionalization": a Historiographical Paradigm in Dispute', *Memoria y Civilización*, 4 (2001), 93–114.
57. Marc Forster, *The Counter-Reformation in the Villages: Religion and Reform in the Bishopric of Speyer, 1560–1720* (Ithaca, 1992).
58. Marc Forster, *Catholic Revival in the Age of the Baroque: Religious Identity in Southwest Germany, 1550–1750* (Cambridge, 2001).
59. R. Po-chia Hsia, *The World of Catholic Renewal 1540–1770* (Cambridge, 1998).
60. D. A. Brading, *The First America* (Cambridge, 1991), p. 104.
61. Robert Ricard, *The Spiritual Conquest of Mexico: an Essay on the Apostolate and the Evangelizing Methods of the Mendicant Orders in New Spain 1523–1572*, trans. Lesley Byrd Simpson (Berkeley, 1966).
62. *Ibid.*, esp. pp. 283–95.
63. Fernando Cervantes, *The Devil in the New World* (New Haven, 1994); Inga Clendinnen, 'Franciscan Missionaries in Sixteenth-Century Mexico', in Jim Obelkevich, Lyndal Roper and Raphael Samuel (eds), *Disciplines of Faith. Studies in Religion, Politics and Patriarchy* (London, 1987).
64. Gauvin Alexander Bailey, *Art on the Jesuit Missions in Asia and Latin America* (Toronto, 1999).
65. Robert Bireley, *The Refashioning of Catholicism, 1450–1700* (Basingstoke, 1999).

10
anabaptism and religious radicalism

michael driedger

scholarship on the early sixteenth century

Many of the concepts we use today to understand the phenomena of religious nonconformity and dissent were also polemical terms of the Reformation era. The largest group of Protestant dissenters quickly became known as 'Anabaptists', and many major theologians of the sixteenth century composed invectives against them. These included Lutheran (Luther, Melanchthon, Menius), Reformed (Zwingli, Bullinger, Calvin), and Catholic (Cochläus, Eck, Faber) writers. Although today 'Anabaptist' has a mostly value-neutral meaning, the term was first used in English by early modern British polemicists, and it was the equivalent of the Continental names *Wiedertäufer* (German) and *wederdooper* (Dutch). Each emphasised the act of rebaptism as an adult, and rebaptism was abhorrent to the great majority of leading Christians of the period, for they considered child baptism to be the only acceptable practice. Many of these leaders believed that dissident baptisers of adults threatened the unity of the Christian community, a unity that the Bible seemed to require in texts like Ephesians 4:5 ('one Lord, one faith, one baptism'). There were also other important polemical terms. Catholics frequently classified all Protestant opponents of the Papacy as 'heretics' (*Ketzer* or *Häretiker*), while Lutherans were fond of the terms 'sects and gangs' (*Sekten und Rotten*) or 'enthusiasts' (*Schwärmer*) to speak about Christian deviants. Because of the diverse character of nonconforming individuals and groups, the polemical attacks amounted to attempts to impose an artificial uniformity on opponents as enemies of God and/or the social order. Polemicists referred frequently to the Peasants' War of 1525 and the period of Anabaptist rule in Münster in 1534–35 as evidence of their charges.

The polemical names had dangerous potential in the early modern period, because they also gained legal significance. The most famous examples are the Imperial decrees of 1528 and 1529, in which rebaptism became a capital crime in the Holy Roman Empire. On this basis hundreds were executed. In response to these and numerous other decrees, some nonconformists sought safety in obscurity, but not all shied away from dangers. Their reasons for putting themselves in danger were varied, but a common unifying feature was their concern to act in conformity with what they understood to be God's will communicated in the Bible or through the Holy Spirit. Some were strong-willed free-thinkers, or participated in short-lived movements of impatient activists who wanted to reform the Christian polity on the model of the early church. Others joined (or, in second or subsequent generations, were born into) groups whose history has become part of the lineage of modern denominations. Most prominent among these are the Mennonites from the Netherlands and northern Germany, the Hutterites from Moravia, and the Amish from Switzerland and the Alsace, all descendants of the first Anabaptists; but there are also Unitarian, Schwenckfeldian and Pietist traditions that have connections with radical nonconformists of the sixteenth century. As each tradition developed its own unique character, it also developed its own interpretation of its heritage. This process of self-definition included refutations of polemical attacks. Important examples of early modern apologetic literature include the Mennonite *Martyrs' Mirror*,[1] the Hutterite *Chronicle*,[2] and the Pietist *Unparteyische Kirchen- und Ketzer-Historie* (Impartial History of Churches and Heretics).[3] One of the common features of this literature is that its protagonists are portrayed as members of a suffering church who were willing to accept high personal costs for the preservation of truth and in obedience to God.

In other words, in the early modern period two opposing traditions of interpreting the significance of Anabaptism and related groups had already emerged: one polemical and one apologetic. These remained the dominant modes of historical writing until a few generations ago.

By the early twentieth century, German social philosophers had introduced new categories for thinking about Reformation-era religious nonconformity. These included the Marxist idea of the 'early bourgeois revolution'. In addition to emphasising social and economic conflict as the root of religious dissent, Marxist historians placed a special emphasis on the role of the preacher and Peasants' War activist Thomas Müntzer as a leading figure in the 'early bourgeois revolution'. While not an Anabaptist in the strict sense, Müntzer was an anti-paedobaptist, an opponent of child baptism. The Marxists' positive portrayal of religious radicals was a departure from the earlier traditions of interpretation, both polemical and apologetic, which saw Müntzer and the struggles of 1525 in merely negative terms.

Another important development in secular historiography was the scholarship of Max Weber and Ernst Troeltsch, both of whom paid close attention to the historical role they claimed Protestant nonconformists played in the emergence of modern economic, social and political relationships. They sought to balance a one-sided materialist approach to socio-historical questions by emphasising how ideas and beliefs shaped social relations. In this connection they gave the old polemical category 'sect' new, value-neutral meaning. For them 'sect' was an ideal type for those groups that organised themselves in self-disciplining, voluntary communities of believers. They distinguished the 'sect'-type from a 'church'-type of socio-religious organisation, and Troeltsch added a further distinction between these two categories and a 'mystic'-type of religious life.

Anabaptist studies were also particularly strong in the early twentieth century in the Netherlands, where toleration of the Mennonites since the late sixteenth century had produced the conditions for congregations to thrive, leaving long institutional legacies and highly trained scholars able and willing to chronicle them. The result was a unique set of Dutch-language debates about the character of early Dutch Anabaptism. On the one side were historians like Karel Vos[4] and A. F. Mellink,[5] who emphasised the radical aspirations of the earliest radical reformers. These men saw close connections between the early Melchiorite Anabaptism that reigned in Münster for a short period and the later Anabaptism of Menno Simons. On the other hand, historians like A. L. E. Verheyden[6] and W. J. Kühler,[7] the heirs of the apologetic tradition of early modern Mennonite writing, argued that it was necessary to make a clear distinction between Münsterite Anabaptists and Menno's followers.

Despite different perspectives and traditions, European and North American Mennonite scholars began meeting in the early twentieth century to plan cooperative projects. Among the fruits of these early meetings were the *Mennonitisches Lexikon* (4 vols, 1913–67) and the *Mennonite Encyclopedia* (5 vols, 1955–90). This was also the period in which two important journals were founded: *Mennonite Quarterly Review* (1927–), and *Mennonitische Geschichtsblätter* (1936–40, 1949–). They joined another already established periodical devoted to Anabaptist and radical history: *Doopsgezinde Bijdragen* (1861–1919; new series, 1975–). While these projects were spearheaded by historians affiliated with Mennonite churches, two important collections of primary sources, the *Täuferakten*

(16 vols, 1930–88)[8] and the *Documenta Anabaptistica Neerlandica* (8 vols, 1975–2002), were the work of scholars from a wide range of backgrounds. All of these journals and sources remain essential for research on sixteenth-century Anabaptists.

By the middle of the twentieth century Goshen College in Indiana had become a centre of Mennonite scholarship, under the leadership of Harold Bender. In 1944 Bender published an essay entitled 'The Anabaptist Vision'[9] which has proven to be one of the most influential works in the development of Anabaptist historiography, for it gave new life to the study of Anabaptism as a positive force in European history. While rejecting Marxist as well as Lutheran and Catholic interpretations which linked Anabaptism to revolutionary moments in the Protestant Reformation, Bender and his supporters received the work of Weber and Troeltsch positively, for they saw in it confirmation that Anabaptist ethics had had a constructive influence on the development of the modern western world, as well as a justification for clearly distinguishing 'sectarian' or separatist Anabaptists from other more 'mystically' oriented Protestant nonconformists.[10]

Ethics rather than academic theology were at the heart of Bender's definition of Anabaptism. He argued that the first Anabaptists, unlike mainstream reformers, had held steadfastly to the original biblical emphasis of the early Reformation. Bender identified three key ethical features shared by these determined evangelicals and their successors: discipleship, brotherhood, and non-resistance. These, he argued, formed the unchanging core of 'Anabaptism proper' or 'Evangelical Anabaptism', which he wanted to distinguish from the 'false' Anabaptism of those who participated in the Peasants' War of 1525 or the Anabaptist kingdom at Münster a decade later. And, in addition to revolutionary Anabaptists, Bender also excluded from his definition those Anabaptists who held unconventional apocalyptic, mystical and spiritualist beliefs. For Bender Zurich was the place where 'Anabaptism proper' had first emerged and from whence it spread.

Not only did Harold Bender die in 1962, but two books published that year also signalled that his arguments about Evangelical Anabaptism were facing challenges. One was Heinold Fast's *Der linke Flügel der Reformation* (The Left Wing of the Reformation).[11] Fast borrowed the metaphor of the 'left wing' from a 1941 essay by Roland Bainton,[12] and his book was a collection of sources grouped according to four categories of authors: Anabaptists, spiritualists, enthusiasts, and anti-Trinitarians. Fast's classification implied family relationships rather than clear distinctions. And in the first group he included not only writers who had been key in Bender's view, but also some like Balthasar Hubmaier and Hans Hut who were suspect for Bender.

The other 1962 book, George H. Williams' *The Radical Reformation*,[13] had a greater impact on historiography than Fast's. The first edition of *The Radical Reformation* is an encyclopaedic monograph of more than nine hundred pages. Williams had introduced the concept of 'Radical Reformation' a few years earlier in a collection of translated sources.[14] In a step that would have probably seemed highly questionable to those in Bender's circle, Williams named the Reformed, anti-Anabaptist polemicist Heinrich Bullinger as the first to recognise the unity of this branch of the Reformation. While Williams accepted Bullinger's argument that those outside of established Protestant and Catholic churches together made up a connected movement, he did not accept Bullinger's polemical interpretations. In effect, he proposed using the formal structure of old polemical arguments, but, like early modern apologists for the nonconformist causes, he inverted anti-sectarian value judgements, so that the marginalised were the heroes of his study.

Williams tried to draw together all the scholarship of recent decades, including the denominationally Mennonite scholarship that was so indebted to Bender, as well as the positive non-Mennonite 'free church' scholarship of the nineteenth and early twentieth centuries. Inspired in part by Troeltsch's typologies, he broke the Radical Reformation into three branches: supporters of adult baptism (Anabaptists), believers in direct inspiration from the Holy Spirit (Spiritualists), and speculators about immortality and the Trinity (Evangelical Rationalists). Williams argued that these groups, despite differences, together formed a tradition distinct from but as historically significant as the Catholic Counter-Reformation and the 'Magisterial Reformation' of mainline Protestants. According to Williams a shared set of ideals were what defined them as radicals. Among these ideals were the restitution of the apostolic church rather than the mere reform of existing institutions, the separation of church and state, a strong missionary impulse, a preference for sanctification rather than the doctrine of salvation by faith alone, an emphasis on freedom of the will, eschatological orientations, and a belief in the sleep of souls until the time of the Resurrection. Compared with Bender, Williams placed more emphasis on theology or ideas as opposed to ethics, and the geographical scope of his study was also much larger. While supporters of the Goshen School of historiography focused on ethical ideals they felt spread from Zurich throughout German-speaking and Dutch lands, Williams' Anabaptists were part of a much more diffuse historical movement which reached from Spain to Eastern Europe and from the British Isles to the Italian peninsula.

The 1960s was a time of transition in Anabaptist historiography. Other scholars interested in new historiographical approaches were starting careers in the 1960s and 1970s, and their work has since come to be known collectively as examples of a 'revisionist' mode of Anabaptist studies – revisionist because they challenged the 'normative' model of Anabaptism proposed by scholars like Bender. The earliest revisionist scholars were a very diverse group who began research careers independently of one another. When more of their work appeared in print in the early 1970s they started to recognise shared conclusions. Early key texts included Hans-Jürgen Goertz' 1967 study of Thomas Müntzer and three books published in 1972: Gottfried Seebass's *Habilitation* on Müntzer's disciple Hans Hut,[15] James M. Stayer's study of Anabaptist attitudes toward government authority and the use of force,[16] and Claus-Peter Clasen's social history of southern German, Swiss and Austrian Anabaptism.[17]

Like Williams, the critics of the Goshen School of historiography understood radical reformers like the anti-paedobaptist preachers Thomas Müntzer and Andreas Karlstadt to be central figures in the prehistory of Anabaptist groups in southern Germany, and also pointed to connections between Anabaptists and spiritualists. These kinds of connections made untenable the clear distinction Bender and his colleagues wanted to draw between 'true' pacifist Anabaptists and 'false' militant ones. For example, in *Anabaptists and the Sword*, James Stayer refuted Bender's claim that non-resistance was a defining feature of the first Anabaptists. Stayer's examination of Anabaptist attitudes toward authority and the use of force showed that these were difficult to reduce to one shared orientation. They ranged from the peaceful apocalyptic hopes of Melchior Hoffman, or the separatist non-resistance of the Swiss Brethren or Menno Simons; through the practical attempts of Balthasar Hubmaier to reform the polities of Waldshut and Nicolsburg on an Anabaptist model; to the militancy of Hans Hut in southern Germany or the Anabaptists of Münster in northern Germany. In other words, while acknowledging the important place of pacifism in early Anabaptist history, Stayer emphasised the diversity of early Anabaptist political orientations.

Among the other features of Stayer's book that have proven significant over the last several decades is his division of Anabaptism into three major regional cultures. Already in 1958 George Williams had argued that rather than having only one point of origin (Zurich), Anabaptism had at least three (Zurich, Nuremberg, and Amsterdam). In *Anabaptists and the Sword*, Stayer devoted individual sections to 'The Swiss Brethren',

'The Upper German Sects', and 'The Melchiorites' of northern Germany and the Netherlands. These regional distinctions for understanding Anabaptism were refined in an essay from 1975 co-authored by Stayer, Werner Packull, and Klaus Deppermann. The title, 'From Monogenesis to Polygenesis',[18] summed up the conclusion. At the time they wrote the essay Deppermann was completing a biography of Melchior Hoffman, and Packull was working on his study of *Mysticism and the Early South German-Austrian Anabaptist Movement, 1525–1531*.[19] Until the late 1990s their tripartite understanding of Anabaptist origins and cultures – including Swiss, southern German and Austrian, and northern German and Dutch branches – dominated revisionist studies. The ideal-typical, tripartite division helped highlight cultural differences between the groups: the Swiss tended to place a premium on congregational discipline on a New Testament model; the southern Germans and Austrians were strongly influenced by the mysticism of Thomas Müntzer; and northern Germans and Dutch believed more than other Anabaptists in the purely divine nature of Christ, a nature untainted by fleshly, human influences.

One of the noteworthy features of this portrait of Anabaptism was that its regional focus was much more narrow than Williams'. While the revisionists argued that spiritualist, mystical, apocalyptic and violent groups were a proper subject for scholars of early Anabaptism, they did tend to share an understanding with denominational Mennonite scholars that it was the Netherlands, Imperial German territories, Swiss cantons, and Austrian Habsburg realms that were the focus for their research. They largely ignored Italian, Spanish, English, and Polish radicals. As a consequence, debates about early Anabaptism became a sub-field of German Reformation studies. For this reason, the publication of excellent studies like *Venice's Hidden Enemies*,[20] which would have found a place in Williams' vision of the Radical Reformation, have received little attention in the mainstream of Anabaptist studies.

It is also noteworthy that the new generation of scholars on Anabaptism was much more cautious, at least initially, than early twentieth-century scholars to make grand claims about the world-historical significance of its subject. Rather than trying to trace the history of the early bourgeois revolution, the emergence of modernity, or the pedigree of a present-day church, the revisionists intended to understand Anabaptism on its own sixteenth-century terms. One study, Claus-Peter Clasen's *Anabaptism: a Social History* (1972), even suggested that the Anabaptists were a small and fractured group whose impact the authorities exaggerated but whose lasting historical significance was minor.

While more cautious in their conclusions, not all revisionists agreed with Clasen's assessment. By the later 1970s some were attempting to develop a fresh understanding of the place of Anabaptists in the larger narrative of the sixteenth century. Hans-Jürgen Goertz, the author of a 1967 study of Müntzer and the editor of *Umstrittenes Täufertum* (Disputed Anabaptism),[21] a collection of essays from 1975 by established and new historians of Anabaptism, was a leader in this regard. In his editor's introduction to *Profiles of Radical Reformers*,[22] which included twenty-one biographies and was originally published in German in 1978, he proposed a new way of thinking about radicalism. In contrast to Williams' definition of 'the Radical Reformation', Goertz did not try to outline an historical movement defined primarily by its members' beliefs. Instead of arguing that particular radical ideas were the defining characteristic of a distinct ecclesiastical tradition (Williams' suggestion), Goertz maintained that a rejection of the status quo in the ecclesiastical, social and political realms was what distinguished radicals from other reformers. In other words, radicalism in his view was a relative phenomenon. This definition avoided a major problem some critics saw in Williams' emphasis on radical doctrines: few if any of Williams' radicals held all the beliefs he listed as typical of the Radical Reformation. For example, rather than advocating the separation of church and state, many early radicals insisted on the total transformation of Christian society on a biblical model. Furthermore, Goertz's redefinition of Williams' category of Radical Reformation provided a way of understanding how figures as different as Thomas Müntzer and Michael Sattler were radical, for Müntzer's militancy and Sattler's pacifism were both active ways of rejecting the status quo.

In subsequent years Goertz continued to elaborate on his conception of radical reform. In *Pfaffenhass und groß Geschrei* (Hate of the Priests and Much Ado),[23] he argued that the concept of anticlericalism or its flip-side, laicism, was the key feature of Reformation radicalism. By anticlericalism, he meant not merely attacks on individual priests or specific institutions, but rather a campaign to remove the social, political and ecclesiastical privileges of the entire clerical estate; the clergy were no longer to be mediators between God and believers. Rather than applying the analytical category to Anabaptists and spiritualists alone, Goertz proposed that anticlericalism was a framework for understanding the original Reformation impulse for radical change. The Protestant rejection of the old Roman order was radical from the very beginning, according to Goertz. In other words, Goertz did not make as strong a distinction as Williams had between 'magisterial' and 'radical' Reformations.

What distinguished those groups and individuals that historians today recognise as the great radical reformers was that they maintained the original radical impulse from the Wittenberg movement of 1521–22, through the Peasants' War, into the 1530s and beyond. By doing so they defied political and ecclesiastical trends toward Protestant reintegration into (re-)established structures of authority. In short, by the late 1980s two competing meanings of Radical Reformation (or the Reformation's radicals) existed side by side. From Goertz's point of view, it makes more sense to speak of religious radicalism, radical reformers (who included Luther in the early 1520s), or radical reforming movements than of the Radical Reformation.

By the late 1980s and early 1990s studies on religious radicalism were plentiful and the field was gaining the attention of a broad group of Reformation historians. In 1989 authors from all over North America and Eastern as well as Western Europe published biographies to mark the 500th anniversary of Thomas Müntzer's birth. Another major development was an international colloquium held in Arizona in 1990 which was devoted to the theme of anticlericalism. The conference proceedings,[24] published in 1993, included contributions from forty scholars.

Even before the collapse of Soviet hegemony in 1989, East German scholars of the Reformation had been involved in an international dialogue with Western colleagues. While Marxist in vocabulary, their work tended not to be strongly ideological in character, and these Marxists also tended to take religion seriously as a factor motivating people's actions. Western European and North American scholars who, since the 1970s, had encouraged East–West discussions on Reformation radicalism included Peter Blickle, Thomas Brady and Robert Scribner. One important record of this dialogue is a small 1988 volume of essays entitled *Radical Tendencies in the Reformation: Divergent Perspectives*.[25]

These historians, Marxist and non-Marxist alike, argued that the Peasants' War of 1525 was an important turning point in the history of the Reformation. Many revisionists – Clasen being an exception – also maintained that the Peasants' War was central in the history of early Anabaptism in southern Germany, Switzerland and Austria. For example, in *The German Peasants' War and Anabaptist Community of Goods* (1991), Stayer highlighted how the Biblical theme of community of goods, which the rebels of 1525 had used in their critique of secular and ecclesiastical authority, also became a common preoccupation for Anabaptists after 1525. Nonetheless, he showed how the cultural peculiarities and historical experiences of groups in each of the three regions posited in the polygenesis model (Swiss, southern German and Austrian, and Dutch and northern German) were reflected in their unique interpretations of community of goods.

Stayer's arguments from 1991 are evidence of a willingness to go beyond the artificially clear typological position outlined in the 1975 essay 'From Monogenesis to Polygenesis'.[26] Another of the co-authors of that essay, Werner Packull, has also added nuance to its stance. In his study of *Hutterite Beginnings* (1995),[27] he detailed how Moravia became a meeting ground for refugee groups from across Swiss, southern German and Austrian regions. As they interacted, their beliefs and community structures took on new and unique forms, the most enduring of which were the Hutterite communes which were later transplanted to North America.

The original revisionists wanted to understand beliefs and actions in their concrete historical settings, and in this regard they shared interpretive concerns with Mennonite-affiliated historians. While many disagreed about specific conclusions, we should not exaggerate the methodological divide between members of both historiographical camps. Nonetheless, an all-too-common assumption among some (usually denominationally affiliated) experts on Anabaptism is that revisionist historians have been and still are unsympathetic to religion and their scholarship is therefore biased by the ideology of secular social history. In a recent study of the first Anabaptists in Zurich and environs,[28] Andrea Strübind, besides disputing arguments about a connection between Anabaptism and the Peasants' War, claims that the revisionist critique of the Bender school is an example of secular social history which does not take theology seriously – and without understanding theology, historians cannot understand religious people. In a long review of Strübind's book, Stayer praises her careful reading of sources, but rejects her characterisation of revisionists as a monolithic and ideologically secularist set of social historians. While critics since the 1970s have rejected the older style of Mennonite and free church historiography, he argues that they by no means have ignored a contextualised account of the role of religious belief in people's lives.[29] In other words, the methodological difference between Strübind's preferred approach to church history and that of the so-called revisionists is not all that great.

There are several further reasons for caution about an all-too-clear portrayal of distinctions between approaches to the study of Anabaptism. For example, readers should be aware of differences on substantial issues among historians labelled as revisionists, such as Clasen and Stayer.[30] Another example is that, just as there has been dialogue between Marxist and non-Marxist historians, so has there been (and continues to be) productive dialogue between non-Mennonite and Mennonite or free church historians. In a 1988 essay entitled 'The Easy Demise of the Normative Vision' Stayer noted the positive reception that he and fellow critics of Bender's style of history received in Mennonite circles.[31] After all, the *Mennonite Quarterly Review*, which Bender had edited until his death, was a welcoming venue for revisionist articles as well as responses to them. Of course, Mennonite and free church historians did not and do not constitute one separate and clearly defined group either. In an historiographical essay from 2002, the current editor of the *Mennonite Quarterly Review*, John Roth, discusses new work from the last two decades by historians such as Arnold Snyder and Denny Weaver.[32] These two scholars, both of whom teach at Mennonite colleges, represent divergent trends in Mennonite historiography of the Anabaptist past. Weaver's scholarship is characterised by an application of postmodernist arguments to early Anabaptist as well as contemporary Mennonite subjects. Snyder's *Anabaptist History and Theology* is an attempt to identify a common core of theological ideas shared by all early Anabaptists, while still acknowledging the importance of revisionist scholarship on religious radicalism.[33]

Among the other issues that stand out in recent discussions in the field of Anabaptist studies are the history of communication and women's history. The first theme, the history of communication, offers promising avenues for further research on early Anabaptism and radical religion. A good introduction to the subject is Arnold Snyder's 1991 essay, 'Orality, Literacy, and the Study of Anabaptism',[34] but numerous studies of martyr songs and polemics, most in German, are also important material. On the second theme, a number of important books have appeared since the early 1990s. These include Marion Kobelt-Groch's *Aufsässige Töchter Gottes* (Uppity Daughters of God);[35] Arnold Snyder and Linda Hubert Hecht's collection on *Profiles of Anabaptist Women*;[36] and Marlies Mattern's *Leben im Abseits* (Life on the Boundaries).[37] One of the insights that women's history has provided is that women were most noticeably active in radical circles in the first years of the Reformation in such roles as martyrs, prophetesses and preachers. Thereafter, they returned to traditional gender roles in their once radical communities. This contrast between early reforming and long-term trends deserves more attention.

a longer view

The early sixteenth century was heroic, tragic, revolutionary, and dynamic. In other words, it is full of drama, and this is one reason why it has been the subject of so many studies. But what happened thereafter? There have until recently been relatively few studies available in English in answer to questions like this.[38] Within the framework of Radical Reformation studies, both Williams and Stayer have suggested that we can speak about the end of radicalism in the late 1570s. For Williams there were a 'cluster of events around 1578 and 1579' which together marked the end point of the Radical Reformation.[39] While Stayer did not disagree about the timing, he placed more emphasis on the gradual transformation inevitable in the dynamics of radical movements. The effervescent energies, as he described them, of early radicalism slowly diminished as rulers and dissidents learned to coexist over the course of several generations.[40]

The historiography of confession-building or confessionalisation is a framework which promises to help us think about the process of gradual transformations that led Anabaptists and descendants of radical reformers to accept and be accepted in mainstream society. In the tradition of Ernst Walter Zeeden, scholars since the 1950s have charted the parallel Lutheran, Catholic and Calvinist strategies of institutionalising separate group identities after the 1530s. Since the late 1970s, Heinz Schilling and Wolfgang Reinhard have argued that this scholarship of confession-building should be expanded to become a model for understanding a pre-absolutist phase in European social, political and ecclesiastical history. Their influential thesis is that rulers used religion as a tool and the clergy as agents in the process of 'social discipline',[41] the process of educating and disciplining obedient populations of subjects and citizens. This definition has led R. Po-chia Hsia to characterise Anabaptists and other minority groups as outsiders in, and victims of, the process of confessionalisation.[42] The tendency in this branch of historiography has been to consider only those groups who were granted rights in the 1648 Treaties of Westphalia to be 'proper' protagonists in the processes of confessionalisation and social discipline. Nonetheless, Heinz Schilling, who has tirelessly promoted confessionalisation as an all-encompassing paradigm for understanding the development of modern forms of social and political organisation in post-Reformation Europe, has also suggested in an essay from 1995 that its application be extended to groups like Anabaptists and Jews.[43]

The pioneer in making the concept of confessionalisation useful to scholars of Anabaptism has been Hans-Jürgen Goertz. It was in essays from 1994 and 1995 that he first wrote explicitly about the new historiography of the post-Reformation era.[44] In his view Mennonites, the dominant group of later Anabaptists in the Netherlands and northern German territories, channelled the strong outward-oriented, anticlerical impulses of their radical forebears to an inward-oriented impulse to maintain strict group discipline – in the vocabulary of the Mennonites, to establish congregations 'without spot or wrinkle' (Ephesians 5:27). Instead of trying to change the world, they tried to control themselves. This shift occurred as mainstream Christian groups were becoming more confessionalised themselves. What was unique and important in Mennonite circles, argued Goertz, was that the later Anabaptist drive toward congregational order amounted to a form of social discipline in which rulers and their agents did not have to become involved. In other words, Goertz's description of Anabaptist social discipline is contrary to the standard model of confessionalisation. Mennonites, in striving to avoid the charge that they were descendants of Thomas Müntzer and the Anabaptists of Münster, became self-policing communities of 'conforming nonconformists'.[45] That is to say, they were willing to act in the interests of the state before being asked or forced to do so. In this way, they could preserve their distinctive character. They gave their religious nonconformity a politically conformist character. In short, Goertz posited a kind of self-directed rather than state-directed social discipline.

While Goertz provided the general outlines of a theory of Anabaptist confessionalisation, he did not suggest how the process may have progressed chronologically or regionally. However, since the mid-1990s a variety of regional studies on later Anabaptist groups have been published. These allow us more detailed insights into the processes of transformation.

The largest study of later Swiss Anabaptism in recent years is Hanspeter Jecker's book from 1994 on Basle and region from 1580 to the end of the seventeenth century. In Protestant Switzerland, Reformed governments maintained repressive measures against Anabaptists[46] even at a time when their Reformed co-religionists in the Netherlands were extending forms of toleration to Mennonites there. As a result, Swiss Anabaptists were forced to the peripheries of society. In Jecker's presentation, Anabaptists around Basle therefore had little chance to develop the clear institutional character that was typical of northern European Mennonites and other mainstream confessional groups in the early modern period. While he did not write about confessionalisation, he did lend qualified support for Goertz's arguments about social discipline. He posited that Swiss Anabaptists' emphasis on leading morally blameless lives had farther-reaching repercussions, for when faced with this example from their ostensible enemies, Reformed governments felt compelled to act in an ethically more rigorous manner, or at least to enforce such a lifestyle on the general Calvinist population. This in effect would be an example not so much of social discipline 'from above' coordinated by the state, but rather pressure for social regulation spurred on 'from below' by subjects of the state, in this case, marginalised Anabaptists.[47]

Among new studies of Mennonites or *doopsgezinden* ('the baptism-minded', a Dutch term Anabaptists in the Netherlands used to describe themselves) is Samme Zijlstra's book from 2000 on the period from the early sixteenth to the later seventeenth century.[48] This long book provides a detailed examination of all aspects of post-Reformation Anabaptist life in a region where the Anabaptist population, although still a minority, was numerically and culturally of great significance. One of Zijlstra's key concerns is to emphasise the diverse institutional, social and religious character of early modern Dutch Anabaptism. By the later sixteenth century, *doopsgezinde* groups also included Waterlanders, Frisians, Flemish, High Germans, and many others. Despite the names, ethnic differences between the groups were not as important as their separate institutions and different attitudes toward ethical rigour and theology. The history of these groups provides good examples of confessionalisation as defined by Ernst Walter Zeeden – a definition focusing on the organisational and dogmatic consolidation of religious communities more than on social discipline.

The factionalism among early modern *doopsgezinden* unsettled many congregational leaders. One common response was for Anabaptist leaders to formulate confessions of faith which codified their groups' attitudes toward important issues such as the nature of God, congregational governance, marriage, church discipline, oath swearing, the use of force, obedience to the authorities, and beliefs about resurrection of souls after death.[49] In fact, Mennonites wrote numerous documents of this sort in the period from the late sixteenth century to the early eighteenth century, and many were published in multiple editions. Both Karl Koop and I have written about Mennonites and their uses of confessions of faith.[50] One of my main conclusions in *Obedient Heretics* (2002) is that the significance of these kinds of identity-defining texts needs to be understood in the context of polemical exchanges, both between opposing Mennonite groups and between Mennonites and their opponents in other churches. I conclude that a key motivation for Mennonite community self-regulation was the maintenance of a positive public image in the face of internal disputes and anti-Anabaptist polemics. To defend their religious nonconformity Mennonite leaders felt obliged to make political conformity a defining feature of their communities.

Another aspect of early modern Anabaptist identity formation was the development of a distinct martyrological tradition, interest in which has been revived by Brad Gregory's *Salvation at Stake*.[51] More recently, Gregory has edited a source collection on the subject: *The Forgotten Writings of the Mennonite Martyrs* (2002).[52] Why did the sources in this collection fall into neglect? Gregory argues that responses on the part of early seventeenth-century martyrology editors to changing historical circumstances in the Netherlands are the main reason. Martyr stories touching on issues which had been at the heart of late sixteenth-century Mennonite schisms were seen as counterproductive amid efforts to heal factionalism. Furthermore, martyrs' songs, which were so popular among Anabaptists in the sixteenth century, fell out of favour in the Dutch Golden Age. They had served the function of unifying believers in times of suffering, but diminished in significance and disappeared from memory in the much more tolerant and economically prosperous circumstances of the seventeenth century. This is a fascinating example of how the character and the use of sources shift with historical circumstances.

Andrea Chudaska's new book from 2003 on Hutterite identity formation is the best long study so far to examine the development toward de-radicalised Anabaptism in the course of the first half of the sixteenth century.[53] The Hutterites emerged as the most successful of groups to unite Anabaptist refugees from Switzerland, southern Germany, and Austria in the early to mid sixteenth century. Chudaska focuses on the career of Pieter Riedemann, a key leader among the Hutterites after the mid-1530s. She describes his early work as a 'theology in transition'. In it he united standard evangelical ideas like *sola gratia* with a spiritualist orientation, an emphasis on believers' baptism and other elements typical of many radical reformers. Chudaska sees in his writing of the 1540s the contours of a uniquely Hutterite theology and group identity. Her main explanation for the development of Hutterite confessional identity focuses on the pressures to establish a clear, de-radicalised group profile in debates and encounters with other Anabaptist groups in changing historical circumstances. She too finds this to have been more important than the government regulation hypothesised in the standard models of confessionalisation and social discipline. Her conclusion about the end of Anabaptist radicalism is important because it locates this development in an earlier period than suggested by Stayer and Williams. Astrid von Schlachta's book from 2003 on Hutterite life and community organisation from the mid-sixteenth to seventeenth century adds further depth to our understanding of confessionalised, institutionalised Anabaptism in east-central Europe.[54]

This summary of recent studies allows a general conclusion about the applicability of confessionalisation to the study of early modern Anabaptism. Especially in northern and central Europe, socio-political conditions provided Anabaptists with an opportunity to create lasting forms of community which show strong parallels with developments in other confessions. At the same time, research suggests that models of social discipline which assume a central place for the state as a leading protagonist do not properly describe the processes involved in the emergence and maintenance of institutionalised and socio-politically obedient forms of Anabaptism. Of greater significance were competition among Anabaptist groups, polemical pressures created by clergymen from other confessions, and ethical impulses inherent in the Anabaptist way of life.

When we look beyond the phenomenon of Anabaptism, one of the most important studies which examines later sixteenth- and seventeenth-century religious radicalism is Leszek Kolakowski's *Chrétiens sans église* (Christians without Churches) (1965).[55] Kolakowski, a Polish expatriate philosopher, examined the competing drives so common among religious free-thinkers: on the one hand to want to break free from the limits set by established ecclesiastical institutions, and on the other to need to re-establish new institutional structures in place of the old ones they had abandoned. Although the radical energies unleashed during the Reformation eventually died away, there was no shortage of continental European material for Kolakowski to study. The Netherlands in particular was populated with religious free-thinkers. Andrew Fix's study of Dutch Collegiants[56] and Jonathan Israel's massive tome on the Radical Enlightenment[57] are two further examples of contributions to this field. Other important monographs are the more than twenty volumes of the *Bibliotheca Dissidentium* (1980–), a series on sixteenth- and seventeenth-century dissenters from all across continental Europe. And, of course, the English Quakers provide another wonderful seventeenth-century example of the transformation of an originally radical religious movement into an established set of communities. Finally, to return to the subject of Anabaptism, it is worth noting that Mennonite preachers were among the leading ideologues and organisers of the republican revolution against the Orange regime in the Netherlands at the end of the eighteenth century.[58] Politically obedient Anabaptism may have been the early modern norm, but there were important exceptions.

A general lesson we can draw from these studies is that radical religion may have diminished as a factor in European history after the middle of the sixteenth century but it was not an entirely spent force. As new orthodoxies established themselves, new forces of renewal and opposition gathered energy. This dynamic deserves more attention in future studies.

further reading

The literature on a subject such as religious radicalism is of course vast. See my essay and notes for references to influential studies such as: Harold Bender's 'The Anabaptist Vision'; George Williams' *The Radical Reformation*; Leszek Kolakowski's *Chrétiens sans église*; James Stayer's *Anabaptists and the Sword* and *The German Peasants' War and Anabaptist Community of Goods*; Deppermann, Packull and Stayer's 'From Monogenesis to Polygenesis'; Werner Packull's *Mysticism and the Early South German-Austrian Anabaptist Movement*; and Hans-Jürgen Goertz's *Pfaffenhass und groß Geschrei*.

For more recent scholarship on the radicalism of the Reformation, an important collection of essays is Hans-Jürgen Goertz and James M. Stayer (eds), *Radikalität und Dissent im 16. Jahrhundert / Radicalism and Dissent in the Sixteenth Century* (*Zeitschrift für historische Forschung*, Beiheft 27: Berlin, 2002).

Also see Euan Cameron, 'Dissent and Heresy', and Hans-Jürgen Goertz, 'Radical Religiosity in the German Reformation', both in R. Po-chia Hsia (ed.), *A Companion to the Reformation World* (Malden, MA, 2004). An older survey of the subject which is exemplary for its long-term perspective, and which also pays greater attention to British as well as continental European subjects, is Michael Mullett, *Radical Religious Movements in Early Modern Europe* (London, 1980). The phenomenon of Reformation-era religious radicalism is best understood not only in its early modern but also in its medieval context, on which see the excellent and influential work on 'textual communities' in Brian Stock, *The Implications of Literacy: Written Language and Models of Interpretation in the Eleventh and Twelfth Centuries* (Princeton, 1983).

For surveys of Anabaptist and Mennonite history, see Hans-Jürgen Goertz, *The Anabaptists*, trans. Trevor Johnson (New York, 1996), and Cornelius J. Dyck, *An Introduction to Mennonite History: a Popular History of the Anabaptists and Mennonites*, 3rd edn (Scottdale, PA, 1993).

One way to get a good sense of the current state of research is to read the essays in *Festschrifts* for three leading scholars: George Williams, James Stayer and Werner Packull. See Rodney L. Petersen and Calvin A. Pater (eds), *The Contentious Triangle: Church, State, and University* (Kirksville, MO, 1999); Werner O. Packull and Geoffrey L. Dipple (eds), *Radical Reformation Studies* (Aldershot: Ashgate, 1999); C. Arnold Snyder (ed.), *Commoners and Community* (Scottdale, PA, and Kitchener, Ontario, 2002).

An important series of sources in translation is *The Classics of the Radical Reformation* (1973–). For selected sources from important German radicals of the 1520s, see Michael Baylor (ed.), *The Radical Reformation* (Cambridge,

1991). Probably the best-known seventeenth-century Mennonite source is Thieleman Jansz van Braght, *The Bloody Theater or Martyrs Mirror of the Defenseless Christians*, trans. Joseph F. Sohm, 11th edn (Scottdale, PA, 1977).

notes

1. Thieleman Jansz van Braght, *The Bloody Theater or Martyrs Mirror of the Defenseless Christians*, trans. Joseph F. Sohm, 11th edn (Scottdale, PA, 1977). It was first published in Dutch in 1660.
2. *The Chronicle of the Hutterian Brethren, 1525–1665*, vol. I (Rifton, NY, 1987).
3. Gottfried Arnold, *Unparteyische Kirchen- und Ketzer-Historie* (1699–1700).
4. Karel Vos, 'Revolutionary Reformation', in J. M. Stayer and W. O. Packull (eds and trans.), *The Anabaptists and Thomas Müntzer* (Dubuque and Toronto, 1980).
5. Albert Fredrik Mellink, *De Wederdopers in de Noordelijke Nederlanden, 1531–1544* (Groningen, 1954).
6. A. L. E. Verheyden, *Anabaptism in Flanders, 1530–1650: a Century of Struggle*, trans. M. Kuitse and J. Matthijssen (Scottdale, PA, 1961).
7. W. J. Kühler, 'Anabaptism in the Netherlands', in Stayer and Packull (eds), *Anabaptists and Thomas Müntzer.*
8. *Quellen zur Geschichte der Täufer* (*der Wiedertäufer* in the first two volumes).
9. Harold S. Bender, 'The Anabaptist Vision', *Church History*, 13 (1944), 3–24; also in *Mennonite Quarterly Review*, 18 (1944), 67–88.
10. Guy Hershberger, 'Introduction', in his *The Recovery of the Anabaptist Vision: a Sixtieth Anniversary Tribute to Harold S. Bender* (Scottdale, PA, 1957), p. 5.
11. Heinold Fast, *Der linke Flügel der Reformation: Glaubenszeugnisse der Täufer, Spiritualisten, Schwärmer und Antitrinitarier* (Bremen, 1962).
12. Roland Bainton, 'The Left Wing of the Reformation', *Journal of Religion*, 21 (1941), 124–34.
13. George Huntston Williams, *The Radical Reformation* (Philadelphia, 1962).
14. George H. Williams and Angel M. Mergal (eds), *Spiritual and Anabaptist Writers: Documents Illustrative of the Radical Reformation* (London, 1957).
15. Although released originally in 1972, the text was not published as a book until recently: Gottfried Seebass, *Müntzers Erbe. Werk, Leben und Theologie des Hans Hut* (*Quellen und Forschungen zur Reformationsgeschichte*, vol. 73: Gütersloh, 2002). This includes a concluding essay in which Seebass reviews literature on the subject published since 1972.
16. James M. Stayer, *Anabaptists and the Sword* (Lawrence, KS, 1972).
17. Claus-Peter Clasen, *Anabaptism: a Social History, 1528–1618* (Ithaca and London, 1972).
18. Klaus Deppermann, Werner Packull and James Stayer, 'From Monogenesis to Polygenesis: the Historical Discussion of Anabaptist Origins', *Mennonite Quarterly Review*, 49 (1975), 83–121.
19. Werner O. Packull, *Mysticism and the Early South German-Austrian Anabaptist Movement, 1525–1531* (Scottdale, PA, 1977).
20. John Martin, *Venice's Hidden Enemies: Italian Heretics in a Renaissance City* (Berkeley, CA, 1993).

21. Hans-Jürgen Goertz (ed.), *Umstrittenes Täufertum*, 2nd edn (Göttingen, 1977).
22. Walter Klaassen, 'Preface', in Hans-Jürgen Goertz (ed.), *Profiles of Radical Reformers: Biographical Sketches from Thomas Müntzer to Paracelsus* (Scottdale, PA, 1982).
23. Hans-Jürgen Goertz, *Pfaffenhass und groß Geschrei. Die reformatorischen Bewegungen in Deutschland (1517–1529)* (Munich, 1987).
24. Peter A. Dykema and Heiko A. Oberman (eds), *Anticlericalism in Late Medieval and Early Modern Europe* (Leiden, 1993).
25. Hans J. Hillerbrand (ed.), *Radical Tendencies in the Reformation: Divergent Perspectives* (Kirksville, MO, 1988). Contributors to the volume included Adolf Laube, Sigrid Looss, Siegfried Hoyer and Günter Vogler from East Germany, as well as Hans Hillerbrand, Eric Gritsch, and James Stayer.
26. Stayer summarises his revised views in 'The Radical Reformation', in Thomas A. Brady, Heiko A. Oberman and James D. Tracy (eds), *Handbook of European History, 1400–1600*, 2 vols (Leiden, New York and Cologne, 1995), vol. II.
27. Werner O. Packull, *Hutterite Beginnings: Communitarian Experiments during the Reformation* (Baltimore and London, 1995).
28. Andrea Strübind, *Eifriger als Zwingli. Die frühe Täuferbewegung in der Schweiz* (Berlin, 2003).
29. James M. Stayer, 'A New Paradigm in Anabaptist-Mennonite Historiography?', *Mennonite Quarterly Review*, 78 (2004), 297–307; cf. Strübind's response, *ibid.*, 308–13.
30. See James M. Stayer, 'Numbers in Anabaptist Research', in C. Arnold Snyder (ed.), *Commoners and Community* (Scottdale, PA, and Kitchener, Ontario, 2002).
31. James M. Stayer, 'The Easy Demise of the Normative Vision of Anabaptism', in Calvin Redekop and Samuel Steiner (eds), *Mennonite Identity: Historical and Contemporary Perspectives* (Waterloo, Ontario, 1988).
32. John D. Roth, 'Recent Currents in the Historiography of the Radical Reformation', *Church History*, 71 (2002), 523–35.
33. C. Arnold Snyder, *Anabaptist History and Theology: an Introduction* (Scottdale, PA, and Kitchener, Ontario, 1995).
34. C. Arnold Snyder, 'Orality, Literacy, and the Study of Anabaptism', *Mennonite Quarterly Review*, 65 (1991), 371–92.
35. Marion Kobelt-Groch, *Aufsässige Töchter Gottes. Frauen im Bauernkrieg und in den Täuferbewegungen* (Frankfurt, 1993).
36. C. Arnold Snyder and Linda Hubert Hecht (eds), *Profiles of Anabaptist Women: Sixteenth-Century Reforming Pioneers* (Waterloo, Ontario, 1996).
37. Marlies Mattern, *Leben im Abseits: Frauen und Männer im Täufertum (1525–1550). Eine Studie zur Alltagsgeschichte* (Frankfurt, 1998).
38. One notable exception is Robert Friedmann, *Mennonite Piety through the Centuries: Its Genius and Its Literature* (Goshen, 1949).
39. George H. Williams, 'The Radical Reformation', in Hans J. Hillerbrand (ed.), *Encyclopedia of the Reformation*, vol. III (Oxford, 1996), p. 383.
40. James M. Stayer, 'The Passing of the Radical Moment in the Radical Reformation', *Mennonite Quarterly Review*, 71 (1997), 147–52.
41. Gerhard Oestreich, 'The Structure of the Absolute State', in Brigitta Oestreich and H. G. Königsberger (eds), *Neostoicism and the Early Modern State*, trans. David McLintock (Cambridge, 1982); Winfried Schulze, 'Gerhard Oestreichs

Begriff "Sozialdisziplinierung in der Frühen Neuzeit"', *Zeitschrift für Historische Forschung*, 14 (1987), 265–302.

42. R. Po-chia Hsia, *Social Discipline in the Reformation: Central Europe, 1550–1750* (London, 1989), p. 168.
43. Heinz Schilling, 'Die Konfessionalisierung von Kirche, Staat und Gesellschaft', in Wolfgang Reinhard and Heinz Schilling (eds), *Die katholische Konfessionalisierung* (Münster, 1995).
44. Hans-Jürgen Goertz, 'Zucht und Ordnung in nonkonformistischer Manier', in his *Antiklerikalismus und Reformation. Sozialgeschichtliche Untersuchungen* (Göttingen, 1995); idem, 'Kleruskritik, Kirchenzucht und Sozialdisziplinierung in den täuferischen Bewegungen der Frühen Neuzeit', in Heinz Schilling (ed.), *Kirchenzucht und Sozialdisziplinierung im frühneuzeitlichen Europa* (Berlin, 1994).
45. The term originates in discussions in the early 1990s between Goertz and myself.
46. See, for example, Mark Furner, 'The Repression and Survival of Anabaptism in the Emmental, 1659–1743' (PhD thesis, University of Cambridge, 1998).
47. Hanspeter Jecker, *Ketzer, Rebellen, Heilige. Das Basler Täufertum von 1580–1700* (Liestal, 1998), esp. pp. 608–12.
48. Samme Zijlstra, *Om de ware gemeente en de oude gronden. Geschiedenis van de dopersen in de Nederlanden 1531–1675* (Hilversum, 2000).
49. Several of these confessions are available in English translation in the preface to the *Martyrs Mirror*.
50. Karl Koop, *Anabaptist-Mennonite Confessions of Faith: the Development of a Tradition* (Scottdale, PA, and Waterloo, Ontario, 2003); Michael Driedger, *Obedient Heretics: Mennonite Identities in Lutheran Hamburg and Altona during the Confessional Age* (Aldershot, 2002).
51. Brad S. Gregory, *Salvation at Stake: Christian Martyrdom in Early Modern Europe* (Cambridge, MA, 1999).
52. Brad S. Gregory (ed.), *The Forgotten Writings of the Mennonite Martyrs* (*Documenta Anabaptistica Neerlandica*, vol. 8: Leiden, 2002).
53. Andrea Chudaska, *Pieter Riedemann: Konfessionsbildendes Täufertum im 16. Jahrhundert* (Gütersloh, 2003).
54. Astrid von Schlachta, *Hutterische Konfession und Tradition: Etabliertes Leben zwischen Ordnung und Ambivalenz* (Mainz, 2003).
55. Leszek Kolakowski, *Chrétiens sans église: la conscience religieuse et le lien confessionnel au XVIIe siècle*, trans. Anna Posner (Paris, 1969) (cf. Polish edition of 1965); cf. Leszek Kolakowski, 'Dutch Seventeenth-Century Anticonfessional Ideas and Rational Religion: the Mennonite, Collegiant and Spinozan Connections', trans. and ed. James Satterwhite, *Mennonite Quarterly Review*, 64 (1990), 259–97, 385–416.
56. Andrew Fix, *Prophecy and Reason: the Dutch Collegiants in the Early Enlightenment* (Princeton, 1991); idem, *Fallen Angels: Balthasar Bekker, Spirit Belief, and Confessionalism in the Seventeenth-Century Dutch Republic* (Dordrecht, 1999).
57. Jonathan Israel, *Radical Enlightenment: Philosophy and the Making of Modernity, 1650–1750* (Oxford, 2001).
58. Michael D. Driedger, 'An Article Missing from the *Mennonite Encyclopedia*: "The Enlightenment in the Netherlands"', in Snyder (ed.), *Commoners and Community*.

11
popular religion

philip m. soergel

Despite recent trends in cultural criticism and theory most historians today still strive to write histories that are faithful to their sources and in which their own contemporary, parochial concerns are kept firmly in check. Historiography, in this regard, is a great tool for keeping scholars honest, because it can often reveal the limits of our abilities to achieve that goal. Historiography fascinates us, in other words, because it shows just how often we have fallen short of the mark of impartial detachment. The study of popular religion, a comparatively young sub-discipline within early modern history, has not been exempt from these problems. Certainly, our knowledge of the ways in which average people practised the Christian religion in the early modern period is far greater today than it was a century ago. Yet a closer examination of the many studies that have treated 'popular religion' since the Second World War reveals all the same that historians have often been influenced by contemporary concerns, rather than by the demand of recreating and analysing the past faithfully.

Seen from hindsight's perspective, several distinct periods are evident in the historiography treating early modern popular religion. Since the 1940s three overlapping, yet distinct generations of scholars have written on this subject. In each of these periods, the questions that historians asked were inspired to a degree by larger social and political realities and intellectual debates in their time. During the 1940s French scholars first began to examine the ways in which Christianity was lived and practised in the late medieval and early modern world. They turned away from what they felt was a long-standing prejudice that had privileged the ideas and actions of theologians and elite churchmen.[1] Yet many of those who were interested in writing a history of what they called *la religion vecue*, 'lived religion', were prompted by fears: fears about the decline of Christianity and about the inexorable course of secularisation in twentieth-century France.[2] Somewhat later, in the more secularised 1960s and 1970s, the *Annales* school's call for a history that was written 'from the bottom up', as well as liberal democratic and Marxist movements, affected debates about popular religion.[3] Where an earlier generation of scholars had looked at 'lived religion' as it was practised within the parish context, scholars of this 'second generation' broadened their focus, concentrating on a range of customs, beliefs, folklore, and rituals that had existed both inside and beyond the confines of the official Church. They reached out to anthropology and other social theory, too, as they hoped to gain a better picture of how popular religion may have been implicated in the great transformations that had dissolved traditional European perceptions and given rise to modernity.[4]

Since 1980, though, there has been a reaction against many of the great synthetic treatments that appeared during this 'second generation' and many scholars of today's 'third generation' have grown increasingly sceptical about the concept of popular religion altogether. Many now doubt that popular religion – as a category of experience separate from the official teachings of the Church or from the beliefs of theological elites – can ever be reliably isolated or described. Yet despite such doubts, the concerns that early modern historians have long had with exploring the contours of normative religious beliefs, customs, and practices in their period persist. If most scholars no longer talk about a 'popular religion' that was distinct from 'official' Church teaching they remain as concerned as ever to explore the ways in which lay people interacted with the clergy, even as they are defining with increasing precision the impact that regionalism, class, ethnicity, and a host of other factors played in conditioning Christian practices in the early modern world. One can see, then, reflected in the current historiography treating early modern religious practices, echoes of broader academic and social debates about cultural diversity. Yet at the same time, the current generation of historiography about Christian practice in the early modern West seems to present us with a paradox. The most recent research of the scholars in the field is eclectic, but also international in flavour. This new internationalism, fuelled by the quickening pace of European integration, means that scholars who treat one part of the Continent are now often asking the same questions of their sources as those in another. Where once historians talked of 'French' theories of acculturation, German notions of 'confessionalisation', and a British 'cultural anthropological' school, early modern religious historians now pose many of the same questions of their sources whether they work with English, German, French, or

Italian documents. This new religious history is, above all, concerned with how lay people displayed and made sense of their devotion and what, if any, permutations of their religion occurred as a consequence of the Protestant and Catholic reformations.

As this brief snapshot suggests, then, debates about the religions of the European peoples are far from resolved, yet the frisson that is evident in scholarly journals and monographs at the present time suggests that the concerns of historians in this sub-discipline, concerns with the ways in which sacraments, rituals, and beliefs informed Europeans' everyday lives, will continue to be debated, theoretically refined, and problematised anew for some time to come. For scholars of the sixteenth- and seventeenth-century Reformations, in particular, the contours of ordinary people's daily religious practices have a special significance, a significance that goes beyond mere antiquarian curiosity. Assessing the character of people's everyday religious practices, in other words, allows us to gauge the effects that dynamic reform movements had when they collided with the workaday world. At this junction, Protestant and Catholic reformers encountered either wellsprings of enthusiasm or walls of indifference. Thus for a moment it might be profitable to consider some of the fertile questions that popular religious historians have posed for scholars of the early modern reformations.

In the years following the Second World War, as French scholars devoted themselves to the study of 'lived religion', they were sometimes puzzled by the low level of religious practice they observed among parishioners in late medieval Europe. The Middle Ages had long been characterised as the pre-eminent Age of Faith. Yet the data being unearthed at this time pointed to a low level of religious practice. Why did most men and women in late medieval Europe usually only attend Mass, perform penance, and receive the Eucharist annually? Did such seeming laxity point to widespread religious indifference or to a failure of indoctrination on the part of the Church? As these scholars examined people's religious beliefs more closely, they found that a bedrock of magic and superstition seemed to underlay every dimension of life in the late medieval and early modern world. Did this general belief in the effectiveness of certain rituals mean that the true religion of the time was magic, and not Christianity? What was at stake, in other words, in the sprinkling of holy water on a farmer's field, the giving of blessed bread to cattle, or a woman's grasp of an amulet in childbirth? Had such beliefs been compatible with Christianity? Or did they reveal that the great mass of Europeans were more magicians than Christians?

As the study of popular religion expanded and British, North American, and German scholars turned to examine ordinary religious beliefs in the premodern world, they were inexorably drawn to consider the sixteenth-century Reformations and their impact on the religious landscape. As they did, new questions multiplied. Were either the Protestant or Catholic Reformations successful in changing the people's long-standing customs and beliefs? Did the reformers, with their decided emphasis on the rote memorisation of catechisms and printed confessional formulae, succeed in weaning many away from the luxuriant world of magic that had flourished in the Middle Ages? Had the reform movements merely widened, rather than closed the gap between the religions of Europe's peoples and its elites?

la religion vecue and *piété populaire*

In the past half-century historians of popular religion have placed these and other questions in the forefront of scholarship treating the Protestant and Catholic Reformations. Yet at the same time we cannot argue that the issues this relatively young field has nourished within early modern studies are completely new. Many of the dilemmas historians of popular religion have recently examined were anticipated in the rich works of late nineteenth- and early twentieth-century scholars. The works of any number of scholars from these years can still be profitably read and mined for their insights. American scholars like Henry Charles Lea (1825–1909) and British historians like George Gordon Coulton (1858–1947) explored the role that magic and superstition had played in lay religion in the medieval and early modern periods. In Germany, innovative scholars like Nikolaus Paulus and Peter Browe, following the lead of the great Catholic scholar Johannes Janssen (1829–1891), integrated the treatment of folklore and social history into their studies of Christian beliefs and practices. And in France as early as the 1920s, Marc Bloch's history of sacral kingship, *The King's Touch*, as well as Lucien Febvre's scholarship on the sixteenth-century Reformation were already laying the foundation for the *Annales* school's 'history of mentalities', an important historical method that has since come to play a vital role in the writing of the history of Christian *praxis*. Yet insofar as the study of popular religion can be said to have a beginning and its own history, its origins can be traced to France and to the years immediately following the Second World War. The interest that developed at that time was qualitatively different from what had come before, as French religious historians came to integrate sociology and its statistical methods into their study of popular piety.

During the war years in France, fears about the decline of Christianity had been nourished by a series of studies that decried the course of the country's secularisation, and in the period following Liberation, sociologists and historians undertook to study France's religious traditions, many with an eye toward stemming the growing tide of religious indifference. From the start many of these scholars were devout believers, while others found the issues surrounding France's de-Christianisation merely an intriguing academic issue. In the late 1940s and 1950s, though, no other country in Europe devoted the same resources to studying the religious beliefs and practices of ordinary people during past epochs as France did. A key figure in these efforts was the sociologist of religion, Gabriel Le Bras (1891–1970), a devout Catholic who had already had an impressive career as a canon law specialist. In the years around 1945 he came to embrace geography and cartography as important tools to examine the lived religion of premodern France and he published a series of studies that explored regional variations in rural Catholic religious practice.[5] In the years following the war he gathered a group of dedicated researchers around him in Paris that would eventually grow to become a separate section, the *Groupe de Sociologie de Religions*, within the *Centre Nationale de Recherches Scientifiques* (CNRS). While initially many of those drawn to Le Bras' circle were practising Christians who were interested in reversing France's religious decline, in time secular-minded academics joined as well. As a result of this steady enlargement the issues that the circle engaged expanded, too. While in the years after 1945 their studies had focused on Catholic France, the group grew by the late 1960s to include scholars that examined Protestant and non-Western religious communities as well.

Jacques Toussaert, Le Bras' student, was one of those who came to associate with the group during its initial burst of creativity in the 1950s and 1960s. Toussaert published important works on the popular religious life of the later Middle Ages, including his massive study, *Le sentiment religieux en Flandres à la fin du moyen-âge* (1963),[6] which exemplifies the techniques of the Le Bras school in this early stage of development. Toussaert confined his analysis to Flanders, and he carefully counted signs of religious practice, focusing, in particular, on compliance with the Church's requirements of baptism, penance, and the Eucharist. Although he relied on numerical data to recreate popular Christianity in the later Middle Ages, he came to mine this data in new ways. His figures concerning the numbers of communicants throughout the year were based on parish wine receipts. Although communion in the later Middle Ages was only given in one species – bread, the Host – unconsecrated wine was often given to communicants after the sacrament to cleanse their palates. As he examined the ebbs and flows in parish wine purchases, he was able to reconstruct the cycle of the traditional liturgical year.

Toussaert's work was the most thorough exploration until that time of the practices of the laity in late medieval Europe, and in making sense of the massive data he collected he came to dismal conclusions about the character of religious life in that era. First, Toussaert found that the practice of the Christian religion in late medieval Flanders had been bifurcated, with an orthodox elite standing in opposition to the broad mass of the people who were religiously indifferent. Most people had received the sacraments of penance and the Eucharist only once each year, in the weeks around Easter. Although everyone conformed to the Church's requirements that they baptise their children, confess in the face of approaching death, and be buried in hallowed ground, most were, all the same, unschooled in the theological, doctrinal, and moral complexities of their faith. Those who were devout, that is, who were labouring to practise the faith in a deeper, more informed way, seemed as well to have frequently had little sympathy for a Church that offered little but uninspiring prohibitions. In this climate heresy loomed as a constant threat, and spiritual anxiety seethed just below society's surface.

Implicit in Toussaert's analysis, then, was a link between the character of late medieval religiosity and the subsequent appearance of sixteenth-century reforms. Later scholars have sometimes sided with him in his assessment that the later Middle Ages suffered from spiritual anxiety. Among contemporary scholars, Steven Ozment has been among the most persistent in promoting a link between the flawed character of late medieval piety and the Reformation.[7] Yet most historians now see Toussaert's discovery of relatively infrequent sacramental practices in the later Middle Ages as a mere fact of religious life, a feature of an age that had very different ideas about the Christian disciplines that should inform daily life. At about the same time another of France's first generation of popular religious historians, Etienne Delaruelle, was to complete the first major synthetic treatment of 'lived religion' in the later Middle Ages, *L'Église au temps du grand schisme et de la crise concilaire, 1378–1449* (1962–64).[8] Where Toussaert's method had been largely adapted from Gabriel Le Bras' sociology, with its hard-hitting quantitative analysis, Delaruelle drew inspiration from the *Annales* school, which was by the 1960s the dominant group of historians within France. Like the *Annales'* founders Marc Bloch and Lucien Febvre, and their disciples, Delaruelle concentrated on a broad range of rituals and practices, instead of merely treating the parochial disciplines and sacraments that Toussaert had examined. He did so in the hope that he might reveal the underlying mental world of late medieval Christians. Where Toussaert's assessments of late medieval religious sentiment had been sombre, Delaruelle's were more optimistic, although he nevertheless sensed a certain spiritual uncertainty lurking below the surface of the popular religious landscape. Delaruelle focused, moreover, on those areas of *praxis* that have since the 1960s become synonymous with the concept of popular religion: processions and pilgrimages, saints' cults, the veneration of relics, and the many strains of late medieval visual piety. Over time, he perceived an increasing interiorisation of religious experience in the later Middle Ages, as people exchanged fondness for crusades and other bellicose embodiments of piety for more personal, subjective forms of devotion. Delaruelle was also acutely aware of the ways in which popular beliefs often veered into magic and superstition, although his work came to be criticised for its absence of a treatment of the differences between 'popular piety' and the official teachings of the Church.

new questions and unprecedented expansion

The earliest studies of popular religion in France from Toussaert, Delaruelle, and others had concentrated on the later Middle Ages and had often paid scant attention to the religious changes the Reformations produced. Yet as scholars from Britain, Germany, and North America responded to this new research, issues about early modern religious reform came to move in the 1970s to the centre of the stage. In that decade a number of synthetic treatments of popular religion appeared, many of them from British historians. In the works of Keith Thomas, Peter Burke, and John Bossy a new agenda for the study of popular religion developed in which the sixteenth-century Reformations were granted an important place in the narrative about popular religion's early modern transformations.

Among these studies, Keith Thomas's *Religion and the Decline of Magic* (1971)[9] stands alongside Lawrence Stone's, *The Family, Sex, and Marriage in England, 1500–1800* (1977) and E. P. Thompson's *The Making of the English Working Class* (1964) as a great achievement of post-war British historiography. While questions about the advent of individualism and liberal values played a key role in Stone's work on aristocratic marriage and a Marxist historiographical perspective informed Thompson's history of the working class, Keith Thomas reached out to embrace sociological and anthropological theory in a bold way. Certainly, many facets of his history have more recently been discredited. Scholars, in particular, have attacked *Religion and the Decline of Magic* for the way in which it set up a too strict dichotomy between religion (which Thomas described as attempts to approach the deity through non-coercive means like prayer and the sacraments) and magic (which he envisioned as the manipulation of supernatural power). Yet at the time in which Thomas wrote, few in the English-speaking world were aware of the full range of 'superstitious' rituals, customs, and beliefs that had informed life in the medieval and early modern world. Importantly, Thomas emblazoned upon the scholarly debates of the 1970s and the following decades an image of the alternative rationalities that had underlain all areas of life in premodern times.

Thomas also linked anthropology, particularly the functionalism of anthropologists like Bronislaw Malinowki, an early twentieth-century Polish émigré to Britain, to the sociology of Max Weber. Even by the time that Thomas wrote, Malinowski's functionalism was an embattled theory, but it had argued that magic played a vital role in primitive societies because it provided people with the illusion they might control harsh, contingent events in the natural world.[10] Thomas applied Malinowski's functionalistic explanation to the magic he saw being practised everywhere in late medieval England, and in the carnival of inquiry that he wrote around that concept, he observed the many ways in which magic had granted an intellectual coherence to English men and women before the Reformation.

Protestantism, Thomas argued, had warred incessantly against this admixture of magic with 'true' religion, and it had tried to separate the wheat of pure belief from the chaff of attempts to manipulate supernatural power. It was England's Puritan divines who engaged most decidedly in this battle, and thus to the functionalism that he took from Malinowski, Thomas also wed the theories of Max Weber, particularly the German sociologist's notions about the link between Calvinism and the 'disenchantment of the world'.[11] This theory – that the transcendental religious teachings of Calvin and his followers had achieved a demystification of the natural order – came to underlie Thomas's treatment of Puritanism. After carefully cataloguing the extent to which religion and magic had mingled uneasily in the later Middle Ages, he proceeded for hundreds of pages to demonstrate the many ways in which Puritans, and Anglicans to a lesser extent, had opposed magic, seeking to establish in its place a religion of faith, prayer, and providence. Yet in his conclusion, Thomas came ultimately to deviate from the traditional Weberian theory of disenchantment. Where Weber had seen the 'spirit' of Calvinism as decisive in producing the modern capitalistic ethic, Thomas argued instead that the Protestant campaign had largely been a failure and that the true sources of England's disenchantment and subsequent modernisation lay, not in religious change, but in science and technology, in developments like insurance and the fire brigade, and in other innovations that ultimately gave people an increasing mastery over the physical environment around 1700.

Thomas's work soon attracted admirers and detractors, but since its publication, most scholars have come to realise that the dividing lines he set up between religion and magic are not as clear as he originally imagined. Many studies have since shown that Protestants continued to embrace magical beliefs, while other scholars have discounted the notion of a premodern magic altogether. This last group has pointed out that many of the rituals late medieval and early modern people practised were not seen as instrumental or coercive, but as signs of devotion that might convince God and the saints to intercede. One person's magic, these scholars caution, is another person's religion.[12] Similarly, Thomas's reliance on the Weberian theory of disenchantment has also been a source of criticism, since many have found that theory wanting in explaining the precise contours of Europe's transformations in the seventeenth and eighteenth centuries.[13]

Yet in the years immediately following Thomas's *Religion and the Decline of Magic*, the way in which he had treated the problem – that is, as a battle between literate, schooled elites on the one side who were opposed to the profound superstitions of the masses on the other – came to be expressed in many studies. In his *Popular Culture in Early Modern Europe* (1978)[14] Peter Burke relied on a similar 'two-tiered' model that juxtaposed the religion of the people against that of elites. Burke insisted, though, that the divisions that had marked Europe's cultural landscape were also conditioned by the profound differences between rural and urban society, between professional groups, and by the 'highland' and 'lowland' cultural spheres common throughout Europe. Like many of the French *Annales* historians of the 1970s, he saw this last category as particularly decisive in fashioning the cultural landscape.[15] Yet Burke's work was not strictly speaking a history of popular religion, but an attempt to throw light upon the 'little' traditions that had informed the lives of all Europeans in the early modern age. Burke saw popular folklore, customs, and beliefs existing in opposition to the 'great' ideas of intellectuals and elites, and over the course of the seventeenth and eighteenth centuries, he noted an increasingly pronounced divorce between these two cultural spheres. Where the scholars, humanists, and writers of the Renaissance had once engaged with the ideas, fables, foibles, and rituals of the people, elites came over the course of the early modern centuries to adopt an increasingly dismissive attitude toward popular culture. Burke's observation was not new, although the way in which he formulated the agenda for future study came to be fruitfully applied in the research that followed.

Two of the book's essays, 'The World of Carnival' and 'The Triumph of Lent', also had important implications for historians interested in the history of religion in the early modern world. In these, Burke traced the gradual establishment throughout society of the disciplines that sixteenth-century reformers had once vigorously promoted and he argued that the new emphasis on social discipline had served to dampen enthusiasm for much of the substance of traditional popular culture. Disarmed in this way, popular folklore was by the end of the eighteenth century poised to make a comeback, that is, to be 'rediscovered' by German intellectuals like Herder and folklorists like the Brothers Grimm.

Throughout these essays, Burke ranged widely, producing evidence from Scotland, Scandinavia, England, the Ukraine, and the Mediterranean world. This breadth of focus was an immediate source of criticism, as some dismissed the work as a premature synthesis that glossed over the many regional variations that had existed in Europe's cultural landscape. Some of Burke's aims, though, were being echoed by other scholars. First, Burke, like Thomas and others, made serious use of anthropological theory, helping to bring theory into the centre of historical debate. In the years that followed, Mary Douglas's studies of taboo and natural symbols[16] and Clifford Geertz's 'thick description' of Balinese rituals[17] were just two of the many anthropological studies that became *de rigeur* reading for anyone hoping to work on popular religion or popular culture in early modern Europe. The techniques of these fields, most notably Geertz's 'thick description', exerted a powerful influence over the ways in which historians came to relate their findings about popular religion. In many studies the religious experience of premodern Europeans was now compared explicitly to that of tribal peoples in other parts of the world.

At the same time many scholars took seriously Burke's call to examine the various religious subcultures that existed within early modern Europe. Increasing specialisation, in other words, became the rule, as religious scholars came to explore the regional permutations that existed beneath the surface of a once seemingly monolithic 'popular religion'. Among the many regional studies from the late 1970s and early 1980s, several are noteworthy. As anthropology and history grew closer together in these years, the anthropologist William Christian published several historical studies of the regional variations of popular piety in sixteenth-century Spain. In his *Local Religion in Renaissance Spain* (1981)[18] Christian echoed Burke's call for a scholarship that might be faithful to regional variations and he showed that the long-standing dynamics of local religion in rural

Spain were far more essential to understanding ordinary people's customs and beliefs than the reform efforts of the sixteenth-century church and state. Christian concentrated on the many devotions that had flourished throughout Spain to the Virgin Mary and the saints, even as Lionel Rothkrug was studying the geographical distribution of saints' cults in late medieval Germany in a way that was strikingly similar. Rothkrug found that the southern part of the Holy Roman Empire had been littered with sacred sites while pilgrimages had been far less popular in the north. In explaining this deep division, he argued that the Reformation and Counter-Reformation had merely formalised long-standing religious divisions within the countryside, with Protestantism finding easy entry in northern Germany because of the pre-existing transcendent character of religious practices there. In the Catholic south, the plethora of saintly devotions had inoculated populations against the Reformation's influence. While Rothkrug's conclusions were controversial, his work presents a still open challenge to historians of the period to explore the underlying ethnographic causes of religious differences.

While the 1970s saw the development of a vigorous school of popular religious historians in the English-speaking world, French scholars continued to make major contributions to the field as well. Many of their histories inspired a deepening debate about the very nature of the concept of popular religion. One of the most important general histories of these years was Jean Delumeau's *Catholicisme entre Luther et Voltaire* (1971).[19] Delumeau's treatment of early modern Catholicism appeared in the *Nouvelle Clio* series of history texts, works intended for use by French university students and which were thus designed to present a state-of-the-art synthesis. Delumeau repeated a view of late medieval popular religion that had by this time become a standard of French scholars. He emphasised that magic and superstition, a fondness for ritual, and a low level of doctrinal knowledge had been the chief characteristics of the religion of the age. He concluded that the people of Europe in 1500 had not been effectively Christianised.[20] This bleak picture thus set the stage for the dramatic efforts at Christianisation he detailed during the Protestant and Catholic Reformations that followed. Delumeau judged the many reform movements of the later sixteenth and seventeenth century successes, since he saw that they established a generally higher level of doctrinal knowledge and discipline among Catholics. But when he came to the later eighteenth century he saw religious indifference and a trend toward de-Christianisation developing in the heat-up to the French Revolution, an observation he drew from other recent research.[21]

Delumeau's thesis, then, interpreted the Catholic Reformation as a great process of acculturation by which ordinary men and women were gradually yet persistently brought under the influence of a doctrinal Christianity and subjected to its discipline. Among French scholars of popular religion, this acculturation thesis acquired great currency in the 1970s, even as it sparked controversy. In a perceptive exchange with Robert Muchembled, one champion of acculturation, Jean Wirth showed that many early modern reform movements had developed from within the laity and had not been imposed from above.[22] Yet despite such cogent rejoinders to the notion of acculturation, the thesis continued to attract adherents. In England, John Bossy has been among those who have adopted elements from the acculturation argument. While he has avoided characterising the religion of the later Middle Ages as 'pre-Christian', as Delumeau, Le Bras and others sometimes did, he has long noted that the social force of Christianity in those centuries derived less from the religion's doctrinal integrity than from its communal nature. And where Delumeau and other French scholars saw a great divide between 'elites' and 'people' in both the late medieval and early modern world, Bossy has stressed that the key differences in both periods were those between the laity and the clergy. In many articles and in *Christianity in the West, 1400–1700* (1985), he has presented a thesis that stresses a process of increasing 'parochialisation' in early modern Europe.[23] The numerous reforms of the post-Tridentine Church, in other words, subjected the laity to new disciplines within their parish, disciplines that were opposed to the communal character of late medieval religion. The impact of the Catholic Reformation, for instance, came to be felt in a new 'technologised' sacrament of penance that heightened the sacrament's inward-looking features, transforming it from a rite that had once healed rifts in towns and villages into a practice of individual self-examination that excavated sins.[24] Baptism, Bossy argues, was similarly shorn of its communal dynamic, as the Church narrowed the numbers of godparents that could be named for an infant child, thus limiting the role that the sacrament had once played in extending Christian community through networks of friends and kin.[25] And finally, Bossy sees that both seventeenth-century Catholicism and Protestantism came to subject their parishioners to a new 'typographical tyranny', in which parishioners were forced to be literally 'on the same page' each week as they celebrated the liturgy, or as they gave their children up for indoctrination in the new printed catechisms.[26] These and other innovative forces in early modern religion spelled, according to Bossy, the end of the rich, communal experience that had flourished in the later Middle Ages. In place of a faith once dominated by vibrant ritual, Christianity became a system of 'hard-edged' intellectualism. In the course of the seventeenth century, in other words, religion was transformed into an arid shadow of its former self, one that ill served the needs, desires, and aspirations of most Europeans. In this way Bossy thesis links the changes of the Protestant and Catholic reformations with the decline of Christian belief in the eighteenth century.

toward a history of christian practice

Since the 1980s Bossy's bleak portrait of early modern religious change, as well as other defences of the acculturation thesis, have been largely dismissed, even as the concept of popular religion as a set of beliefs and practices that existed in isolation from, and in opposition to the official teaching of 'elites' has crumbled as well. Yet while historians have increasingly found these theories wanting, efforts to reconstruct the history of religion as it was lived and practised in the later Middle Ages and the early modern world have continued, and even intensified. While the past two decades have produced little in the way of grand syntheses to match those of Le Bras, Delumeau, Thomas, and others, these years have still seen great creative energy. Much of this energy, though, has been spent in an effort to understand the regional mutations of European piety, that is, in a grassroots effort to determine how specific local historical and ethnographic realities, as well as the political and confessional dynamics of the Reformations, shaped people's religious experience.

Of the many scholars who have explored lay religious practices in this 'third generation', Robert Scribner has been among the most active and imaginative. Although much of Scribner's considerable output of articles and texts concentrated on the history of the Reformation in Germany, his studies have come to have important implications for historians of all early modern European cultures. Scribner began his career as a political historian, examining the dynamics of the Reformation within Germany's cities.[27] In the early 1980s, he shifted focus away from the political history of the urban Reformation and instead began to explore the role traditional religious mentalities had played in shaping early sixteenth-century reform movements. While many in these years were criticising the concept of popular religion as unwieldy and untenable, Scribner proved to be its most vigorous proponent, although he came to jettison the long-standing two-tiered model that had juxtaposed 'popular' religion against the learned prescriptions of elites. By contrast, he insisted that the study of popular religion could retain its usefulness as a category of analysis precisely because its mentalities pervaded all strata of early modern society, from university classrooms, to city streets, and peasant villages. Armed with this and other insights about the importance of rituals, traditional customs, and folklore, Scribner developed a highly imaginative, sometimes controversial set of theories about the Reformation and religious change.

In his path-breaking study of the visual propaganda of the early Reformation, *For the Sake of Simple Folk* (1981), he demonstrated the ways in which the overwhelmingly oral world of the sixteenth-century had shaped the dynamics of religious reform.[28] In that work he deployed communication theory, semiotics, and the more traditional discipline of iconography to read early Reformation visual propaganda. Throughout he demonstrated the ways in which the Reformers and their adherents and critics had alternately popularised and denigrated the doctrines of reform before a society that was largely preliterate. Somewhat later, Scribner turned to examine the role that traditional rituals had played in promoting as well as limiting the appeal of the early Reformation movement.[29] In a series of articles published in the later 1980s and early 1990s, Scribner explored the deployment of violence directed at the clergy and at elements of traditional Catholic cult. In contrast to previous scholarship on iconoclasm that has stressed its revolutionary dynamic, Scribner demonstrated that the impulses that reposed within these early Reformation collective actions were not so much new, but the product of deeply traditional societies that had long 'tried' relics and saintly images. Other case studies from these years brought to light much about the early history of Protestantism that had long been overlooked. In articles with provocative titles like 'Incombustible Luther', Scribner detailed the development of a 'Luther cult' within the evangelical tradition that in many ways mirrored the popular devotions medieval Europeans had once entertained for saints like Francis.[30]

Although it was always Scribner's intention to write a general history of the Reformation that might take account of popular culture, his premature death prevented him from completing that work. A series of essays published in the 1990s, though, suggests the direction that study may have taken.[31] In these years Scribner came to present a forceful response to those who had argued in Weberian fashion that the Reformation represented an important breakthrough along the way toward modern rational thought. Scribner opposed such conclusions, insisting instead that there was no evidence of Reformation-era disenchantment. Although Protestants had attacked the 'magic' and 'superstitions' of the medieval Church, Scribner observed, they came very quickly to create their own forms of magic that were little different from those of the medieval world. And as Scribner's studies to this point had also detailed, the Reformation arose within a highly traditional world where the impulses that fed its popularity were not so much revolutionary, as conservative.

The long shadow of tradition also played an important role in Eamon Duffy's *The Stripping of the Altars* (1992). In contrast to previous works on 'popular religion', Duffy characterised his study as an effort to recreate 'traditional' English Catholicism, and much of his book was given over to a careful recounting of that religion's vigorous liturgical culture as well as the annual undulations of its devotional and festive life. It is fruitful to read *The Stripping of the Altars* in terms of a 'third-generation' response to the older works of Keith Thomas, Peter Burke, Jean Delumeau, and other older popular religious scholars. Where those earlier studies often drew dismal conclusions about the character of religious knowledge among the people of late medieval Europe, Duffy is far more optimistic. Instead of detailing a world of popular magic and superstition, he carefully reads a broad range of prayer books, sermons, and devotional texts, even as he also relies on literary and poetic texts and artistic evidence to bring to life late medieval Catholicism. His analysis of ecclesiastical architecture and art are particularly suggestive, since they detail the vitality of religion as it was experienced in England immediately before the Reformation.

Duffy's work consequently reveals the pitfalls inherent in many older studies of 'popular religion', which frequently concentrated on ritual behaviour and oral, folkloric traditions at the expense of textual evidence. The fascination of previous generations of researchers with affective forms of behaviour, in particular, often proved problematic precisely because rituals were capable of being significantly misread by modern historians anxious to discover worlds very different from their own. These scholars often invested the many customs early modern Europeans practised with 'magical' significance. Yet behind those rituals, as Duffy's works show, often reposed some quite sophisticated religious assumptions. While some have found Duffy's portrayal of the Reformation's violent extinguishing of a vibrant Catholicism unduly nostalgic, the evidence that he has amassed – evidence that suggests that England's traditional religion was intellectually coherent, theologically respectable, and enthusiastically embraced – continues to challenge historians of England, as well as those of the Continent, to reformulate their narratives of the Reformation. Where one or two generations ago, late medieval English Catholicism was often treated as a decaying system for the ignorant swept away by the triumphant rise of Protestantism, Duffy's *Stripping of the Altars*, as well as other recent works like Robert Whiting's *The Blind Devotion of the People* (1991), dispel the underlying Whiggery of that narrative.[32]

Other studies of early modern religion in England, too, similarly point to the increasingly international influences that have made their impact felt upon all Reformation historiography. England, to be sure, was a country long seen to have had an exceptional religious, political, and social history, and undertaking study of its sixteenth- and seventeenth-century history was seen as requiring training very different from historians of the Continent. This notion of English exceptionalism was still alive thirty years ago when Keith Thomas and others first explored the history of the country's popular religion. Yet in recent years the historiography of English religion in the late medieval and early modern periods has come to mirror many of the same issues, and rely on the same methods as those that are current among historians of France, Germany, and Italy. Tessa Watt's recent study of ballads and religious broadsides, for example, addresses larger issues about the diffusion of print and its impact on the patterns of cultural transmission in England in ways that echo and expand upon Robert Scribner's study of the visual propaganda of the German Reformation.[33] In her study of the Reformation in early sixteenth-century London, Susan Brigden raises and answers questions about the nature of late medieval religious practice, the piety of the laity, and the broader context of sixteenth-century religious reform movements, issues that have been treated in similar ways in France by scholars like Philip Hoffman and Natalie Davis.[34] Yet this exchange has not occurred in one direction only, for studies of English Reformation history like Eamon Duffy's offer historians of continental Europe excellent models of ways to treat medieval and early modern religious practices free from the modernist, cultural bias that was inherent in older, popular religious models.

The past twenty years, then, may have seen the gaze of historians shorten as they have come to explore with increasing precision regional mutations in lay piety. Yet an increasing methodological sophistication and cross-fertilisation has emerged in these years to compensate for the loss of a larger synthetic vision. For the student of Christian piety, this situation often presents a curious irony. While on the one hand religious historians are concerned more than ever before with remaining faithful to their specific local contexts, on the other, the questions they often ask of their sources appear strikingly similar across cultures. In the coming years historians will continue to locate their research within specific national and regional contexts, but the issues that they treat will likely also be being examined by many others who are working in geographical settings distant from their own. Scholars who are interested in early modern religious practice will likely continue to examine the changes that the Protestant and Catholic reformations produced in pious bequests, saints' cults, pilgrimages, and the broad range of customs, folklore, and devotions that informed European faith in premodern times. Yet the quickening pace of the Continent's political integration seems poised to make its influence felt on the writing of history, and as scholarly exchanges between scholars working in disparate parts of Europe will likely grow even more common, integration's effects promise to be as profound as any of the other political or social movements of the past fifty years were on the scholarship of their day. It remains unclear, though, what models will emerge to replace the increasingly outdated notion of 'popular religion', or of theories of change like the 'acculturation thesis'.

Still, some studies of the past few years suggest the direction that historians may take as they seek to synthesise current research, and as they move to forge a new paradigm for understanding Christian practice in the early modern period. In his recent work on 'village religion' in the German south-west, for example, Marc Forster has shown that the vigour of traditional religious practices proved a significant source of Catholicism's strength in the region. As the Counter-Reformation found its way into these provinces, it did not war against customs like pilgrimage or the widely held love for processions, as early scholars who relied on the notion of an elite 'reform of popular culture' imagined. Instead the clerical elite appropriated these long-standing practices, and relied upon them as the source for a widespread, truly popular religious renewal; both the clergy and laity thus came to mould these pilgrimages into symbols of a new Roman Catholic identity.[35] To the north, in Brandenburg, an enduring and widespread conservatism among the people developed widespread reserves of affection for the essentially traditional currents of Lutheran reforms. As Bodo Nischan has demonstrated,[36] the Electors of the Mark Brandenburg failed in their efforts to introduce harsher, more disciplined Calvinist reforms throughout their lands. Faced with such innovations their subjects threatened revolt. Until the nineteenth century, Brandenburg remained a land with a thin elite composed of Calvinist ministerial and military officials who cautiously refused to impose their religious assumptions on a populace that retained its Lutheranism.

Farther afield in Italy, recent scholarship by Simon Ditchfield on Counter-Reformation hagiography has challenged us to rethink the very nature of the many innovations that occurred in historical writing at this time. Ditchfield has shown that the works of Caesar Baronius and other great Counter-Reforming hagiographers, figures who were once thought to epitomise the very nature of the Counter-Reformation's 'top-down' cultural strategy, were inspired by altogether different realities. The audience for their lives of the saints lay in scores of communities, who relied upon their source-driven hagiographies to protect local religious customs and practices against the encroachment of bishops and liturgical reformers anxious to prune away older 'superstitions'.[37] These and other studies remind us that early modern religion was shaped, not by binary opposites of 'elite' and 'popular' religion, but by complex exchanges, a long dialogue between interests that might often be working at cross-purposes, but which could just as easily identify considerable areas of consensus. As historians thus come to make sense of the mounting evidence their local studies present concerning these complex relationships – between laity and clergy, between villagers and the central state and the ecclesiastical hierarchy, and between competing professions, sodalities, and social orders – they will need to listen carefully to this conversation and to take account of the many skirmishes, rapprochements, and sheer indifferences that marked the communication between all the various spheres that comprised the early modern world.

further reading

On early investigations of popular religion, see Pierre Boglioni, 'Some Methodological Reflections on the Study of Medieval Popular Religion', *Journal of Popular Culture*, 11/3 (1977), 696–705; and John Van Engen, 'The Christian Middle Ages as an Historiographical Problem', *American Historical Review*, 91/3 (1986), 519–52. Steven Ozment, *The Reformation in the Cities: the Appeal of Protestantism to Sixteenth-Century Germany and Switzerland* (New Haven, 1975) takes a (now) unfashionably negative view of late medieval religion but still repays reading.

Keith Thomas's *Religion and the Decline of Magic* (London, 1971) is dated in places but remains an impressively magisterial survey of the subject. On critiques of Thomas, see Philip M. Soergel, 'Miracle, Magic, and Disenchantment in Early Modern Germany', in Peter Schäfer and Hans G. Kippenberg (eds), *Envisioning Magic: a Princeton Seminar and Symposium* (Leiden, 1997), and especially Robert Scribner, 'The Reformation, Popular Magic and the Disenchantment of the World', *Journal of Interdisciplinary History*, 23 (1993), 475–94.

Peter Burke, *Popular Culture in Early Modern Europe* (New York, 1978) is another ambitious and enduring study, placing religion in its broadest cultural context, drawing on the immensely influential methodology of Clifford Geertz, *The Interpretation of Cultures* (New York, 1973). Comparable but less wide-ranging studies include Robert Muchembled, *Popular Culture and Elite Culture in Early Modern France*, trans. Lydia G. Cochrane (Baton Rouge, LA, 1985); William Christian, *Local Religion in Renaissance Spain* (Princeton, 1981).

The clearest statement of the 'acculturation' thesis is Jean Delumeau, *Catholicism between Luther and Voltaire*, trans. Jeremy Moiser (London, 1977). John Bossy has refined these ideas further: see especially his *Christianity in the West, 1400–1700* (Oxford, 1985).

Robert Scribner pioneered more positive views of 'popular religion': see his *For the Sake of Simple Folk* (Cambridge, 1981; revised edn, Oxford, 1994), and *Popular Culture and Popular Movements in Reformation Germany* (London, 1987). Another pioneer was Natalie Zemon Davis: see *Society and Culture in Early Modern France* (Stanford, CA, 1975). An even more vibrant portrait of English 'traditional religion' is painted in Eamon Duffy, *The Stripping of the Altars* (New Haven and London, 1992). Other key regional studies of the 'third generation' of popular-religion scholarship include Tessa Watt, *Cheap Print and Popular Piety, 1540–1650* (Cambridge, 1991); Marc Forster, *Catholic Revival in the Age of the Baroque: Religious Identity in Southwest Germany, 1550–1750* (Cambridge, 2001); Simon Ditchfield, *Liturgy, Sanctity, and History in Tridentine Italy* (Cambridge, 1995).

notes

1. William J. Swatos (ed.), *Encyclopedia of Religion and Society* (Lanham, MD, 1998), s.v. 'Sociology of Religion'.
2. Although several works appeared treating this theme, the most influential was likely by two Jesuits, Henri Godin and Yvan, *La France, pays de mission?* (Lyons, 1943).
3. Peter Burke, *The French Historical Revolution: the Annales School, 1928–1989* (Cambridge, 1990).
4. The early historiography of the investigation of popular religion is reviewed in Pierre Boglioni, 'Some Methodological Reflections on the Study of Medieval Popular Religion', *Journal of Popular Culture*, 11/3 (1977), 696–705; and John Van Engen, 'The Christian Middle Ages as an Historiographical Problem', *American Historical Review*, 91/3 (1986), 519–52.
5. The development of Le Bras' methods is reviewed in F.-A. Isambert, 'Développement et dépasssement de l'étude de la pratique religieuse chez Gabriel Le Bras', *Cahiers Internationaux de Sociologie*, 20 (1956), 149–69; and in his own *Études de sociologie religieuse* (Paris, 1955–56).
6. (Paris, 1963).
7. Steven Ozment, *The Reformation in the Cities: the Appeal of Protestantism to Sixteenth-Century Germany and Switzerland* (New Haven, 1975).

8. Published in two parts with the collaboration of E.-R. Labande and Paul Ourliacas as volume XIV of A. Fliché and V. Martin (eds), *Histoire de l'Église des origines jusqu'à nos jours* (Paris, 1962–64).
9. (London, 1971).
10. Bronislaw Malinowski, 'Magic, Science and Religion', in J. Needham (ed.), *Science, Religion, and Reality* (New York, 1925).
11. This dimension of his theory of religious sociology is outlined in M. Weber, *The Protestant Ethic and the Spirit of Capitalism*, trans. Steven Kalberg (Chicago, IL, 2001); cf. German edition, 1904.
12. For a survey of these studies, see Philip M. Soergel, 'Miracle, Magic, and Disenchantment in Early Modern Germany', in Peter Schäfer and Hans G. Kippenberg (eds), *Envisioning Magic: a Princeton Seminar and Symposium* (Leiden, 1997).
13. A point made shortly after the work appeared in Natalie Davis, 'Some Tasks and Themes in the Study of Popular Religion', in Charles E. Trinkaus and Heiko A. Oberman (eds), *The Pursuit of Holiness in Late Medieval and Renaissance Religion: Papers from the University of Michigan Conference* (Leiden, 1976). The criticism was echoed in other reviews. See especially Hildred Geertz, 'An Anthropology of Religion and Magic, I' and Keith Thomas, 'An Anthropology of Religion and Magic, II', *Journal of Interdisciplinary History*, 6 (1975), 71–89, 91–109. Robert Scribner also examined the notion of Weberian disenchantment trenchantly in his 'The Reformation, Popular Magic and the Disenchantment of the World', *Journal of Interdisciplinary History*, 23 (1993), 475–94.
14. (New York, 1978).
15. An observation that was at the heart of Robert Muchembled's roughly contemporaneous *Culture populaire et culture des élites dans la France moderne: XVe–XVIIIe siècles* (Paris, 1978); translated as *Popular Culture and Elite Culture in Early Modern France*, trans. Lydia G. Cochrane (Baton Rouge, LA, 1985).
16. Mary Douglas, *Purity and Danger: an Analysis of Concepts of Pollution and Taboo* (London, 1966); idem, *Natural Symbols: Explorations in Cosmology* (New York, 1970).
17. Clifford Geertz, *The Interpretation of Cultures* (New York, 1973).
18. (Princeton, 1981).
19. *Nouvelle Clio* 30 (Paris, 1971); translated as *Catholicism between Luther and Voltaire*, trans. Jeremy Moiser (London, 1977).
20. A conclusion that Gabriel Le Bras and others had also come to over the years. Van Engen, 'The Christian Middle Ages as an Historiographical Problem'.
21. Particularly in the research of Michel and Gabrielle Vovelle, *Vision de la mort et de l'au-delà en Provence d'après les autels des âmes du purgatoire, XVe–XXe siècles* (Paris, 1970). Michel Vovelle long continued this line of inquiry in his numerous monographs, most notably *Piété baroque et déchristianisation en Provence au XVIIIe siècle; les attitudes devant la mort d'après les clauses des testaments* (Paris, 1973); and *Religion et révolution: la déchristianisation de l'an II* (Paris, 1976).
22. Robert Muchembled, 'Lay Judges and the Acculturation of the Masses' and Jean Wirth, 'Against the Acculturation Thesis', in Kaspar von Greyerz (ed.), *Religion and Society in Early Modern Europe 1500–1800* (London, 1984).
23. John Bossy, 'The Counter-Reformation and the Catholic People of Europe', *Past and Present*, 46 (1970), 51–70; idem, 'The Mass as a Social Institution',

Past and Present, 100 (1983), 29–61; idem, 'Blood and Baptism: Kinship, Community and Christianity in Western Europe from the Fourteenth to the Seventeenth Centuries', *Studies in Church History*, 10 (1973), 129–43; idem, *Christianity in the West, 1400–1700* (Oxford, 1985).
24. John Bossy, 'The Social History of Confession in the Age of the Reformation', *Transactions of the Royal Historical Society: 5th Series*, 25 (1975), 21–38.
25. Bossy, 'Blood and Baptism.'
26. Bossy, *Christianity in the West.*
27. Scribner's doctoral dissertation treated the political dynamics of reform at Erfurt; parts of it were published in article form, including 'Civic Unity and the Reformation in Erfurt', *Past and Present*, 66 (1975), 29–60.
28. (Cambridge, 1981; revised edn, Oxford, 1994).
29. Many of these studies were reprinted in Robert Scribner, *Popular Culture and Popular Movements in Reformation Germany* (London, 1987).
30. Scribner, *For the Sake of Simple Folk*; idem, 'Incombustible Luther: the Image of the Reformer in Reformation Germany', *Past and Present*, 110 (1986), 38–68.
31. See the posthumous collection edited by Lyndal Roper, *Religion and Culture in Germany (1400–1800)* (Leiden, 2004).
32. Robert Whiting, *The Blind Devotion of the People* (Cambridge, 1991).
33. Tessa Watt, *Cheap Print and Popular Piety, 1540–1650* (Cambridge, 1991); Scribner, *For the Sake of Simple Folk.*
34. Susan Brigden, *London and the Reformation* (Oxford, 1989); Philip Hoffman, *Church and Community in the Diocese of Lyon, 1500–1789* (New Haven, 1984); Natalie Zemon Davis, *Society and Culture in Early Modern France* (Stanford, CA, 1975).
35. Marc Forster, *The Counter-Reformation in the Villages* (Ithaca, 1992); idem, *Catholic Revival in the Age of the Baroque: Religious Identity in Southwest Germany, 1550–1750* (Cambridge, 2001).
36. Bodo Nischan, *Prince, People, and Confession: the Second Reformation in Brandenburg* (Philadelphia, 1994).
37. Simon Ditchfield, *Liturgy, Sanctity, and History in Tridentine Italy* (Cambridge, 1995).

12
women, gender and sexuality

merry wiesner-hanks

The study of gender in the past began as investigations into the history of women, who had largely been left out or kept out of what was defined as 'history' since the time of the ancient Greeks. Though a few historical studies in the Middle Ages and early modern period, such as convent chronicles and group biographies, had been written by or about women, most history was written by men, and, not surprisingly, focused on their lives and achievements. This what we might now call 'gender bias' became even more pronounced when history became a profession in the nineteenth century, and real historians were defined as those who participated in university seminars, which were closed to women.[1] A few women were inspired by the women's rights movement (now termed the 'first wave' of the feminist movement) to write about women's lives in the past, but this was not viewed as serious history.

The second wave of the feminist movement in the late 1960s brought an even more dramatic upsurge of interest in women's history, and it also brought many more women into universities. Students in history programmes in North America and Western Europe in the late 1960s and early 1970s, most (though not all) of them women, began to focus on women, asserting that any investigation of past oppression or power relationships had to include information on both sexes. Initially these studies were often met with derision or scepticism, but this criticism did not quell interest in women's history, and may in fact have stimulated it. By the late 1970s, hundreds of colleges and universities in the United States and Canada offered courses in women's history; universities in Britain, Australia, and the Netherlands began offering courses somewhat later, and the rest of continental Europe later still.

Courses in women's history were accompanied by research, and over the last thirty years, historians of women have demonstrated that there is really no historical change that does not affect the lives of women in some way, though often very differently than it affects the lives of men of the same class or social group. This scholarship has called into question many basic historical categories – class, modernity, capitalism, and even how historical periods are divided and designated. It has also become increasingly self-critical, putting greater emphasis on differences among women. Women's experience differed according to categories already set out based on male experience – social class, geographic location, rural or urban setting – but also categories that had previously not been taken into account: marital status, health, number of children. Historians are thus much less comfortable talking about the 'status of women' in general without sharply qualifying exactly what type of women, or in what type of sources; conclusions can be made about, for example, the legal status of women in theoretical treatises, or the role of widows in a particular organisation, but statements about the status of women rising or falling are too vague to have any meaning.

Building on studies of women, some historians during the 1980s shifted their focus somewhat to ask questions about gender itself, that is, about how past societies fashioned their notions of what it means to be male or female. They distinguished between sex – physical or biological differences between men and women – and gender – socially constructed differences. Historians studying gender often used and continue to use theories and methodology drawn from sociology, anthropology, and literary studies, and emphasise that gender structures are often contradictory, unstable, and frequently changing.

The study of gender has not completely replaced the study of women, because there is still much more basic information available about the lives of men than the lives of women, but it has resulted in new types of questions about the lives of men and relations between the sexes. Historians attuned to gender now study the construction of masculinity and men's experiences in history *as men*, rather than simply as 'the history of man' without noticing that their subjects were men. Gender roles were just as prescribed for men as they were for women, and history that ignores the effects of 'private' factors such as marital status, sexuality, and friendships in men's lives is incomplete. This new 'men's history' is only in its infancy, and has been criticised both by traditional historians who see it as trivialising 'great men and their ideas' and by some women's historians who view it as simply a way to refocus historical concern back where it has always been, on men.

Despite the criticism, studying men as men is an important part of the understanding that people's notions of gender shaped not only the way they thought about men and women, but about their society in general. In the words of the historian Joan Scott, 'gender is a primary way of signifying relationships of power', not simply relationships between men and women, but between any superior and subordinate group.[2] These ideas affected the way women and men acted, but explicit and symbolic ideas of gender could also conflict with the way people chose or were forced to operate in the world. Thus the status of women *and* men was at once more varied and more shaped by their gender than most historians would have concluded thirty years ago.

Along with a focus on the gendered nature of both women's and men's experiences, some historians have turned their attention more fully to the history of sexuality. Just as interest in women's history has been part of feminist political movements, interest in the history of sexuality has been part of the gay liberation movement that began in the 1970s. The gay liberation movement encouraged the study of homosexuality in the past and present and the development of gay and lesbian studies programmes, and also made both public and academic discussions of sexual matters more acceptable. Historians have attempted to trace the history of men's and women's sexual experiences – both homosexual and heterosexual – in the past, and, as in women's history, to find new sources that will allow fuller understanding. The history of sexuality has contributed to a new interest in the history of the body, with historians investigating how cultural understandings of the body shaped people's experiences of their own bodies and also studying the ways in which religious, medical, and political authorities exerted control over those bodies. This in turn has often led back to a focus on women, as female sexuality and the female body have generally been of greater concern to authorities than male sexuality.

Scholarship focusing on the Protestant and Catholic Reformations has been an important part of the study of women, gender, and sexuality over the last thirty years. Like the field in general, it has stressed diversity and complexity. It has emphasised the differences between the ideas and ideals of the reformers and the institutions that were established and ended, highlighted women's actions and decisions (what is often termed 'agency') and the actions of men supporting and restricting that agency, discussed the great differences between northern and southern Europe, rural and urban, rich and poor. Intensive archival research in many parts of Europe has meant that most scholars are less willing to make general conclusions about the impact of the Protestant or Catholic Reformations on *all* women or on ideas about gender in *all* of Europe than they were several decades ago.

The enormous amount of scholarship over the last several decades has made generalisations more difficult, but it has also changed the scholarly view of sixteenth-century women and, to a somewhat lesser degree, of the Reformations. It has developed new theoretical and methodological directions, and also presented new ways to look at more 'old-fashioned topics', such as the lives of great women and the ideas of great men. This chapter will take a closer look at this breadth of scholarship and discuss its impact on more traditional areas of Reformation research.[3]

Most considerations of women and the Reformations go off in one of two directions. The first explores women's actions in support of or in opposition to the Protestant and Catholic Reformations and looks more broadly at women's spiritual practices during this period. The second focuses on the ideas of the reformers and the effects of the Reformations on women and on structures that are important to women, such as the family. In surveying recent scholarship it will be useful to consider these separately, as they tend to focus on different issues and incorporate different theoretical perspectives.

The line of inquiry analysing women's actions and religious practices began in the 1970s, and tends to emphasise women's agency. Studies have looked at women active in iconoclastic riots and religious wars, women defending convent life in word and deed, women preaching in the early years of the Protestant Reformation, pastors' wives creating a new ideal for women, women defying their husbands in the name of their faith, women converting their husbands or other household members, women writing and translating religious literature.[4]

Research exploring women's actions in institutionalising Protestant or Catholic reforms originally focused primarily on female rulers, including Mary and Elizabeth in England, various queen mothers in France, and the female rulers of states within the Holy Roman Empire.[5] Within the last ten years convents have received more and more attention. In part this is a function of sources, for convents housed literate women, controlled property and people, and were often linked with powerful families, all of which means they have frequently left extensive records. These records were often transferred as a body to some larger archives, but remain a manageable group of sources for a dissertation project. In Italy, Gabriella Zarri at the University of Bologna has directed teams of researchers exploring convents, holy women, and hagiographical texts, and sponsored regular conferences on these topics.[6]

In Germany, convents in areas becoming Protestant often fought the Reformation through letter writing, family influence, physical bravery, and what one might call sheer cussedness, stuffing wax in their ears so as not to hear Protestant sermons and refusing to leave their houses unless they were physically dragged out; in some cases authorities finally gave up and the convents remained islands of Catholicism for centuries. Others supported the Protestant Reformation theologically on some issues, but ignored its message about the value of convent life and remade themselves into institutions that were acceptable to Protestant authorities, educating girls and providing an honourable place for women who could not or chose not to follow the Protestant injunction to marry.[7]

Research on convents and religious women in Catholic Europe has also focused on those who challenged boundaries. Not surprisingly, the prominent Spanish nun and reformer Teresa of Ávila has been explored from the most angles: her milieu, her political influence, her spirituality, and her sense of authorship and of self.[8] These studies have been joined recently by several that focus on her predecessors and contemporaries and which make it clear that though Teresa is unique, she also followed a pattern found in other Spanish and Italian women: close relationship with a confessor, physical manifestations of her piety, doubts about her own self-worth, effective alliances with local and sometimes national political leaders.[9] Some of these women were able to retain reputations as holy women – *beatas* – throughout their lives or even, like Teresa, make it somewhere on the ladder to sanctity. Others were judged to be false saints, accused of faking their stigmata or ability to live without food and exerting malicious influence over their followers and confessors, and ultimately ending up before an Inquisition or other type of religious court, which is how their stories became known.[10]

Along with Catholic women who walked the (narrow) boundary between sanctity and heresy, there were also those who challenged the boundaries between lay and religious life. Mary Ward (the eventual founder of the English Ladies), the Ursulines and Daughters of Charity in Italy and France, and other so-called 'Jesuitesses' have all received scholarly attention in the last decade for their attempts to create an active religious vocation for women out in the world.[11] The older opinion about such efforts is that they were either a failure or insignificant – what are a few hospitals or a few schools for girls? – but there has been increasing recognition that their actions and those of women religious in Protestant areas disrupt standard narratives about the Reformations. It is difficult to make the claim that 'the Reformation led to the closing of the convents in Protestant areas' (which is standard in textbooks and more specialised studies) when they survived in the German province of Saxony and Brunswick, the city of Strasburg and who knows yet where else; it is difficult to say that 'convents in Catholic Europe after the Council of

Trent were all enclosed' (another standard statement) when the convent walls were permeable in so many places. The activities of early modern women religious also disrupt standard narratives in women's history. It is difficult to say, for example, that 'Florence Nightingale was the first female nurse' or even that 'Florence Nightingale made nursing respectable for women' when completely respectable middle-class French women had been nursing as members of the Ursuline religious order for several centuries.

Studies of women religious have not only examined those who broke boundaries, they have also broken boundaries themselves, particularly those between disciplines. Art historians have explored how convents acted as patrons of the visual arts, ordering paintings and sculpture with specific subjects and particular styles for their own buildings and those of the male religious institutions they supported, thus shaping the religious images seen by men as well as women. Music historians have shown how women sang, composed, and played musical instruments, with their sounds sometimes reaching far beyond convent walls. Religious historians have examined the ways in which women circumvented, subverted, opposed, and occasionally followed the wishes of church authorities. Social historians have explored the ways in which women behind convent walls shaped family dynamics and thus political life. More importantly, scholars in all these fields have thought about the ways their stories intersect, as art and music both shape devotional practices and are shaped by them, as family chapels and tombs – often built by women – represent and reinforce power hierarchies, as artistic, literary, political, and intellectual patronage relationships influence and are influenced by doctrinal and institutional changes in the church.[12] This scholarship has thus changed the narrative of the Catholic Reformation, and also provided a good example of how problematic the notion of a clear public/private dichotomy can be in women's history; even after the Council of Trent, convents and their residents were very much part of the public realm of power politics and culture.

Many abbesses and other female religious wrote extensively, and their works are beginning to see modern editions and translations, or in some cases the first appearance of their words in print.[13] These texts have deepened our understanding of convent life and female spirituality, showing plays that nuns wrote for their sisters to perform, letters telling of attempts at converting Protestants in the street and supporting Catholics in prison, polemics praising convent life for its richness and others attacking it for its shallowness. These newly-discovered or newly-made-available sources have provided excellent examples of women's religious opinions and spiritual creativity, but they further increase the complexity of the story rather than making generalisations easier.

Editing, analysing, and translating women's religious writings has not been limited to Catholic women or convent residents. Though the names of a few Protestant women writers have been known for decades, only within the last several years have their whole works finally been made available. The model of such scholarship is Elsie McKee's magisterial two-volume work on Katherina Schütz Zell, a woman who used to be described as 'the wife of the reformer Matthias Zell' but for whom a better description, following McKee, would be 'a Strasburg reformer'.[14] Most of this work has so far concentrated on previously known works and figures, such as the German noblewoman Argula von Grumbach, the Genevan abbess Marie Dentière, or the English martyr Anne Askew, but the works of lesser-known writers, such as Dutch Anabaptist women who wrote hymns and songs, are beginning to see the light of day.[15]

No matter how many Protestant women are eventually discovered, however, research into women's actions and spirituality in the Reformation era will no doubt continue to favour Catholics and convents. This contrasts with the earliest years of research, when the focus was almost solely on Protestant women, but it should actually come as no surprise, given Reformation historians' dependence on institutional records; there simply are many more sources about women religious than their lay sisters, particularly the type of specific sources that lend themselves well to a dissertation or other focused study. In terms of sources *by* women about religious issues, those from educated nuns in the sixteenth century far outweigh those from lay women, Protestant or Catholic. By the seventeenth century, Protestant groups such as the Quakers or English Puritans often called for a written examination of conscience as part of the process of conversion; such reflections have survived for quite average women and men as well as the spiritual and social elite, so they allow for more study of lay women's spirituality and more direct comparisons between men's and women's experiences. For the sixteenth century, however, the survival of more sources contributes to the sense that Catholic women had more options than Protestant women during this period: they had, as long recognised, *maritus aut murus* (marriage or the convent) but also various in-between forms, limited and criticised though these were. Those studying Teresa, or convents, or women striving to be Jesuits do not argue this explicitly, as they are very careful to talk about the many constraints within which women acted. But the sheer number of studies – whatever their tone or conclusions

– makes it appear as if Catholic women were more likely than Protestant to have lives and ideas worthy of note.

If the Protestant/Catholic balance in the work emphasising women's agency implicitly suggests comparisons between Catholic and Protestant women, much of the work in the second line of inquiry – studies of the ideas of the reformers and the effects of the Reformations – has made these comparisons explicitly. This is particularly true for scholarship focusing on Germany, which began even earlier than studies investigating women's actions with works on Luther's notions of home and family. Older studies of Protestant ideas about marriage and the family were largely positive, seeing the Protestant Reformation as rescuing women, especially married women, from the misogyny of late medieval monastic culture. This positive evaluation has continued in the work of some German scholars and a few intellectual historians who write in English, including Luise Schorn-Schütte and Steven Ozment.[16]

Other scholars, primarily social historians or literary scholars trained outside of Germany, such as Lyndal Roper, Susan Karant-Nunn, and Sigrid Brauner, tend to view the impact of Protestant ideas more negatively.[17] They stress that viewing *marriage* in a positive light, as the Protestant reformers did, is not the same as viewing *women* positively, and may, in fact, have contributed to suspicion of unmarried women – and, to a lesser degree, unmarried men – as deviant and dangerous. Laws were passed prohibiting unmarried women to move into cities or live on their own, and ordering unmarried female servants to take positions only in households headed by men. In some cases grown, unmarried daughters were ordered to leave the household of their widowed mothers to find a position in a male-headed household.[18] Renate Dürr has traced the demonisation of female servants in the writings of Protestant moralists and the authors of urban law codes, who denounced them for immorality, sloth, laziness and disobedience.[19]

Scholarship on the radical Reformation has also been divided about whether the ideas and practices of Anabaptists, spiritualists and other radicals were positive or negative toward women. G. H. Williams argues that the radical groups offered women more opportunities and Claus-Peter Clasen that they were more restrictive and patriarchal.[20] Jane Dempsey Douglass and John Thompson have disagreed sharply about Calvin's ideas, though their dispute has centred on the degree to which Calvin broke with his predecessors and contemporaries, with only implicit references to the issue of whether the impact of Calvinism on women was positive or negative.[21]

The question of whether Protestantism was good or bad for women – a question a friend of mine calls the Glinda-test, from the good witch in *The Wizard of Oz* who asks Dorothy whether she is a good witch or a bad witch – has generally fallen out of favour as a research issue in the last decade. This is due in part to the general stress on difference and diversity in women's history – which women? Where? When? Married or single? Old or young? Urban or rural? Mothers or childless? In part this is also because many scholars have de-emphasised the role of the Protestant Reformation alone in changing women's lives or gender structures. Examination of the ideas of Christian humanists on such issues as marriage, spousal relations, and proper family life has led historians of England such as John Yost and Margo Todd to question whether the Protestants or Puritans were saying anything new.[22] Puritans might have emphasised spousal affection and the importance of stable families to the social order more loudly and at greater length than their predecessors, but these were hardly new ideas. More recently, Christine Peters finds continuities in patterns of piety centring on Christ and ideals for women extending from the fifteenth through the seventeenth centuries, and Kathleen Crowther-Heyck finds not much difference between Protestants and Catholics in terms of ideas about reproduction and childbirth.[23]

Doubts about the novelty of the Protestant message also emerge in the work of several historians of Germany, though they interpret this differently than do Yost and Todd. Heide Wunder sees changes in family life and ideas about marriage as a result not of changes in religious ideology, but of social and economic changes which allowed a wider spectrum of the population to marry and made the marital pair the basic production and consumption unit. This 'familialisation of work and life' happened, in her view, in the High Middle Ages, which means Reformation ideas about the family did not create the bourgeois family, but resulted from it.[24] This was one of the reasons that Protestant arguments in favour of marriage as the 'natural' vocation of women were accepted so readily by Catholics, and that the suspicion of unmarried women, unless they were safely in a convent, could be found in Catholic as well as Protestant areas. I have also examined the social context of 'Protestant' ideas about familial relations, and have argued that gendered ideas about men as workers and women as 'helpmeets' emerged in craft and journeymen's guilds before they become part of Reformation ideology.[25] Beate Schuster similarly sees hostility to unmarried women and praise of the male-headed bourgeois household as part of a new 'morality of settledness' that emerged out of an urban context before the Reformation.[26]

Isabel Hull traces the further development of the link between men's position as heads of household and notions of citizenship and civil society. She establishes how both the public and private spheres were male ones, with the latter created explicitly for the benefit of married men as a realm of life outside government interference. Only through a lack of state coercion in sexual matters could a man exercise the qualities expected of a citizen – independence, self-actualisation, energy – in all aspects of his life. Potent male sexuality, trained to be self-restraining rather than externally coerced, was the key quality for fitness for the new civil society; those who lacked this – women and lower-class men who could not restrain themselves – could be neither members of civil society nor citizens, and were still viewed as needing the institutions enforcing social discipline that had developed in the sixteenth century. Theories undergirding these ideas and laws that put them in place developed in both Protestant and Catholic areas of Germany.[27]

These historians of early modern England and Germany are not the only ones to question whether the Reformation or other events around 1500 brought any dramatic changes in the lives of women. As one might expect, medieval historians, most prominently Judith Bennett, have also questioned the significance of this great divide, and called for a greater focus on continuity.[28] In response to this, some scholars have begun using the term 'premodern' to refer to the long period from the tenth century through the eighteenth.

The term 'premodern' also has its critics, however. The literary critic Lee Patterson sharply attacks this 'crude binarism that locates modernity ("us") on one side and premodernity ("them") on the other'.[29] Historians of areas other than Europe, such as Barbara and Leonard Andaya, have wondered whether 'especially in light of subaltern writings that reject the notion of modernity as a universal', both premodern and early modern, 'implicitly set a "modern Europe" against a "yet-to-be modernised" non Europe'.[30]

These doubts about a premodern/modern dichotomy have emerged at just the same time that this dichotomy has been reinforced in an exploding area of historical research, the history of sexuality. Because of the influence of the French philosopher Michel Foucault, or better said, because of the influence of a particular way of reading Foucault's work on sex, much research has explored the development of 'modern' sexuality, or simply taken 'modern' sexuality as its topic.[31] Foucault, in the first volume of *History of Sexuality*, locates the beginning of the 'transformation of sex into discourse' – which he sees as the core of modern sexuality – with the practice of confession. He recognises that this expanded after the Reformations, as Catholics required more extensive and frequent confession and Protestants substituted the personal examination of conscience for oral confession to a priest.[32] This discourse about sexuality was later taken over by medical, political, and educational authorities, and it is this point that most of Foucault's successors (and disciples) see as the beginning of what interests them, going on to explore the mechanisms that define and regulate sexuality and investigating the ways in which individuals and groups described and understood their sexual lives. Even those who do look at earlier periods, and who stress the socially-constructed and historically-variable nature of all things sexual, tend to accept this notion of one clear break, terming their work the study of 'premodern' or even 'before' sexuality.[33]

This notion of a decisive break sometime in the eighteenth century works fine for medievalists, who have always been before the break anyway; this situation, combined with the significant and highly-respected early work on sexuality by James Brundage and Vern Bullough, may explain why there is so much good research on medieval sexuality going on right now.[34] For Reformation scholars, however, this is more problematic; the Reformation has always been on *this* side of the great break, part of what made the early modern period modern. So how is it that the modern family emerged from the Reformation (or, following Heide Wunder, even earlier and then was strengthened by the Reformation), but modern sexuality did not develop until several centuries later? If 1500 is dethroned but there is a break later, does that make Reformation scholars all medievalists?

There are two lines of research related to women and the Reformations that suggest it might still be valid to view the period around 1500 as a time of significant change. The first of these are studies of the ways in which after the Reformations, Protestant and Catholic religious authorities worked with rulers and other secular political officials to make people's behaviour more orderly and 'moral'. This process, termed 'social discipline' or 'the reform of popular culture', has been studied intensively over the last several decades in many parts of Europe. Scholarship on social discipline recognises the medieval roots of such processes as the restriction of sexuality to marriage, the encouragement of moral discipline and sexual decorum, the glorification of heterosexual married love, and the establishing of institutions for regulating and regularising behaviour, but it also emphasises that all of these processes were strengthened in the sixteenth century, and that this had a different effect on women than on men. Laws regarding such issues as adultery, divorce, 'lascivious carriage' (flirting), enclosure of members of religious orders, inter-denominational and inter-racial marriage were never gender-neutral. The enforcement of such laws was even more discriminatory, of course, for though undisciplined sexuality and immoral behaviour were portrayed from the pulpit or press as a threat to Christian order, it was women's lack of discipline that was most often punished. The newer scholarship on marriage, divorce, the family, and sexuality in the context of the Reformations makes clear that the roots of this strengthening are complex, and include much more than theology; though most studies are not directly comparative, as a whole they make it clear that these changes occurred in Catholic areas as well as Protestant, and that they involved church bodies such as the Inquisition as well as secular courts and other institutions.[35]

This complexity leads to an unwillingness to draw up grand schema and timetables – no-one is currently setting specific dates for developments like the 'rise of the restricted patriarchal nuclear family', as Lawrence Stone once did – but this scholarship does suggest that there was enough of a break with a pre-1500 past to retain this date as significant in the history of gender. It also makes clear that changes in gender ideology and structures are processes that take many decades or even centuries, but that they are just as significant as and also tightly interwoven with more familiar processes from this era such as 'the rise of the nation-state' and 'the growth of proto-industrial capitalism'.[36]

The second line of research is one that is only beginning to be related to the Reformations, but which will clearly be an important area of research in the coming decades – the expansion of Christianity beyond Europe with the first wave of colonialism. The Andayas' reference to subaltern theory noted above reminds us that viewing 1500 as a great gulf did not come solely from the chain of events that began in 1517, but also from the events that began in 1492.

Viewing the religious activities of early modern women and the creation of new intellectual and social structures of gender and sexuality in a global context has begun, not surprisingly, for scholars of Iberia, who now regularly hold conferences, publish article collections, and carry out their own research from a trans-oceanic perspective.[37] This has been joined by research on French Catholic women in Canada, some of whom continued the fight for a non-cloistered life on the other side of the Atlantic, and by studies of the Jesuits that analyse the way their experiences outside of Europe shaped their mission of social discipline within Europe.[38]

This wider geographic perspective should not be limited to those countries that actually founded colonies in the early modern era, however.

Like their French counterparts, German Ursulines thought about the world beyond Europe as in need of their services. Luther linked not only popes but also Turks with the Whore of Babylon, viewing his non-Christian foes in highly gendered and sexualised terms. German, Italian, and other reformers of all confessions considered the dangers of mixed-blood relationships, whether it was Anabaptists marrying Lutherans, Catholic wetnurses suckling Calvinist babies, or, as in a statute from the early colonial period of Virginia, a 'christian' (that is, English) woman 'committ[ing] fornication with a negro man'.[39] Studies of women, gender, and sexuality have reshaped the way we view the origins and impact of the Protestant and Catholic Reformations in Europe over the last thirty years, and will no doubt be key to understanding the role of religion as Europeans established colonial empires in the centuries after Luther. Studies that explore the intersection of gender and empire have overwhelmingly focused on the later British experience, but it is clear that analysis of this intersection in the early modern period has much to offer both Reformation studies and wider world history.

further reading

General overviews of women and/or gender in early modern Europe are a good place to start in providing a broader context for the issues discussed in this chapter. Those that cover many parts of Europe are: Merry E. Wiesner, *Women and Gender in Early Modern Europe*, 2nd edn (Cambridge, 2000); Olwen Hufton, *The Prospect Before Her: a History of Women in Western Europe, 1500–1800* (London, 1995); Margaret King, *Women of the Renaissance* (Chicago, 1991). For England, see Sara Mendelson and Patricia Crawford, *Women in Early Modern England* (Oxford, 1998) and Anthony Fletcher, *Gender, Sex, and Subordination in England 1500–1800* (New Haven and London, 1995). For Germany, see Heide Wunder, *He is the Sun, She is the Moon: Women in Early Modern Germany*, trans. Thomas Dunlap (Cambridge, MA, 1998).

The endnotes provide a number of suggestions for specialised studies of particular geographic areas and topics. More general works that explore women and the Reformations include: Sherrin Marshall (ed.), *Women in Reformation and Counter-Reformation Europe: Public and Private Worlds* (Bloomington, 1989); Patricia Crawford, *Women and Religion in England, 1500–1750* (London, 1993); Patricia Ranft, *Women and the Religious Life in Premodern Europe* (New York, 1996); Barbara Diefendorf, *From Penitence to Charity: Pious Women and the Catholic Reformation in Paris* (Oxford, 2004). For an overview of the Reformations and sexuality in a global context, see

Merry E. Wiesner-Hanks, *Christianity and the Regulation of Sexuality in the Early Modern World: Regulating Desire, Reforming Practice* (London, 2000). For other perspectives on recent research, see the historiographical 'focal point' in *Archiv für Reformationsgeschichte* 92 (2001), 274–320, with articles by Lyndal Roper, Heide Wunder, and Susanna Peyronel Rambaldi.

Issues of gender figure prominently in studies of the Reformation and marriage, such as Eric Josef Carlson, *Marriage and the English Reformation* (Oxford, 1994) and Joel F. Harrington, *Reordering Marriage and Society in Reformation Germany* (Cambridge, 1995). The rituals that marked transitions from one stage of life to another were also highly gendered, as discussed in Susan C. Karant-Nunn, *The Reformation of Ritual: an Interpretation of Early Modern Germany* (London, 1997) and David Cressy, *Birth, Marriage and Death: Ritual, Religion, and the Life-Cycle in Tudor and Stuart England* (Oxford, 1997).

The writings of many of the reformers are available in English translations. For a book that brings together many of Luther's writings on women and gender, see Susan C. Karant-Nunn and Merry E. Wiesner-Hanks, *Luther on Women: a Sourcebook* (Cambridge, 2003).

notes

1. Bonnie Smith, *The Gender of History: Men, Women, and Historical Practice* (Cambridge, MA, 1998).
2. Joan Scott, 'Gender: a Useful Category of Historical Analysis', *American Historical Review*, 91 (1986), 1053–75.
3. Some of the following material was presented as a paper in February of 2000 for the Burdick-Vary Symposium honouring Robert Kingdon at the University of Wisconsin-Madison, and published as 'Reflections on a Quarter-Century of Research on Women' in Lee Palmer Wandel (ed.), *History Has Many Voices* (Kirksville, MO, 2003). Material from that paper is reprinted here by permission of the Sixteenth Century Journal Publishers.
4. See, for example: Marian Kobelt-Groch, *Aufsässige Töchter Gottes: Frauen im Bauernkrieg und in den Taüferbewegungen* (Frankfurt, 1993); C. Arnold Snyder and Linda A. Huebert Hecht (eds), *Profiles of Anabaptist Women: Sixteenth-Century Reforming Pioneers* (Waterloo, Ontario, 1996); Diane Watt, *Secretaries of God: Women Prophets in Late Medieval and Early Modern England* (Rochester, NY, 1997).
5. The debate about female rulers, which became particularly intense in the sixteenth century as dynastic circumstances led to many women serving as rulers in their own right or as advisers to child kings, has been well-studied in the last several decades. A female ruler's power over religious matters generally figured in these debates, especially in the writings of those who opposed female rule, such as John Knox, who used Biblical injunctions against women having authority in church as one of many reasons to oppose women's rule.

A useful survey of the English part of this debate is Amanda Shepherd, *Gender and Authority in Sixteenth-Century England: the Knox Debate* (Keele, 1994).

6. Zarri has also edited the proceedings of many of these conferences. See Gabriella Zarri (ed.), *Donna, disciplina, creanza cristiana dal XV al XVII secolo: studi e testi a stampa* (Rome, 1996), including a very large bibliography of early modern works by and about religious women; *Ordini religiosi, santità e culti: prospettive di ricerca tra Europa e America Latina: atti del Seminario di Roma, 21–22 giugno 2001* (Galatina, 2003).
7. Merry E. Wiesner, 'Ideology Meets the Empire: Reformed Convents and the Reformation', in Susan Karant-Nunn and Andrew Fix (eds), *Germania Illustrata: Essays Presented to Gerald Strauss* (Kirksville, MO, 1991); D. Jonathan Grieser, 'A Tale of Two Convents: Nuns and Anabaptists in Münster, 1533–1535', *Sixteenth Century Journal*, 26 (1995), 31–48; Paula S. Datsko Barker, 'Charitas Pirckheimer: a Female Humanist Confronts the Reformation', *Sixteenth Century Journal*, 26 (1995), 259–72; Amy Leonard, *Nails in the Wall: Catholic Nuns in Reformation Germany* (Chicago, forthcoming); Charlotte Woodford, *Nuns as Historians in Early Modern Germany* (Oxford, 2003).
8. Jodi Bilinkoff, *The Avila of St. Theresa* (Ithaca, 1989); Alison Weber, *Teresa of Avila and the Rhetoric of Femininity* (Princeton, 1990); Carole Slade, *Saint Teresa of Avila: Author of a Heroic Life* (Berkeley, 1995); Gillian T. W. Ahlgren, *Teresa of Avila and the Politics of Sanctity* (Ithaca, 1996).
9. Electa Arenal and Stacey Schlau, *Untold Sisters: Hispanic Nuns in their Own Works* (Albuquerque, 1989); Craig Harline, *The Burdens of Sister Margaret: Private Lives in a Seventeenth-Century Convent* (New York, 1994); Ronald E. Surtz, *Writing Women in Late Medieval and Early Modern Spain: the Mothers of Saint Teresa of Avila* (Philadelphia, 1995); Sherry M. Velasco, *Demons, Nausea and Resistance in the Autobiography of Isabel de Jesús 1611–1682* (Albuquerque, 1996). Studies of the relationships between female penitents and their confessors include: Rudolph M. Bell, 'Telling her Sins: Male Confessors and Female Penitents in Catholic Reformation Italy', in Lynda L. Coon et al. (eds), *That Gentle Strength: Historical Perspectives on Women in Christianity* (Charlottesville, 1990); Jodi Bilinkoff, 'Confessors, Penitents, and the Construction of Identities in Early Modern Avila', in Barbara B. Diefendorf and Carla Hesse (eds), *Culture and Identity in Early Modern Europe (1500–1800): Essays in Honor of Natalie Zemon Davis* (Ann Arbor, 1993); Patricia Ranft, 'A Key to Counter-Reformation Women's Activism: the Confessor-Spiritual Director', *Journal of Feminist Studies in Religion*, 10 (1994), 7–26; Stephen Haliczer, *Sexuality in the Confessional: a Sacrament Profaned* (New York, 1996); Cordula van Wyhe, 'Court and Convent: the Infanta Isabella and her Franciscan Confessor Andres de Soto', *Sixteenth Century Journal* (forthcoming 2005).
10. Recent studies of *beatas* and 'living saints' in southern Europe include Gabriella Zarri, *Le sante vive: Profezie di corte e devozione feminille tra '400 e '500* (Torino, 1990) and her 'Living Saints: a Typology of Female Sanctity in the Early Sixteenth Century', in Daniel Bornstein and Roberto Rusconi (eds), *Women and Religion in Medieval and Renaissance Italy* (Chicago, 1996); Richard Kagan, *Lucrecia's Dreams: Politics and Prophecy in Sixteenth Century Spain* (Berkeley, 1990); Jodi Bilinkoff, 'A Spanish Prophetess and her Patrons: the Case of María de Santo Domingo', *Sixteenth Century Journal*, 23 (1992), 21–34; Fulvio Tomizza, *Heavenly Supper: the Story of Maria Janis*, trans. Anne

Jacobson Schutte (Chicago, 1993); Anne Jacobson Schutte, *Aspiring Saints: Pretense of Holiness, Inquisition, and Gender in the Republic of Venice, 1618–1750* (Baltimore, 2001).

11. Ruth Liebowitz, 'Virgins in the Service of Christ: the Dispute over an Active Apostolate for Women During the Counter-Reformation', in Rosemary Radford Ruether and Eleanore McLaughlin (eds), *Women of Spirit: Female Leadership in the Jewish and Christian Traditions* (New York, 1979); Elizabeth Rapley, *The Dévotes: Women and Church in Seventeenth-Century France* (Montreal and Kingston, 1990); Anne Conrad, *Zwischen Closter und Welt: Ursulinen und Jesuitinnen in der Katholischen Reformbewegung des 16./17. Jahrhunderts* (Mainz, 1991); Jeanne Cover, *Love, The Driving Force: Mary Ward's Spirituality: its Significance for Moral Theology* (Milwaukee, 1997); Susan Dinan, 'An Ambiguous Sphere: the Daughters of Charity between a Confraternity and a Religious Order', in John Patrick Donnelly and Michael Maher (eds), *Confraternities and Catholic Reform in Italy, France, and Spain* (Kirksville, MO, 1999).
12. For convent residents and the arts, see Craig Monson (ed.), *The Crannied Wall: Women, Religion and the Arts in Early Modern Europe* (Ann Arbor, 1992); idem, *Disembodied Voices: Music and Culture in an Early Modern Italian Convent* (Berkeley, 1995); Ann Matter and John Coakley (eds), *Creative Women in Medieval and Early Modern Italy* (Philadelphia, 1997); Robert Kendrick, *Celestial Sirens: Nuns and Their Music in Early Modern Italy* (Oxford, 1996); Cynthia Lawrence (ed.), *Women and Art in Early Modern Europe: Patrons, Collectors, and Connoisseurs* (University Park, PA, 1997); Elissa B. Weaver, *Convent Theatre in Early Modern Italy: Spiritual Fun and Learning for Women* (Cambridge, 2002); K. J. P. Lowe, *Nuns' Chronicles and Convent Culture in Renaissance and Counter-Reformation Italy* (Cambridge, 2003); Helen Hills, *Architecture and the Politics of Gender in Early Modern Europe* (London, 2004). For the economic and political patronage of religious women, see María Leticia Sánchez Hernández, *Patronato, regio y órdenes religiosas femeninas en el Madrid de los Austrias* (Madrid, 1997); Monica Chojnacka, 'Women, Charity and Community in Early Modern Venice: the Casa dell Zitelle', *Renaissance Quarterly*, 51 (1998), 68–79; Jodi Bilinkoff, 'Elite Widows and Religious Expression in Early Modern Spain: the View from Avila', in Sandra Cavallo and Lyndan Warner (eds), *Widowhood in Medieval and Early Modern Europe* (London, 1999); Elizabeth A. Lehfeldt, 'Discipline, Vocation, and Patronage: Spanish Religious Women in a Tridentine Microclimate', *Sixteenth Century Journal*, 30 (1999), 1009–30; Sharon T. Strocchia, 'Sisters in Spirit: the Nuns of Sant'Ambrogio and their Consorority in Early Sixteenth-Century Florence', *Sixteenth Century Journal*, 33 (2002), 735–68; Silvia Evangelisti, '"We do not have it, and we do not want it": Women, Power, and Convent Reform in Florence', *Sixteenth Century Journal*, 34 (2003), 677–700. The relationship between family politics and women's entry into convents has been explored in Elizabeth Rapley, 'Women and the Religious Vocation in Seventeenth-Century France', *French Historical Studies*, 18 (1994), 613–31; P. Renée Baernstein, *A Convent Tale: a Century of Sisterhood in Spanish Milan* (New York, 2002); Barbara Diefendorf, 'Give Us Back our Children: Patriarchal Authority and Parental Consent to Religious Vocation in Early Counter-Reformation France', *Journal of Modern History*, 68 (1996), 265–307; Joanne Baker, 'Female Monasticism and Family Strategy: the Guises and Saint Pierre de Reims', *Sixteenth Century Journal*, 28 (1997), 1091–108; Thomas Worcester, '"Neither Married nor Cloistered": Blessed

Isabelle in Catholic Reformation France', *Sixteenth Century Journal*, 30 (1999), 457–76; Elizabeth Lehfeldt, 'Gender, Order, and the Meaning of Monasticism in the Reign of Ferdinand and Isabella', *Archiv für Reformationsgeschichte*, 93 (2002), 145–71.

13. Many of these texts have been published by the University of Chicago Press, in the series *The Other Voice in Early Modern Europe*, ed. Margaret King and Albert Rabil, Jr. The works by nuns included in this series are: Cecelia Ferrazzi, *Autobiography of an Aspiring Saint*, ed. and trans. Anne Jacobson Schutte (1996); Antonia Pulci, *Florentine Drama for Convent and Festival*, ed. and trans. James Wyatt Cook and Barbara Collier Cook (1996); Bartolomea Riccoboni, *Life and Death in a Venetian Convent: the Chronicle and Necrology of Corpus Domini, 1395–1436*, ed. and trans. Daniel Bornstein (2000); Lucrezia Tornabuoni, *Sacred Narratives*, ed. and trans. Jane Tylus (2001); Maria de San José, *Book of Recreations*, ed. and trans. Amanda Powell and Alison Weber (2002); Jacqueline Pascal, *A Rule for Children and Other Writings*, ed. and trans. John Conley (2003). In addition, Marquette University Press's *Reformation Texts with Translation (1350–1600): Women of the Reformation* Series, ed. Kenneth Hagen and Merry Wiesner-Hanks, has issued: Merry Wiesner-Hanks and Joan Skocir (ed. and trans.), *Convents Confront the Reformation: Catholic and Protestant Nuns in Germany* (1996); Elizabeth Rhodes (ed. and trans.), *'This Tight Embrace': Luise de Carvajal y Mendoza* (2000); Linda Lierheimer (ed. and trans.), *Spiritual Autobiography and the Construction of Self: the Mémoires of Antoinette Micolon* (2004).
14. Elsie Anne McKee, *Katharina Schütz Zell*, 2 vols (Leiden, 1999); idem, *Reforming Popular Piety in 16th Century Strasburg: Katherine Schütz Zell and Her Hymnbook*, Studies in Reformed Theology and History, Princeton Theological Seminary 2/4 (1994); idem, 'Katherina Schütz Zell: Protestant Reformer', in Timothy J. Wengert and Charles W. Brockwell (eds), *Telling the Churches' Story: Ecumenical Perspectives on Writing Christian History* (Grand Rapids, 1995); Merry E. Wiesner, 'Katherine Zell's "Answer to Ludwig Rabus" as Autobiography and Theology', *Colloquia Germanica*, 28 (1995), 245–54.
15. Peter Matheson, *Argula von Grumbach: a Woman's Voice in the Reformation* (Edinburgh, 1995); idem, 'Breaking the Silence: Women, Censorship and the Reformation', *Sixteenth Century Journal*, 27 (1996), 97–109; idem, 'A Reformation for Women? Sin, Grace, and Gender in Argula von Grumbach', *Scottish Journal of Theology*, 49 (1996), 1–17; Susannah Brietz Monta, 'The Inheritance of Anne Askew, English Protestant Martyr', *Archiv für Reformationsgeschichte*, 94 (2003), 134–60; Thomas Head, 'The Religion of the "Femmelettes": Ideals and Experiences Among Women in Fifteenth and Sixteenth-Century France', in Coon et al. (eds), *That Gentle Strength*; Hermoine Joldersma and Louis Grijp (eds and trans.), *'Elisabeth's Manly Courage': Testimonials and Songs by and about Martyred Anabaptist Women* (Milwaukee, 2001).
16. Steven Ozment, *When Fathers Ruled: Family Life in Reformation Europe* (Cambridge, MA, 1983); Luise Schorn-Schütte, '"Gefährtin" und "Mitregentin": Zur Sozialgeschichte der evangelischen Pfarrfrau in der Frühen Neuzeit', in Heide Wunder and Christina Vanja (eds), *Wandel der Geschlechterbeziehungen zu Beginn der Neuzeit* (Frankfurt, 1991). For a more recent, and less celebratory interpretation, see Scott Hendrix, 'Luther on marriage', *Lutheran Quarterly*, 14 (2000), 335–50.

17. Lyndal Roper '"The Common Man", "the common good", "common women": Reflections on Gender and Meaning in the Reformation German Commune', *Social History*, 12 (1987), 1–21; idem, *The Holy Household: Women and Morals in Reformation Augsburg* (Oxford, 1989); Susan C. Karant-Nunn, 'Continuity and Change: Some Effects of the Reformation on the Women of Zwickau', *Sixteenth Century Journal*, 13 (1982), 17–42; idem, 'The Transmission of Luther's Teachings on Women and Matrimony: the Case of Zwickau', *Archiv für Reformationsgeschichte*, 77 (1986), 31–46; idem, 'The Reformation of Women', in Renate Bridenthal, Susan Mosher Stuard and Merry E. Wiesner (eds), *Becoming Visible: Women in European History* (Boston, 1998); Sigrid Brauner, *Fearless Wives and Frightened Shrews: the Construction of the Witch in Early Modern Germany* (Amherst, 1994).
18. Merry Wiesner-Hanks, 'Having Her Own Smoke: Employment and Independence for Unmarried Women in Germany, 1400–1700', in Judith Bennett and Amy Froide (eds), *Singlewomen in the European Past* (Philadelphia, 1999).
19. Renate Dürr, *Mägde in der Stadt: Das Beispiel Schwäbisch-Hall in der Frühen Neuzeit* (Frankfurt, 1995).
20. G. H. Williams, *The Radical Reformation* (Philadelphia, 1962), pp. 506–7; Claus-Peter Clasen, *Anabaptism: a Social History, 1525–1618* (Ithaca, 1972), p. 207. For more recent scholarship on gender in the Radical Reformation, see Marion Kobelt-Groch, 'Why did Petronella Leave her Husband? Reflections on Marital Avoidance among the Halberstadt Anabaptists', *Mennonite Quarterly Review*, 62 (1988), 26–41; Wes Harrison, 'The Role of Women in Anabaptist Thought and Practice: the Hutterite Experience of the Sixteeenth and Seventeenth Centuries', *Sixteenth Century Journal*, 23 (1992), 49–70; Snyder and Hecht, *Profiles*; Craig D. Atwood, 'Sleeping in the Arms of Christ: Sanctifying Sexuality in the Eighteenth-Century Moravian Church', *Journal of the History of Sexuality*, 8 (1997), 25–47; Stephen Boyd, 'Theological Roots of Gender Reconciliation in Sixteenth Century Anabaptism: a Prolegomenon', *Journal of Mennonite Studies*, 17 (1999), 34–51; Linda Heubert Hecht, 'A Brief Moment in Time: Informal Leadership and Shared Authority among Sixteenth-Century Anabaptist Women', *Journal of Mennonite Studies*, 17 (1999), 52–74; Werner O. Packull, '"We are Born to Work Like Birds to Fly": the Anabaptist-Hutterite Ideal Woman', *Mennonite Quarterly Review*, 74 (2000), 51–85; Jeni Hiett Umble and Linda Huebert Hecht (eds), *Strangers at Home: Amish and Mennonite Women in History* (Baltimore, 2002).
21. Jane Dempsey Douglass, *Women, Freedom and Calvin* (Philadelphia, 1985); John Lee Thompson, *John Calvin and the Daughters of Sarah: Women in Regular and Exceptional Roles in the Exegesis of Calvin, His Predecessors and His Contemporaries* (Geneva, 1992).
22. John K. Yost, 'Changing Attitudes Towards Married Life in Civic and Christian Humanists', *American Society for Reformation Research, Occasional Papers*, 1 (1977), 151–66; Margo Todd, 'Humanists, Puritans and the Spiritualized Household', *Church History*, 49 (1980), 18–34; idem, *Christian Humanism and the Puritan Social Order* (Cambridge, 1987). See also Diane Willen, 'Godly Women in Early Modern England: Puritanism and Gender', *Journal of Ecclesiastical History*, 43 (1992), 561–80.
23. Christine Peters, *Patterns of Piety: Women, Gender, and Religion in Late Medieval and Reformation England* (Cambridge, 2003); Kathleen Crowther-Heyck, '"Be

Fruitful and Multiply": Genesis and Generation in Reformation Germany', *Renaissance Quarterly*, 55 (2002), 904–35.

24. Heide Wunder, *'Er ist die Sonn, sie ist der Mond': Frauen in der frühen Neuzeit* (Munich, 1992), translated as *He is the Sun, She is the Moon: Women in Early Modern Germany*, trans. Thomas Dunlap (Cambridge, MA, 1998).
25. Merry Wiesner-Hanks, 'Guilds, Male Bonding and Women's Work in Early Modern Germany', *Gender and History*, 1 (1989), 125–37; idem, 'The Religious Dimensions of Guild Notions of Honor in Reformation Germany' in Sibylle Backmann et al. (eds), *Ehrkonzepte in der Frühen Neuzeit: Identitäten und Abgrenzungen*, Colloquia Augustana 8 (Berlin, 1998).
26. Beate Schuster, *Die freie Frauen: Dirnen und Frauenhäuser in 15. und 16. Jahrhundert* (Frankfurt, 1995).
27. Isabel Hull, *Sexuality, State, and Civil Society in Germany, 1700–1815* (Ithaca, 1996).
28. Judith Bennett, 'Medieval Women, Modern Women: Across the Great Divide', in David Aers (ed.), *Culture and History 1350–1600: Essays on English Communities, Identities and Writing* (London, 1992).
29. Lee Patterson, 'On the Margin: Postmodernism, Ironic History, and Medieval Studies', *Speculum*, 65 (1990), 107.
30. Leonard Y. Andaya and Barbara Watson Andaya, 'Southeast Asia in the Early Modern Period: Twenty-Five Years On', *Journal of Southeast Asian Studies*, 26/1 (1995), 92.
31. For a good survey of modern Western sexuality, see Carolyn Dean, *Sexuality and Modern Western Culture* (New York, 1996).
32. Michel Foucault, *L'Histoire de la sexualité 1: La Volonté de savoir* (Paris, 1976).
33. Jacqueline Murray and Konrad Eisenbichler (eds), *Desire and Discipline: Sex and Sexuality in the Premodern West* (Toronto, 1996); Louise Fradenburg and Carla Freccero (eds), *Premodern Sexualities* (New York, 1996); Anne L. McLanan and Karen Rosoff Encarnación (eds), *The Material Culture of Sex, Procreation, and Marriage in Premodern Europe* (London, 2002); David M. Halperin, John J. Winkler and Froma I. Zeitlin (eds), *Before Sexuality: the Construction of Erotic Experience in the Ancient Greek World* (Princeton, 1990).
34. The best introductions to the scholarship on medieval sexuality are Joyce E. Salisbury, *Medieval Sexuality: a Research Guide* (New York, 1990) and Vern L. Bullough and James A. Brundage, *Handbook of Medieval Sexuality* (New York, 1996).
35. The scholarship on social discipline and the Reformations is huge. Monographs alone include: Thomas Max Safley, *Let No Man Put Asunder: the Control of Marriage in the German Southwest* (Kirksville, MO, 1984); Martin Ingram, *Church Courts, Sex and Marriage in England 1570–1640* (Cambridge, 1987); James R. Farr, *Authority and Sexuality in Early Modern Burgundy* (New York, 1995); Jeffrey R. Watt, *The Making of Modern Marriage: Matrimonial Control and the Rise of Sentiment in Neuchâtel, 1550–1800* (Ithaca, 1992); Robert Kingdon, *Adultery and Divorce in Calvin's Geneva* (Cambridge, MA, 1995); Michael F. Graham, *The Uses of Reform: 'Godly Discipline' and Popular Behavior in Scotland and Beyond 1560–1610* (Leiden, 1996); Helen Parish, *Clerical Marriage and the English Reformation: Precedent, Policy and Practice* (London, 2000); Georgina Dopico Black, *Perfect Wives, Other Women: Adultery and Inquisition in Early Modern Spain* (Durham, 2001); Helmut Puff, *Sodomy in Reformation Germany and Switzerland, 1400–1600* (Chicago, 2003); Renato Barahona, *Sex Crimes, Honour, and the Law*

in Early Modern Spain, Vizcaya, 1500–1750 (Toronto, 2003); Daniela Hacke, *Women, Sex, and Marriage in Early Modern Venice* (London, 2004). For a recent theoretical discussion, see the 'Focal Point: Confessionalization and Social Discipline in France, Italy, and Spain', with articles by James R. Farr, Wietse de Boer, and Allyson Poska, in *Archiv für Reformationsgeschichte*, 94 (2003), 276–319.

36. There is a growing body of literature linking gender and the nation-state in the early modern period, but most of this does not pay much attention to religion. Three recent exceptions are Luisa Accati, 'L'importanza della diversità nella definizione de sé e nella definizione dello stato moderno: Appunti sulla controriforma', *Archiv für Reformationsgeschichte*, 92 (2001), 264–73; Anne Mclaren, 'Gender, Religion, and Early Modern Nationalism: Elizabeth I, Mary Queen of Scots, and the Genesis of English anti-Catholicism', *American Historical Review*, 107 (2002), 739–67; Ulrike Strasser, *State of Virginity: Gender, Religion, and Politics in an Early Modern Catholic State* (Ann Arbor, 2003).
37. *I Congreso Internacional del monacato femenino en España, Portugal y América, 1492–1992*, vol. 2 (León, 1993); Mary G. Giles (ed.), *Women in the Inquisition: Spain and the New World* (Baltimore, 1998); Kenneth J. Adrien and Rolena Adorno (eds), *Transatlantic Encounters: Europeans and Andeans in the Sixteenth Century* (Berkeley, 1991); Josiah Blackmore and Gregory S. Hutcheson (eds), *Queer Iberia* (Durham, 1999); Charlene Villaseñor Black, 'Love and Marriage in the Spanish Empire: Depictions of Holy Matrimony and Gender Discourses in the Seventeenth Century', *Sixteenth Century Journal*, 32 (2001), 637–68; idem, *Creating the Cult of St. Joseph: Art and Gender in the Spanish Empire* (Princeton, 2003); Kathleen Ann Myers, *Neither Saints nor Sinners: Writing the Lives of Women in Spanish America* (Oxford, 2003); Carolyn Brewer, *Shamanism, Catholicism, and Gender Relations in Colonial Philippines, 1521–1695* (London, 2004); Nora E. Jaffary, *False Mystics: Deviant Orthodoxy in Colonial Mexico* (Lincoln, NE, 2004). Scholars of Iberia have also led the way in studying the religious activities of early modern Muslim women. See, for example: Mary Elizabeth Perry, *The Handless Maiden: Moriscos and the Politics of Religion in Early Modern Spain* (Princeton, 2005) and Ronald E. Surtz, 'Morisco Women, Written Texts, and the Valencia Inquisition', *Sixteenth Century Journal*, 32 (2001), 421–34.
38. Natalie Zemon Davis provides an extensive analysis of Marie of the Incarnation, the founder of the first women's convent in Quebec, in *Women on the Margins: Three Seventeenth-century Lives* (Cambridge, MA, 1995); Patricia Simpson discusses the early life of Marguerite Bourgeoys, who founded a community of lay missionary teaching sisters in Montreal in *Marquerite Bourgeoys and Montreal* (Montreal and Kingston, 1997); Allan Greer discusses the first Native American to make a Christian vow of virginity and perpetual spirituality in *Mohawk Saint: Catherine Tekakwitha and the Jesuits* (New York, 2004). Susan E. Dinan and Debra Meters (eds), *Women and Religion in Old and New Worlds* (London, 2001), and Allan Greer and Jodi Bilinkoff (eds), *Colonial Saints: Discovering the Holy in the Americas, 1500–1800* (New York, 2003) include material from several areas. For the Jesuits, see Jennifer D. Selwyn, *A Paradise Inhabited by Devils: the Jesuits' Civilising Mission in Early Modern Naples* (Aldershot, 2004).
39. William Waller Hening (ed.), *The Statutes at Large; Being a Collection of All the Laws of Virginia*, vol. II (Charlottesville, 1823), p. 170.

13
religious persecution and warfare

william monter

chronological overview, 1520–1648

To oversimplify matters: a phase of religious persecution in Reformation Europe (c.1520–60) rapidly generated a theory of political resistance which was used to justify a phase of religious warfare (c.1560–1630). The first Lutheran martyrs died only a few months after Luther's famous defiance of papal authority at the Diet of Worms. Nevertheless, the movement spread with amazing rapidity in the 1520s. Its political leaders soon formed a self-defence league, calling themselves 'Protestants' after filing an official protest to the Imperial Diet in 1529 (their enemies simply called them 'heretics' or used even less polite terms). However, with one brief exception, religious conflict between Protestant and Catholic states began only after Luther's death in 1547, and was neither prolonged nor severe until the 1560s. By then, the bulk of state-sponsored religious persecution had already ended in Latin Christendom.

Although the subject has never been investigated with proper thoroughness,[1] we can be sure that the repression of early forms of Protestantism was both rapid and extensive. By the time the term 'Protestantism' emerged, more than a hundred of its adherents had already been tried and executed, mainly in a few of the movement's early centres in the Holy Roman Empire and the Low Countries. In the Austrian Alpine province of Tyrol, several hundred Protestants died in the aftermath of a (non-Protestant) peasant uprising of 1526. Several hundred more died in the Low Countries in the mid-1530s. Twenty years later, the English persecutions under Mary Tudor added nearly three hundred more – Europe's third highest total in any region in any five-year period. Europe's largest kingdom, France, never experienced this level of intensity, but nevertheless publicly executed nearly five hundred Protestants before

1560. However, with the partial exception of the 'Council of Troubles' in the Low Countries (which executed over a thousand people for both sedition and heresy between 1567 and 1572), Reformation Europe experienced no large-scale religious persecutions after 1560.

Of course, western Europe's well-known wars of religion broke out precisely in the 1560s; among other things, they provoked the creation of the 'Council of Troubles'. These wars lasted until the century's end in France and even well beyond in the Low Countries. However, for most early modern historians, the Thirty Years War in the Empire (1618–48) marks the crescendo of religious warfare in Reformation Europe. It also marks its conventional end – even if the Swiss fought a short Protestant–Catholic civil war as late as the 1840s and occasional atavistic vestiges still afflict a few places like Belfast. Although the Thirty Years War clearly began as a Protestant–Catholic conflict in the Holy Roman Empire, it progressively lost its religious coloration as it dragged on and spread geographically. Most historians agree that it ceased to be primarily a war of religion after 1630, when Cardinal Richelieu allied France with the great Lutheran 'Lion of the North', Gustavus Adolphus of Sweden. After 1635 its major protagonists were two Catholic states, France and Spain, who would remain at war until 1659.

protestant theories of resistance

Although it rapidly evolved a theory of political resistance to 'idolatrous' Catholic rulers, Reformation Europe managed to avoid prolonged religious warfare for a full generation. Martin Luther was an extremely reluctant rebel, for obvious reasons: his Biblical principle of *sola scriptura* and his primary reliance on the New Testament taught him that Jesus fully accepted the legitimacy of Roman rule over the Holy Land, and therefore that duly constituted secular authority should be obeyed. Among leading Protestant theologians, Scriptural authority was not easily bent and twisted into justifications for rebellion, which however seemed necessary for this emerging movement. Since purely political alliances preceded its elaboration, the history of Protestant resistance theory was largely an *ex post facto* search for satisfactory exegeses of generally unpromising Biblical texts. Luther avoided this problem and found a different way to overcome his aversion to physical resistance on behalf of the Gospel. Luther experts like Mark Edwards agree with such major historians of political thought as Quentin Skinner that Luther reversed his earlier positions in October 1530, by accepting that resistance to the Emperor was a matter reserved exclusively to civil lawyers rather than theologians. In a treatise composed shortly afterwards, Luther drily noted that rebellion 'should have another name, which the law will surely find'.[2]

Others found different solutions. Because of their enormous heterogeneity, one cannot synthesise the political theory of Luther's radical enemies of the 1520s, the people he called *Schwärmer* and modern historians call Anabaptists (James Stayer's *Anabaptists and the Sword* remains the best survey). A few of their more notorious apocalyptical leaders, from Thomas Müntzer during the great popular revolt of 1525 to the rulers of Münster's 'New Jerusalem' in 1534, ignored New Testament pacifism in favour of a war of annihilation against all enemies of the Gospel. But after they were defeated and eliminated, nearly all the remnants of Anabaptism either remained or became pacifists; as Stayer argues, their political theory continued to insist on the necessarily corrupt nature of *all* secular governments.[3] In this way, Anabaptists bear an odd resemblance to modern anarchists; but their conclusion was to avoid political authority as far as possible rather than overturn it. They remained impaled on Protestantism's Scriptural dilemma, and unlike Luther (who was steadfastly protected by a ruler of major importance), they suffered enormously in the Reformation era.

Jehovah's protection of the Israelites in the Old Testament offered more promising material for Protestant resistance theory. Under the strain of military defeat in 1547 and subsequent attempts to reimpose Catholicism by force, German Lutherans discovered the usefulness of Old Testament texts supporting the argument that even lesser magistrates had the right to oppose idolatrous rulers. But these texts and arguments were taken over after 1560 and soon amplified by theologians and propagandists of the Reformed churches, the final and most bellicose variety of 'magisterial' Protestantism. If Jean Calvin, the outstanding exponent of the Reformed tradition, extended the doctrine of resistance to political authorities without sovereign authority, some of his most important followers stretched the most serviceable Old Testament texts to the point of justifying the assassination of 'idolatrous' rulers. Living in an age of Christian ecumenism, we cannot hope to understand Reformation Europe's wars of religion without understanding the sometimes complicated justifications offered by rebelling Protestants, replete with legal technicalities but resting ultimately on Scriptural references.[4]

persecution in reformation europe

Probably even more difficult to grasp today than political theory justified by Scripture was the extent and severity of religious persecution in

Reformation Europe. Secular courts conducted the vast majority of these heresy trials in northern Europe; even in Mediterranean Europe, they were usually managed by state-controlled inquisitions. Together, they provided a bloody prelude to the subsequent political risings, and recent memories from it stiffened the resolve of Protestant rebels.

Largely because of the printing press, Reformation Europe produced the first successful celebrations of martyrs since the Roman Empire. Everyone accepted St Augustine's definition that 'not the punishment, but the cause, makes the martyr', but everyone interpreted it selectively. Following the useful division proposed in Brad Gregory's magisterial survey of this phenomenon,[5] Reformation-era martyrs can be split into 'magisterial Protestants', combining three confessions (Lutherans, Anglicans, and Reformed), each of which produced separate printed martyrologies; Anabaptists, who had to use novel forms of commemoration because they had extremely few printing presses in the sixteenth century; and Roman Catholics, who faced quite different theological problems in commemorating their numerous sixteenth-century martyrs.

So-called 'magisterial' Protestants produced several early and remarkably successful printed martyrologies; 'between 1552 and 1559, four major martyrologies appeared in a milieu of threatened international Protestantism'.[6] John Foxe's *Acts and Monuments*, now being reproduced in a state-of-the-art technological version, clearly became a major prop of the established Protestant Church of England, and was placed in some parish churches alongside approved versions of the Bible.[7] Although it never acquired such quasi-official status, Jean Crespin's *Histoire des Martyrs*, commemorating the martyrs of the Reformed church, was begun even earlier and revised more often. German Lutherans had the earliest Protestant martyrs, but embarrassingly few of them; Luther had envisaged his own martyrdom at the Diet of Worms in 1521, but he died peacefully in bed over twenty-five years later. Nevertheless, not wanting to be left behind, Ludwig Rabus produced a multi-volume Lutheran collection.[8]

Confessional purity remained a constant and major concern of all major Protestant martyrologists; Augustine's maxim meant that excluding 'confessionally incorrect' martyrs was just as important as finding every documentable case where someone's life had been sacrificed for the true Gospel faith. At the same time, there was considerable overlap among Protestants; everyone, for example, could include John Hus. Augustine's dictum also implied that quality was as important as quantity, so martyrologists gave enormous importance to the recorded testimony of people facing death for the truths of the Gospel. Therefore this popular Reformation literary genre recycled large amounts of confessionally-oriented discourse from 'authentic' martyrs, while ignoring others. The most important shared feature of 'magisterial' Protestant martyrologists was to exclude anyone who died as an Anabaptist or other type of sectarian radical; therefore, all of them excluded the overwhelming majority of people burned to death for religious reasons during the early Protestant Reformation.

Anabaptists faced the reverse problem: huge numbers of martyrs, but almost no printers. Consequently, their methods of commemoration were often original, but probably effective in reinforcing confessional loyalty. However, this movement was also far more fragmented than magisterial Protestantism; consequently, its martyrological commemorations became 'micro-confessional'.[9] The most innovative among them were Dutch Mennonites, who soon developed elaborate forms of musical commemorations for their martyrs. The second volume of Dutch Anabaptist martyrology, *The Sacrifice to the Lord*, published in 1560, consisted entirely of lyrics. Europe's other major permanent Anabaptist sect, the Hutterites, first printed their early martyrs' songs in 1914.

Overall, as Judith Pollmann notes, 'we know far more about Protestant than about Catholic uses of vernacular song; there have been few attempts to compare the use of song in different parts of Reformation Europe, and no attempts to compare the messages of songs with those of pamphlets and sermons'.[10] Nevertheless, certain features seem clear. Luther's very first song, *Ein Neues Lied Wir Heben An*, glorifying two Augustinians who died for their faith in Brussels in 1523, deliberately echoed the wording of traditional songs about saints and martyrs. However, Anabaptist martyrs' songs did not just glorify the martyrs' steadfastness and condemn their cruel oppressors, but also taught people for what reasons these martyrs had died. 'By singing details of interrogations,' argues Pollmann, 'they also taught believers how to articulate, defend and provide biblical evidence for their beliefs.' If Anabaptists had most reason to invent musical commemorations of their martyrs, they had no monopoly in these matters. As early as 1574, a song lamenting the fate of fourteen Catholic priests martyred by Dutch rebels at Gorinchem was set to the tune of the already-famous Dutch Protestant *Wilhelmus*.

As this incident suggests, Catholics also suffered hundreds of martyrdoms at the hands of Protestants, but they had different ways of commemorating them and experienced different problems in establishing their status. Dozens were killed in France and the Low Countries, but the largest number perished in the British Isles. An important common feature, first pointed out thirty years ago by Natalie Zemon Davis in a pathbreaking article on the 'rites of violence' in sixteenth-century France, is that everywhere it was the ordained Catholic clergy who suffered most from Protestant attacks. For example, a list of Irish martyrs from 1537 to 1600, based on a compilation made at Salamanca around 1590, included eight bishops, eighteen secular clergy and 119 regular clergy, but only twenty-five laymen and one woman.[11] However, few of them were ever officially honoured at Rome, simply because most of them (unlike the Jesuits and other Catholics who were hanged in Elizabethan England) were murdered in cold blood, giving them little opportunity to testify to their steadfastness in the faith. Moreover, few Irish martyrs could produce properly attested miracles, another prerequisite for Roman Catholic sainthood since the High Middle Ages.

In Reformation Europe, recorded martyrdoms represent only the tip of a statistical iceberg of repression. A larger but submerged layer includes thousands of religious dissidents who were caught, interrogated, and given comparatively minor punishments, usually after recanting unorthodox beliefs. If Luther's song celebrated two martyrs, a third Augustinian, imprisoned with them, saved his life by recanting. England's Archbishop Thomas Cranmer provides a much more famous and significant example: he went through five partial recantations before finally dying as a martyr.[12] Numbers of recantations remain far more difficult to establish than numbers of martyrs, precisely because wafflers and especially turncoats were never commemorated. For example, we know that an informer broke the underground network which had distributed a radically anti-Catholic placard all over Paris one night in 1534, thereby triggering a bloody repression from furious authorities; a generation later, the martyrologist Crespin mentioned his nickname, but his actual name remains unknown.

The iceberg metaphor applies with full force to what was by far its largest and most deeply submerged layer, the one composed of so-called 'Nicodemites', people who hid their unorthodox religious beliefs in the name of prudence and expediency. If many recanters can be traced in judicial records, although far less easily than martyrs, these people remain almost completely invisible. As its name implies, this practice had Biblical precedents extending beyond the tale of Nicodemus, who concealed his allegiance to Christ by visiting him under cover of darkness. In 1970, Carlo Ginzburg traced the intellectual genealogy of a cluster of Reformation-era artful dodgers known by such various names as 'Nicodemites', '*temporiseurs*', or 'dissembling sects' to a 1527 Biblical commentary which first offered a bundle of Scriptural texts useful for justifying dissimulation in religious questions.[13] Its author, Otto Brunfels, was a former monk and Anabaptist sympathiser who probably practised what he preached. At the opposite confessional extreme from him stands the extreme subtlety and discretion of English Catholic apologists for religious dissimulation, admirably sketched by Perez Zagorin under an ironic title.[14] Although the very nature of their behaviour makes individual 'Nicodemites' all but impossible to detect in Reformation Europe, the old metaphor of smoke and fire suggests that religious dissimulation was widespread. Vehement authors from many confessional camps denounced such tepid and moderate Christians; in France, for example, Calvin's blasts against 'Nicodemites' were followed a generation later by French Catholic zealots denouncing tepid *politiques* – those who supported religious toleration for pragmatic reasons. Similarly, religious moderation was excoriated by all confessional camps in Reformation Europe. But since relatively few of parish clergy usually had to be replaced whenever official religion changed (Tudor England provides Europe's largest and most important examples), moderation must have been widespread even among 'professionals' as well as the less-instructed laity. A recent collection has tried to anatomise this 'moderation' and its limits.[15]

It is worth remembering that sixteenth-century apologists of religious toleration, who have been studied intensively by scholars for over a century, were almost as rare as Anabaptist authors, and that both groups often published anonymously. However, our ecumenical age has produced few sympathetic portraits of religious compromisers, with the notable exception of Mario Turchetti's sympathetic treatment of the French *moyenneurs* whose failure immediately preceded the kingdom's wars of religion.[16]

effects of persecution

Persecution was especially severe in France, Europe's largest kingdom, and even worse in the Low Countries, the commercial heart of sixteenth-century northern Europe. However, even though these same places experienced sustained and confessionally-driven warfare after 1560, it seems impossible to establish any causal relationship between these phenomena. Persecution never provided a sufficient cause for the outbreak of confessional warfare, and it was not even a necessary provocation; in Germany, heresy trials against Lutherans were extremely rare and played no part when Lutheran governments formed a military league. Instead, we propose to assess the significance of religious persecution in Reformation Europe by asking whether overall it was a success or a failure. The answer depends on where one looks. Although persecution completely failed to discourage organised religious dissent in transalpine

Europe, it probably succeeded in eliminating Protestantism throughout Mediterranean Europe.

Why were the outcomes so different in northern and southern Europe? First, although historians can trace elements of sympathy for the early Protestant Reformation almost everywhere in Latin Christendom, receptivity to Luther's Reformation proposals varied enormously in different parts of Europe, and early public support was clearly strongest among northern Europeans speaking languages closely related to Martin Luther's native German; elsewhere, most Protestant sympathisers followed different drummers.

However, what we call public opinion usually played a minor role in determining whether or not Protestantism eventually triumphed. Nearly everywhere outside the Swiss Confederation, early sixteenth-century Europe's most powerful republic, dynastic interests played decisive roles. Tudor England, with its rapid changes of official faith after 1534 and the insistence of its schismatic rulers on using the papal title 'Defender of the Faith' (awarded for Henry VIII's refutation of Luther), provides Europe's most extreme example of the German legal maxim *cuius regio eius religio* (the religion of the prince is the religion of his subjects). Many other parts of Protestant Europe, the Scandinavian kingdoms and their dependencies as well as many principalities within the Holy Roman Empire, also followed the *cuius regio* principle. More importantly, Europe's Catholic rulers, most of whom already enjoyed satisfactory legal concordats with the papacy, enabling them to control appointments to the ecclesiastical hierarchy, also followed the *cuius regio* rule. Of course, there were some important exceptions, mainly in north-western Europe: in the British Isles, Protestant Scotland removed a Catholic ruling queen, while Catholic Ireland stubbornly resisted its Protestant English queen. And nearly half of the Habsburg Low Countries (seven of its seventeen provinces) successfully repudiated their absentee hard-line Catholic sovereign after prolonged struggle.

However, most European rulers had little incentive to adopt some form of Protestantism, and desired to crush support for these novelties with sufficient use of force. This appears to be exactly what happened in the Iberian peninsula, through a vigorous repression carried out by the state-run inquisitions of Spain and Portugal against an embryonic and 'pre-confessional' Protestant movement. In politically fragmented Renaissance Italy, organised repression developed more slowly: the Roman Inquisition was not created until 1547 and never covered all of Italy. Led by Adriano Prosperi, current historiography argues that post-Tridentine Catholicism rapidly and effectively suffocated Italy's philo-Protestant movements by co-opting parish confessors to supplement the work of inquisitors and of new religious orders.[17]

The most ironic aspect of the repression of Protestantism in Mediterranean Europe is its chronology. Long after indigenous Protestant movements had been extirpated throughout Mediterranean Europe, so-called 'Lutheran' heretics, mostly Reformed and always foreigners, were still being burned: at Rome itself as late as 1610, but even later (1618) at Goa in Portuguese India and long after that (1640) in Spanish-ruled Sicily. In Italy, it seems certain that most of the executions ordered by the Vatican's Holy Office occurred only *after* any authentic Protestant movement had been eliminated and that the largest totals occurred during the pontificate of the man later canonised as St Pius V.[18] Mediterranean Europe thus illustrates what might be called the 'inverse theory of persecution' in Reformation Europe. Secular authorities burned a few 'outsiders' – not in order to frighten waverers, but literally as burnt sacrifices designed to reinforce militant confessional solidarity.

Meanwhile, the history of persecution in northern and western Europe suggests a totally different dynamic: an ultimately complete failure to intimidate religious dissenters into conformity with the established faith, or even to prevent the survivors from organising themselves into counter-churches. These persecutors did not fail for lack of trying; before 1560, 95 per cent of Europe's heresy executions occurred north of the Alps. Moreover, as we have seen, the bulk of them occurred in France and the Low Countries, the very regions which rose in rebellion soon after 1560. Under Francis I and Henry II, Valois France burned over 425 heretics,[19] but it was statistically overshadowed by the Habsburg Low Countries, whose reputedly Erasmian municipal governments ordered over three times as many heretics executed during the same years, in a population only one-sixth as large. However, the vast majority of those executed were pacifist Anabaptists, who were absent in France. The number of Reformed Protestants executed in both places seems far less disproportionate to their incidence in the mid-sixteenth century – and these were the martyrs who were commemorated in print both before and during the rebellions.

The only other western European country which experienced hundreds of heresy executions in a short period was Marian England. And England serves as a useful counter-example to any attempt to link persecution with religious warfare in Reformation Europe. England did become a Protestant nation, but overwhelmingly because of the dynastic accident that brought Elizabeth I to the throne in 1558 and kept her there, despite numerous assassination plots and marriage projects, for forty-five years.

I have argued elsewhere that the very weakness of England's Protestant movement vis-à-vis France required England's re-established Protestant state religion to remain unusually latitudinarian in confessional terms.[20] Coupled with dynastic loyalty among most English Catholics, it also helps explain why the English managed to avoid the prolonged religious warfare that ravaged their French neighbours.

confessional warfare, 1560–1650

We have now reached the major chronological divide within the political history of Reformation Europe, separating a phase of intensive persecution of nascent dissenting movements in several countries (punctuated in 1531 by a brief civil war in Switzerland and by equally brief outbreaks of hostilities in Germany between 1547 and 1552) from an age of confessional warfare (featuring innumerable executions for sedition during the Dutch and French revolts, and some very late public executions for heresy in Mediterranean countries). A fundamental sixteenth-century religious development, related to the *cuius regio* principle, further separates these periods. Although official 'Protestant' declarations of religious doctrine date from the Augsburg and Tetrapolitan confessions of 1530, current wisdom locates the consolidation of Europe's 'confessional century' around 1560, with several Reformed confessions of faith at national levels, the articles of the Church of England, and of course the decrees of the Council of Trent. Before 1560, Protestant movements and Catholic repression; afterwards, rival public churches confronting each other.

It seems worth asking where wars of religion did *not* occur in confessionally-divided western Europe. The Swiss Confederation, Europe's first confessionally-divided republic, long avoided overt religious conflicts after Zwingli's defeat and death on the battlefield in 1531; however, they held brief confessionally-driven civil wars as late as 1712. The two obvious major exceptions are the Holy Roman Empire, where major conflicts were postponed until 1618, and Britain, where (depending on one's point of view) they were either avoided altogether or postponed until the 1630s or later. Conventional wisdom insists that each place avoided religious conflicts through very different tactics: the Empire through relatively early legislative coexistence between Lutherans and Catholics, and England primarily through a relatively undogmatic established church. It is worth noting that Germany's Lutherans had fought Emperor Charles V twice before the religious Peace of Augsburg was negotiated in 1555 and put into effect during the brief interregnum between Charles' resignation and his brother's accession. Numerous critics have pointed out that it was ultimately unable to prevent a devastating civil war in 1618 and that its articles granted rights to only two groups, Catholics and Lutherans. Nevertheless, the terms of this understudied compromise were unprecedented in some progressive ways, including the *jus emigrandi* or right of subjects to sell their assets at a reasonable price and move to another overlord in order to worship as they pleased. And Germany's Reformed Protestants – a confession controlling no major territory within the Empire in 1555 – became the 'outsiders' primarily responsible for rupturing this long-established status quo.

At the opposite extreme from England and the German Empire stood France and the Low Countries, sites of almost interminable confessionally-driven military conflicts by the 1560s. Scholars agree that no direct causal links connected the extensive heresy persecutions of previous decades in either place to these acts of political subversion. At the same time, seminal articles by H. G. Koenigsberger and R. M. Kingdon back in the 1950s demonstrated how Reformed Protestants organised and financed revolutionary uprisings against legitimate dynastic rulers in both places.[21] They argued that these revolts required only two necessary triggers: a sufficiently well-organised Reformed church, and a sufficiently well-placed dissident prince claiming to share their views and willing to serve their cause.

What made these civil wars both so long and so indecisive? Their length is undeniable. Dozens of Dutch historians have proudly recounted their region's eighty-year struggle for political independence. And although conventional wisdom ends the French wars of religion with the promulgation of the Edict of Nantes in 1598, Mack Holt makes an excellent case for continuing them until Richelieu crushed the last Huguenot military rebellion thirty years later.[22] The indecisive eventual outcomes seem equally undeniable. After eighty years of chronic warfare, the Dutch Republic controlled only seven of the original seventeen provinces comprising the Low Countries in 1560, including only one of the three largest; and the rebels came to adopt a *de facto* religious toleration. After sixty-five years of intermittent wars, the French monarchy remained in nearly the same bi-confessional situation as when they originally began.

One often overlooked explanation for the extreme length and indecisiveness of these religious wars emphasises the ongoing rivalry between Europe's greatest sixteenth-century monarchies, France and Spain. Huguenot rebels opposed kings of France enjoying the papal title of 'Very Christian King' (*roi très chretien*); Low Countries rebels opposed kings of Spain, recognised by the Vatican as 'Catholic Kings' (*reyes católicos*). But with rare and brief exceptions, these two pre-eminent Catholic monarchs never cooperated against a common enemy during these religious wars. They had been locked in a titanic struggle for control of Renaissance Italy ever since Luther was a child, and remained rivals rather than allies even after officially making peace in 1559. If Catholic solidarity usually (but not always) prevented the French king from overtly supporting Dutch Protestant rebels militarily during their prolonged sixteenth-century rebellion, both France and Spain saw each other's difficulties as opportunities. Their perennial hostility gave both Dutch and French rebels some much-needed manoeuvring space before the 'Catholic King' and the 'Very Christian King' again resumed their struggle for dynastic supremacy during the later phases of the Thirty Years War.

The indecisive results of western Europe's wars of religion also reflected the strengths and weaknesses of the revolutionary confessional churches which began and sustained them. In both France and the Low Countries, the rebellious Reformed church endured; it enjoyed official status in the Dutch Republic and full legal recognition in France. However, for different reasons, it never acquired majority status in either place and therefore stood no chance of becoming a *Landeskirche* on the German or Scottish model. Estimates of their actual size vary greatly, but the leading authority on the subject, Philip Benedict, suggests that the French Reformed church began the civil wars with the hearts and minds of perhaps one-fifth of the kingdom's population. Battered by defeat and massacres, it fell to about one-tenth of French subjects by the 1570s and dropped very slowly thereafter.[23] In the Low Countries, this trend was reversed: when the revolt began, the Reformed church had proportionately fewer adherents than in France, especially in the northern provinces which became the modern Netherlands. As the wars continued, the geographical core of the resistance moved north into exactly these regions. When the first Dutch political union was formed in 1579, membership in Reformed churches remained below 10 per cent of the population of any city within its boundaries.[24] Even after the Dutch Republic received *de facto* recognition in 1609, historians agree that its official church remained a minority almost everywhere. Although Dutch Catholicism remains a redoubtable political force, it was undeniably the Dutch Reformed church that provided the backbone for a successful revolution. And it also seems obvious that its minority status had much to do with the much-celebrated *de facto* toleration of the seventeenth-century Dutch Republic.

Regional geography became another critical factor helping to prolong the wars of religion. Although rebel armies lost the vast majority of sixteenth-century battles, these revolts endured by consolidating themselves in well-fortified regions beyond the effective reach of central authorities. This process was clearest in the Low Countries, where the rebels faced the formidable Spanish infantry of their overlord Philip II. A series of military and political events in the 1570s – the capture of an island base by rebels, lifting the siege of Leiden after rebels flooded the area, and the rebels' failure to unite anti-Spanish opposition throughout the Low Countries following the Pacification of Gent in 1576 – forced them into a low-lying but militarily defensible north-western base. In the French case, Huguenot rebels maintained clusters of mountainous strongholds forming a loose arc across southern France from Dauphiné and Languedoc to Guyenne, from which sixteenth-century royal armies never dislodged them.

How important was religion in these wars? It seems evident that without a huge measure of unwavering devotion to a cause whose foundations were confessional churches, these rebellions could never have lasted more than a few years. Nevertheless, sceptical French contemporaries like Montaigne doubted the sincerity of leaders on both sides, and a nation which has become a fortress of *laicité* finds the religious fanaticism of their ancestors embarrassing, even while scholars like Denis Crouzet dissect its minutiae in major studies.[25] Himself a Catholic, Montaigne needed only point to a prince he knew well: Henry of Navarre, official leader of the French Reformed cause for many years, converted to Catholicism twice – once to save his life, and again in order to pacify the kingdom he had unexpectedly inherited.

Meanwhile, Dutch historians, looking back with understandable pride at their extraordinarily tolerant and prosperous seventeenth-century 'Golden Age', have even more reason to doubt the religious fanaticism of their sixteenth-century rebels and every reason to emphasise the undeniable fanaticism of their Spanish opponents. They have a point. The contrast becomes evident when one compares the four wives of the rebel leader William of Orange (ruler of a small principality in southern France, with most of his estates outside the Low Countries) and the four wives of his archenemy Philip II of Spain. Philip's wives came from several royal dynasties (Portugal, England, France, Austria), but all were staunch Catholics. William's wives came from the upper aristocracy, but as his revolt began and continued, they became successively more radical in religion, moving from a traditional Catholic through a German Lutheran and ending with two French Reformed princesses.

Politics and confessionalism remained inextricably mixed during these wars, which were themselves inextricably interconnected. The modern Dutch national anthem, the *Wilhelmus*, provides an excellent example of both. As Pollmann notes, it was originally a Dutch Protestant *contrafactum* or burlesque of a French Catholic song mocking the defeat of a Huguenot prince, whose sixteen stanzas simultaneously justified William of Orange's revolt against a 'King he had honoured always' and emphasised the godliness of his cause. First printed in German and known as far away as Switzerland, the *Wilhelmus* soon became a real anthem for the Dutch revolt and the international Protestant cause. It was also immediately parodied by Catholics; by 1600, a printed Catholic songbook already included dozens of examples.

During the wars of religion, confessionalism usually hampered Protestant efforts because their two main branches, Lutherans and Reformed, never cooperated politically, especially within the Holy Roman Empire. One can point to occasional ecumenical Protestant efforts in raising money; for instance, the leading patron of the new Protestant church erected in 1610 in Prague's Old City was James I of England, followed by almost all of Germany's leading Protestant princes, both Calvinist and Lutheran. Yet ever since Luther quarrelled with Zwingli at Marburg in 1529, all attempts at political alliances between Lutherans and Reformed had failed. Moreover, the same English king who patronised the Prague church so lavishly refused military help to his German Calvinist son-in-law nine years later, and soon married his own heir to a Catholic princess. But if Protestant princes avoided military commitments across confessional boundaries, the bitter rivalry between Europe's greatest Catholic rulers proved even more serious for the other side.

The Thirty Years War (1618–48) not only culminated Europe's confessional warfare, but also ultimately ended it. Begun when a Reformed prince tried to seize Bohemia from the Catholic Habsburgs and was crushingly defeated, this conflict continued through successive interventions by foreign rulers claiming to uphold German 'liberties' threatened by victorious German Catholics. Scandinavian Lutherans came first; after 1630, they were subsidised by Catholic France; finally, after 1635 both France and Spain entered the German conflict directly. After a quarter-century of chronic warfare and unparalleled suffering, every major European state except England and the Ottoman Empire had become involved. Stalemate ensued, and diplomats finally had to sort things out.

Diplomatic history has long remained unfashionable – understandably so after Europe collapsed into a second World War in 1940, but it remains unfashionable even after the multilateral successes of the European Community during the past generation. Perhaps it is time for a renaissance; in any event, the history of diplomacy offers some penetrating insights into confessional Europe during this prolonged phase of religious warfare. As Garrett Mattingly told us in a well-written survey fifty years ago, modern Europe's system of reciprocal resident ambassadors, withdrawn only during wars, was basically put into place during Martin Luther's lifetime.[26] Rome was then the most important diplomatic centre of Europe. Because few Protestant princes appointed ambassadors (even major German princes like Luther's overlord could not), the Reformation took a long time to have a significant impact on European diplomacy. Resident ambassadors always brought their own chaplains and their rights of extra-territoriality had been well established. But ultimately, confessionalism affected diplomacy precisely during the most intensive phase of the wars of religion. As Mattingly acutely noted, diplomatic relations between Protestant and Catholic states ceased completely after England and Spain finally broke diplomatic ties; the schism lasted for nearly two decades (1585–1604).

It seems fitting to conclude with the creation of a 'post-confessional' European diplomacy. Originally published in 1959 and still untranslated, Fritz Dickmann's classic account of the marathon conferences which culminated in the famous Peace of Westphalia in 1648 remains an indispensable guide.[27] It is worth noting that all other major European treaties, including some before 1600 (Cateau-Cambrésis, Vervins), bear the name of the city in which they were signed. 'Westphalia', however, is a region, and these negotiations were actually held in two separate cities, Münster and Osnabrück, over a period of several years. Because every German state represented in the Imperial Diet had the right to attend, the official size of this peace congress parallels that of the current United Nations: 176 plenipotentiaries, with the largest embassy (that of France) approaching two hundred people. It soon broke into separate gatherings along religious lines. The seventy-three Protestant delegates of the *Corpus Evangelicorum* met at Osnabrück, with a Danish mediator; meanwhile, the seventy-two Catholic diplomats of the *Corpus Catholicorum* met at Münster, mediated by a papal legate. Couriers shuttled incessantly between them as rival proposals were gradually hammered into compromises. However, in order to ensure that their interests were adequately represented, the three major players (France, Austria, and Sweden) found it necessary to name both a Protestant to represent them at Osnabrück and a Catholic to negotiate at Münster. Despite confessional divisions, the Protestant lobby rather surprisingly held together better than their Catholic counterparts and gained most of the disputed religious issues. In the end, literally hundreds of diplomats negotiated an agreement which has remained a landmark of European history. Its recent 450th anniversary was lavishly commemorated. In a telling illustration of the end of confessional diplomacy, the Vatican, after brokering the Münster negotiations, immediately (and futilely) repudiated the agreements signed there.

further reading

The outstanding survey of sixteenth-century political thought is Quentin Skinner, *The Foundations of Modern Political Thought, vol. 2: The Age of the Reformation* (Cambridge, 1978). It can be supplemented by a sympathetic work on Anabaptist political theory: James Stayer, *Anabaptists and the Sword*, 2nd edn (Lawrence, KS, 1976) and the detailed exposition of 'maximalist' Protestant resistance theory by Robert M. Kingdon, *Myths About the St. Bartholomew's Day Massacre, 1572–1576* (Cambridge, MA, 1988).

The masterful cross-confessional study of martyrdom in Reformation Europe by Brad Gregory, *Salvation at Stake* (Harvard University Press, 1999), has set the agenda for twenty-first-century research. On religious songs, see a forthcoming article by Judith Pollmann, 'Singing for Reformation in the Sixteenth Century', in H. Schilling and I. Toth (eds), *Religious Identities, Confessional Formation, and Cultural Diffusion in Early Modern Europe* (Cambridge, 2005); and W. Stanford Reid's classic essay, 'The Battle Hymns of the Lord: Calvinist Psalmody of the Sixteenth Century', *Sixteenth Century Essays and Studies*, 2 (1971), 36–54. Oddly enough, the most interesting work in English about 'Nicodemism' deals primarily with English Catholics: Perez Zagorin, *Ways of Lying: Simulation and Dissimulation in Early Modern Europe* (Harvard University Press, 1990); see also Andrew Pettegree, *Marian Protestantism: Six Studies* (Aldershot, 1996), chapter 4. On sixteenth-century religious toleration, one must still begin with the massive survey by the French Jesuit Joseph Lecler; originally published in 1955, it had an English translation as *Toleration and the Reformation* in 1960. See also the essays in O. P. Grell and Bob Scribner (eds), *Tolerance and Intolerance in the European Reformation* (Cambridge, 1994).

For a convenient summary of confessionalism, see Heinz Schilling, 'Confessional Europe', in T. A. Brady, H. A. Oberman and J. D. Tracy (eds), *Handbook of European History 1400–1600*, 2 vols (Leiden, 1995), vol. II.

There are numerous reliable accounts of both the French and Dutch religious wars. For France, the best brief survey in English is Mack P. Holt, *The French Wars of Religion, 1562–1629* (Cambridge, 1995); for the Dutch revolt, it is Geoffrey Parker, *The Dutch Revolt* (London, revised edn 1985). If little has been done to supplement the old articles of Koenigsberger and

Kingdon comparing the origins of both conflicts, we have an outstanding analysis of religious violence during the French wars by Natalie Zemon Davis, *Society and Culture and Society in Early Modern France* (Stanford, 1975). On the Thirty Years War, by far the best general account is Geoffrey Parker et al., *The Thirty Years War* (London, 1984).

The best survey of sixteenth-century diplomacy remains Garrett Mattingly, *Renaissance Diplomacy* (Boston, 1955), but we still lack a first-rate English-language investigation of the Westphalian peace congresses.

notes

1. For a preliminary survey, see William Monter, 'Heresy Executions in Europe, 1520–1565', in O. P. Grell and Bob Scribner (eds), *Tolerance and Intolerance in the European Reformation* (Cambridge, 1994).
2. See Mark U. Edwards, *Luther's Last Battles: Politics and Polemics, 1531–1546* (Ithaca, 1983), chapter 2.
3. James Stayer, *Anabaptists and the Sword*, 2nd edn (Lawrence, KS, 1976).
4. Quentin Skinner, *The Foundations of Modern Political Thought. Volume 2: The Age of the Reformation* (Cambridge, 1978). For some of the theorists, see Donald R. Kelley, *The Beginning of Ideology: Consciousness and Society in the French Reformation* (Cambridge, 1981); John Knox, *On Rebellion*, ed. Roger Mason (Cambridge, 1994).
5. Brad Gregory, *Salvation at Stake* (Harvard University Press, 1999).
6. *Ibid.*, p. 165.
7. On Foxe see above, pp. 139–40.
8. On Rabus see Robert Kolb, *For All the Saints: Changing Perceptions of Martyrdom and Sainthood in the Lutheran Reformation* (Macon, GA, 1987).
9. Gregory, *Salvation at Stake*, pp. 232–5.
10. Judith Pollmann, 'Singing for Reformation in the Sixteenth Century', in H. Schilling and I. Toth (eds), *Religious Identities, Confessional Formation, and Cultural Diffusion in Early Modern Europe* (Cambridge, 2005).
11. Samantha Meigs, *The Reformations in Ireland: Tradition and Confessionalism, 1400–1690* (London, 1997), p. 180.
12. Diarmaid MacCulloch, *Thomas Cranmer: a Life* (New Haven and London, 1996).
13. Carlo Ginzburg, *Il Nicodemismo: Simulazione e dissimulazione religiosa nell'Europa del '500* (Turin, 1970).
14. Perez Zagorin, *Ways of Lying: Simulation and Dissimulation in Early Modern Europe* (Cambridge, MA, 1990).
15. Luc Racaut and Alec Ryrie (eds), *Moderate Voices in the European Reformation* (Aldershot, 2005).
16. Mario Turchetti, *Concordia o toilleranza? Franois Baudoin ed i 'moyenneurs'* (Geneva, 1986).
17. Adriano Prosperi, *Tribunali della coscienza. Inquisitori, confessori, missionari* (Turin, 1996).
18. William Monter, 'The Roman Inquisition and Protestant Heresy Executions in Sixteenth-Century Europe', in A. Borromeo (ed.), *L'Inquisizione: Atti des Simposio internazionale .. 1998* (Vatican City, 2003).

19. William Monter, *Judging the French Reformation* (Cambridge, MA, 1999).
20. William Monter, 'The Fate of the English and French Reformations, 1554–1563', *Bibliothèque d'Humanisme et Renaissance*, 64 (2002), 7–19.
21. H. G. Koenigsberger, 'The Organization of Revolutionary Parties in France and the Netherlands during the Sixteenth Century', *Journal of Modern History*, 37 (1955), 335–51; and Robert M. Kingdon, 'The Political Resistance of the Calvinists in France and the Low Countries', *Church History*, 27 (1958), 3–16.
22. Mack P. Holt, *The French Wars of Religion, 1562–1629* (Cambridge, 1995).
23. Benedict's previous work on this subject has been conveniently condensed in his *Christ's Churches Truly Reformed: a Social History of Calvinism* (New Haven and London, 2002), pp. 137, 148, 372–3.
24. Jonathan Israel, *The Dutch Republic: Its Rise, Greatness, and Fall, 1477–1806* (Oxford: Clarendon Press, 1995), pp. 363ff; see above, pp. 92–5.
25. See especially his massive two-volume survey, *Les Guerriers de Dieu* (Seyssel, 1990).
26. Garrett Mattingly, *Renaissance Diplomacy* (Boston, 1955), chapter 20.
27. Fritz Dickmann, *Der Westfälische Frieden*, 5th edn (Münster, 1985).

index